Library of Congress Cataloging-in-Publication Data

DNA probes for infectious diseases.

Includes bibliographies and index.
1. DNA probes — Diagnostic use. 2. Nucleic acid hybridization. 3. Diagnostic microbiology — Technique. I. Tenover, Fred C., 1954- [DNLM: 1. Communicable Diseases — diagnosis. 2. Nucleic Acid Hybridization. WC 100 D629]
QR69.D54D54 1989 616.9'047583 88-9526
ISBN 0-8493-6176-1

Direct all inquiries to CRC Press, Inc., 2000 Corporate Blvd., N.W., Boca Raton, Florida, 33431.

International Standard Book Number 0-8493-6176-1

Library of Congress Card Number 88-9526
Printed in the United States

INTRODUCTION

"...In the 1880s medicine was turned topsy-turvy by convincing proofs that the world of microorganisms, known to exist for several centuries by scientists, was responsible for many of the major epidemic diseases afflicting people. These revelations converted the medical laboratory, which had been developed to advance basic understanding of biologic process, also into a place where diagnostic analyses of the problems of patients were conducted. Perhaps more than any other institution the laboratory symbolized to doctor and patient the arrival of science in medicine..."*

Clinical microbiology laboratories have not changed much over the decades. Automation has been introduced to improve turn-around times, and computers now speed results to anxious clinicians, but the basic bench routines are not so dissimilar from those that Koch and Pasteur instituted in the early years of bacteriology. These techniques have stood the test of time because they are sensitive and effective. But, there is room for improvement. The time-honored techniques are often too time consuming to impact patient management. New pathogens are emerging. Some can be propagated in vitro, while others remain elusive. Thus, there is a need for faster, more accurate methods of detecting and identifying infectious agents. DNA probes may be the answer.

The use of hybridization technology for the identification of pathogenic microorganisms is still a novel idea and is far from being an accepted modality by clinical microbiologists, pathologists, or infectious disease specialists. Yet, fast and accurate recognition of pathogenic microorganisms directly in tissues or secretions is an important goal if infectious diseases are to be controlled and ultimately eliminated.

With this volume, I hope to impart a broad view of what has happened and is happening with the application of DNA probe technology to the field of infectious disease. Some chapters deal specifically with the application of DNA probes to the clinical laboratory; others explore the use of probes to answer questions concerning the natural history of several viral pathogens of humans. One chapter deals with the application of DNA probes to the quality control of food products. The applications of this technology are indeed broad.

As with any major undertaking of this kind, there are several individuals to whom I am indebted for their valuable contributions. In particular, I would like to thank Jim Kaper, Steve Moseley, Marie Coyle, Carol Spiegel, and Dick Tilton for taking the time to thoughtfully review several of the manuscripts in this volume. I also thank the contributors for the time and effort they each put into preparing their manuscripts. I also would like to thank Stan Falkow and Frank Young, my former mentors, for their encouragement in my pursuit of this project. Finally, I would like to thank my wife, Lisa, for patiently reading and correcting all of my chapters and for her support during the preparation of this volume.

* Reiser, S. J., Technology and the eclipse of individualism in medicine, in *In Search of the Modern Hippocrates*, Bulger, R. J., Ed., University of Iowa Press, Iowa City, 1987, 215.

THE AUTHOR

Fred Carron Tenover, Ph.D. received his Baccalaureate degrees in Biology and Chemistry from the University of Dayton where he was awarded the John E. Dlugos, Jr. memorial award of excellence in biology. While pursuing his Masters and Doctoral degrees in Microbiology at the University of Rochester in New York under the supervision of Dr. Frank Young, the author was awarded a President's Fellowship by the American Society for Microbiology. After completing his doctoral degree, Dr. Tenover continued his training as a post-doctoral Fellow in Clinical Microbiology and Public Health at the University of Washington in Seattle. During this period, he worked on the cloning of aminoglycoside resistance genes in the laboratory of Dr. Stanley Falkow. After completing his fellowship, Dr. Tenover accepted the position as Associate Chief, and more recently, Chief of Microbiology at the Seattle Veterans Administration Medical Center. He is also Associate Professor of Laboratory Medicine and Adjunct Associate Professor of Microbiology at the University of Washington. Dr. Tenover serves on the editorial boards of the Journal of Clinical Microbiology and Antimicrobial Agents and Chemotherapy, is a diplomate of the American Board of Medical Microbiology, and is a member of the American Society for Microbiology, the Infectious Diseases Society of America, the American Association for the Advancement of Science, and the Academy of Clinical Laboratory Physicians and Scientists. Besides his interest in diagnostic DNA probes and antimicrobial resistance, Dr. Tenover is interested in the pathogenesis of campylobacter infections.

Dedication

To my wife and best friend, Lisa.

CONTRIBUTORS

Bernadino Albuquerque
Institute of Medicine
Tropical de Manaus
Manaus, Amazonas, Brazil

Anna Marie Beckmann, Ph.D.
Assistant Member
Department of Experimental Pathology
Fred Hutchinson Cancer Research Center
Seattle, Washington

Hubert E. Blum, M.D.
Associate Professor
Department of Medicine
University of Freiburg
Freiburg, West Germany

Michael P. Busch, Ph.D.
Assistant Scientific Director
Scientific Services
Irwin Memorial Blood Bank
San Francisco, California

Jonathan P. Clewley, Ph.D.
Virus Reference Laboratory
Public Health Laboratory Service
London, England

Kenneth D. Croen
Laboratory of Clinical Investigation
National Institute of Allergy and Infectious Diseases
National Institutes of Health
Bethesda, Maryland

Wayne M. Dankner, M.D.
Associate Physician
Department of Pediatrics/Infectious Diseases
University of California
San Diego, California

Frank Denaro, Ph.D.
Research Virologist
Department of Pediatrics/Infectious Diseases
University of California
San Diego, California

Heitor Dourado
Institute of Medicine
Tropica de Manaus
Manasu, Amazonas, Brazil

Mary Kolb Estes, Ph.D.
Professor
Department of Virology
Baylor College of Medicine
Houston, Texas

Walter E. Hill, Ph.D.
Research Microbiologist
Division of Microbiology
Food and Drug Administration
Washington, D.C.

Timo Hyypiä, Ph.D.
Department of Virology
University of Turku
Turku, Finland

Garret M. Ihler, M.D., Ph.D.
Professor and Head
Department of Medical Biochemistry and Genetics
Texas A&M College of Medicine
College Station, Texas

Edward G. Jablonski, Ph.D.
Director of Research
Molecular Biosystems, Inc.
San Diego, California

James B. Kaper, Ph.D.
Associate Professor
Department of Medicine
University of Maryland School of Medicine
Baltimore, Maryland

Myron J. Levin, M.D.
Professor
Departments of Pediatrics and Medicine
University of Colorado School of Medicine
Denver, Colorado

J. Glenn Morris, Jr., M.D.
Associate Professor
Department of Medicine
University of Maryland School of Medicine
Baltimore, Maryland

David Myerson, Ph.D.
Assistant Member
Department of Clinical Research
Fred Hutchinson Cancer Research Center
Seattle, Washington

Mitsuaki Nishibuchi, Ph.D.
Assistant Professor
Department of Microbiology
Faculty of Medicine
Kyoto University
Kyoto, Japan

Jeffrey M. Ostrove
Laboratory of Clinical Investigation
National Institute of Allergy and Infectious Diseases
National Institutes of Health
Bethesda, Maryland

Airi Palva, Ph.D.
Researcher
Recombinant DNA Laboratory
University of Helsinki
Helsinki, Finland

Ulf Pettersson, M.D.
Professor
Department of Medical Genetics
Uppsala University
Uppsala, Sweden

Marjut Ranki, Ph.D.
Head
Molecular Biology Department
Orion Corporation, Limited
Helsinki, Finland

Allison C. Rice-Ficht, Ph.D.
Assistant Professor
Department of Medical Biochemistry and Genetics
Texas A&M College of Medicine
College Station, Texas

William O. Rogers
Department of Tropical Public Health
School of Public Health
Harvard University
Boston, Massachusetts

Harley A. Rotbart, M.D.
Assistant Professor
Departments of Pediatrics and Microbiology/Immunology
University of Colorado School of Medicine
Denver, Colorado

Sydney B. Shaw, Ph.D.
Research Manager
Research and Development Department
Gen-Probe, Inc.
San Diego, California

Pamela S. Shirley
Research Specialist
Department of Infectious Disease
St. Jude Children's Research Hospital
Memphis, Tennessee

John W. Sixbey, M.D.
Associate Member
Departments of Infectious Disease and Virology and Molecular Biology
St. Jude Children's Research Hospital
Memphis, Tennessee

Deborah H. Spector, Ph.D.
Professor
Department of Biology
University of California
San Diego, California

Stephen A. Spector, M.D.
Department of Pediatrics
Co-Chief
Division of Infectious Diseases
University of California
San Diego, California

Stephen E. Straus, M.D.
Head, Medical Virology Section
Laboratory of Clinical Investigation
National Institute of Allergy and Infectious Diseases
National Institutes of Health
Bethesda, Maryland

Tomoyuki Tanaka, M.D.
Visiting Associate Professor
Department of Virology and Epidemiology
Baylor College of Medicine
Houston, Texas

Fred C. Tenover, Ph.D.
Associate Professor
Departments of Laboratory Medicine and Microbiology
University of Washington
Seattle, Washington

Paul P. Ulrich, M.D.
Postdoctoral Fellow
Department of Laboratory Medicine
University of California
San Francisco, California

Girish N. Vyas, Ph.D.
Professor
Department of Laboratory Medicine
University of California
San Francisco, California

Dyann F. Wirth, Ph.D.
Department of Tropical Public Health
School of Public Health
Harvard University
Boston, Massachusetts

TABLE OF CONTENTS

PART I: PROBE DEVELOPMENT

PART II: IMPACT OF PROBE TECHNOLOGY ON DIAGNOSTIC LABORATORIES

PART III: NUCLEIC ACID PROBES FOR DIARRHEAL PATHOGENS

PART IV: NUCLEIC ACID PROBES FOR RESPIRATORY PATHOGENS

PART V: NUCLEIC ACID PROBES FOR THE DETECTION AND IDENTIFICATION OF VIRAL PATHOGENS AND *CHLAMYDIA*

PART VI: NUCLEIC ACID PROBES FOR IDENTIFICATION OF PARASITIC AGENTS

PART VII: FUTURE DIRECTIONS

Part I: Probe Development

Chapter 1

CONSTRUCTING DNA PROBES FOR INFECTIOUS AGENTS

Fred C. Tenover

TABLE OF CONTENTS

I. INTRODUCTION

The age of molecular biology is now becoming the age of molecular medicine. Clinical problems, both diagnostic and therapeutic, are being solved using novel approaches made possible by recombinant DNA and monoclonal antibody technology.[1] Although the technology itself is now a decade old, many laboratory directors and clinicians are, as of yet, unaware of the potential impact of this technology on our ability to diagnose diseases rapidly and accurately. Within 5 years, however, these techniques will become the backbone of diagnostic laboratory medicine. It is imperative that those directing laboratories and those that use diagnostic laboratories understand the promises and pitfalls of this technological revolution.

In this chapter, the basic principles of molecular biology, as well as several approaches to the development of DNA probes for infectious agents, will be reviewed. While few laboratorians or clinicians will be involved in the design or construction of DNA probes, a basic review of probe development should help illuminate the strengths and weaknesses of this approach to diagnostic microbiology.

A. Basic Vocabulary of Molecular Biology

Like any branch of science, molecular biology has it own vocabulary. To understand the process of constructing probes, it is important to know something about the tools that are used. In this case, the tools are plasmids, cloning vectors, restriction endonucleases, ligase, reverse transcriptase, and the process of DNA-mediated transformation.

1. DNA, Plasmids, Cloning Vectors, and Restriction Endonucleases

The heart of molecular biology is deoxyribonucleic acid or DNA. DNA is a double-stranded, helical molecule composed of pairs of nucleotide bases (adenine, guanine, thymine, and cytosine) located between two sugar-phosphate backbones (Figure 1). The DNA molecule, which resembles a twisted ladder with the adenine and thymine molecules and guanine and cytosine molecules linked together by hydrogen bonds, can be separated into two single strands by heat (thermal denaturation) or by raising the pH and lowering the ionic strength of the DNA solution. The DNA molecule is most stable in its native double-stranded state. Thus, when single-stranded DNA is placed in solution under appropriate temperature and salt conditions, the complementary strands will recombine to form a duplex molecule. DNA molecules will recombine only when the two strands are composed of complementary sequences. It is this property of DNA that is the basis of the DNA probe reaction.

Besides the large, circular molecules of DNA that compose the chromosomes of the bacterial cell, there are much smaller segments of DNA, known as plasmids, that are often found within the bacterial cell. Plasmids are covalently closed circular pieces of DNA that replicate independently of the bacterial chromosome. They are not required for cell replication, but often give the host cell some selective advantage when it finds itself in a hostile environment. These extrachromosomal elements often encode antimicrobial resistance genes, adhesion factors, hemolysins, bacteriocins, and the genes that orchestrate transfer of the plasmid to other bacteria during conjugation. Plasmids range in size from a few thousand base pairs (bp) to as large as 400,000 bp (400 kilobases or 400 kb). One subset of plasmids, often used in recombinant DNA technology, is called cloning vectors. Cloning vectors are small plasmids, often just 2 to 5 kb, that contain a selectable marker such as ampicillin resistance, and a stretch of DNA that can be cleaved by many different types of restriction endonucleases (see below). These specialized plasmids are often constructed in the laboratory and combine different functional regions of several extrachromosomal elements. Examples of commonly used cloning vectors are pBR322, a 4.6-kb plasmid encoding ampicillin and tetracycline resistance;[2] pUC18 and pUC19, a pair of plasmids that contain (1) a region with multiple restriction sites that facilitate insertion of foreign DNA, (2) an ampicillin resistance gene that indicates the presence of the plasmid in the recipient

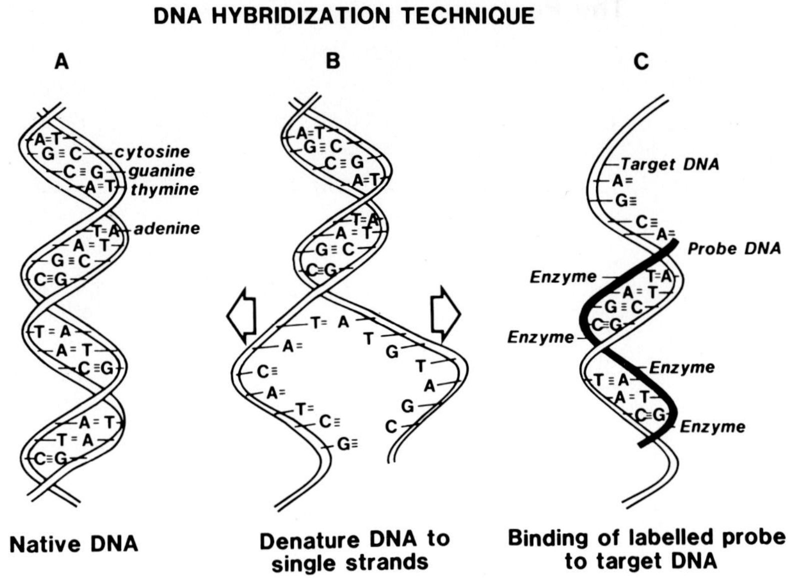

FIGURE 1. Structure of deoxyribonucleic acid, effects of denaturation and binding of probe to target DNA. (From Tenover, F. C., *Clin. Microbiol. Rev.*, 1, 83, 1988. With permission.)

bacterium after transformation, and (3) a clever indicator system utilizing a β-galactosidase gene where plasmids containing DNA inserts produce colorless colonies, while those organisms carrying plasmids without inserts appear as blue colonies on special media,[3] and M13mp18/19, two similar bacteriophage with the same cloning region as the pUC18 and pUC19 plasmids that facilitate DNA sequencing.[3] Other vectors can be found in several of the cloning manuals now available.[4-6] Assembling these chimeric plasmids is accomplished through the use of restriction endonucleases, which are enzymes that cleave double-stranded DNA at specific nucleotide sequences, often leaving stretches of single-stranded DNA on both ends, known as "sticky ends". These sticky ends allow two pieces of DNA from diverse sources (such as the cloning vector pBR322 and DNA from a malaria parasite), that have both been cleaved with the same restriction endonuclease, to join together as a single molecule. An example of the restriction endonuclease *Eco*RI is shown in Figure 2. Restriction endonucleases are named by combining the first letter of the bacterial genus from which they are isolated with the first two letters of the species, adding any strain designations, and, finally, indicating whether the enzyme was the first, second, or third enzyme discovered from that strain. Thus, *Hin*dIII is the third restriction endonuclease isolated from *Haemophilus influenzae* strain d.

2. DNA Ligase and DNA-Mediated Transformation

After cleaving the vector DNA and the DNA to be inserted with the same enzyme, the two are mixed in the presence of DNA ligase, a sort of molecular glue that seals the ends of the two strands of DNA into a continuous circle. After sealing the ends, the new chimeric plasmid is introduced into *Escherichia coli* by a process called transformation. Once inside the bacterial cell, the plasmid will replicate, making hundreds of identical copies of the plasmid containing the heterologous DNA (i.e., the cloned fragment). The plasmid, now greatly amplified in copy number, can be purified from the bacterial cell, and the cloned fragment can be isolated away from the cloning vector by digesting the DNA with the same restriction endonuclease used to construct the chimera. The purified fragment, once labeled with a radioisotope, enzyme, or some other method of detection, is now ready to be used as a diagnostic probe. This process is illustrated in Figure 3.

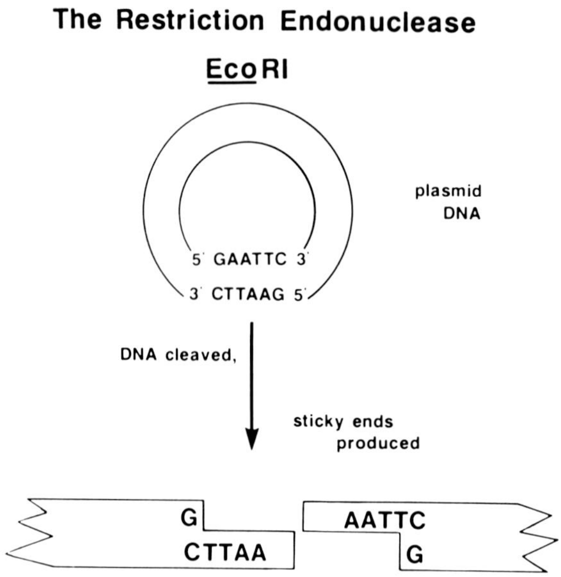

FIGURE 2. The restriction endonuclease *Eco*RI. (From Tenover, F. C., *Clin. Microbiol. Rev.*, 1 , 83, 1988. With permission.)

3. Reverse Transcriptase

For some organisms, such as RNA viruses or complex eukaryotic organisms, it is easier to isolate the work with RNA rather than DNA. In such cases, an enzyme known as reverse transcriptase is used to make a DNA copy of the desired RNA molecule. A second complementary DNA strand is then synthesized, this time using the DNA copy as a template, and the new double-stranded DNA molecule is inserted in the cloning vector. Reverse transcriptase allows one to clone virtually any segment of RNA, including messenger and ribosomal RNA, with relative ease.

4. Probes, Targets, Reporter Molecules, and Stringency

In general, a probe is a piece of nucleic acid, either DNA or RNA, labeled in some fashion, that can seek out and bind itself to DNA or RNA of complementary sequences in a test sample. (Since the processes for DNA and RNA are essentially the same, DNA will be used for convenience to refer to both DNA and RNA molecules for the remainder of this chapter unless otherwise designated.) The DNA in the sample (i.e., a clinical specimen such as sputum or feces) is referred to as the target nucleic acid, and the label on the probe (often a radioactive isotope, biotin, or an enzyme complex such as alkaline phosphatase) is referred to as the signal generator or reporter molecule. Both the probe DNA and the target DNA must be single stranded to initiate the hybridization reaction. During hybridization, the labeled probe DNA seeks its complementary target sequences in the sample and binds to them, resulting in a labeled DNA duplex. Since this is the natural state of DNA, it is also the most stable, especially if there is a perfect match between each of the paired nucleotide bases. The temperature, salt conditions, and the concentration of formamide in the reaction (if used) are chosen so that only perfect or nearly perfect matches of the two nucleic acid sequences will stay together during the wash phases that follow the hybridization reaction, during which excess probe is removed from the reaction mixture. These conditions are referred to as the stringency of the reaction. If reaction conditions are chosen such that only exact matches of DNA strands will stay together, the stringency of the reaction is said to be "high". As the salt concentration of the hybridization buffers is raised, or the temperature of the washes after hybridization is lowered, or the formamide concentration is

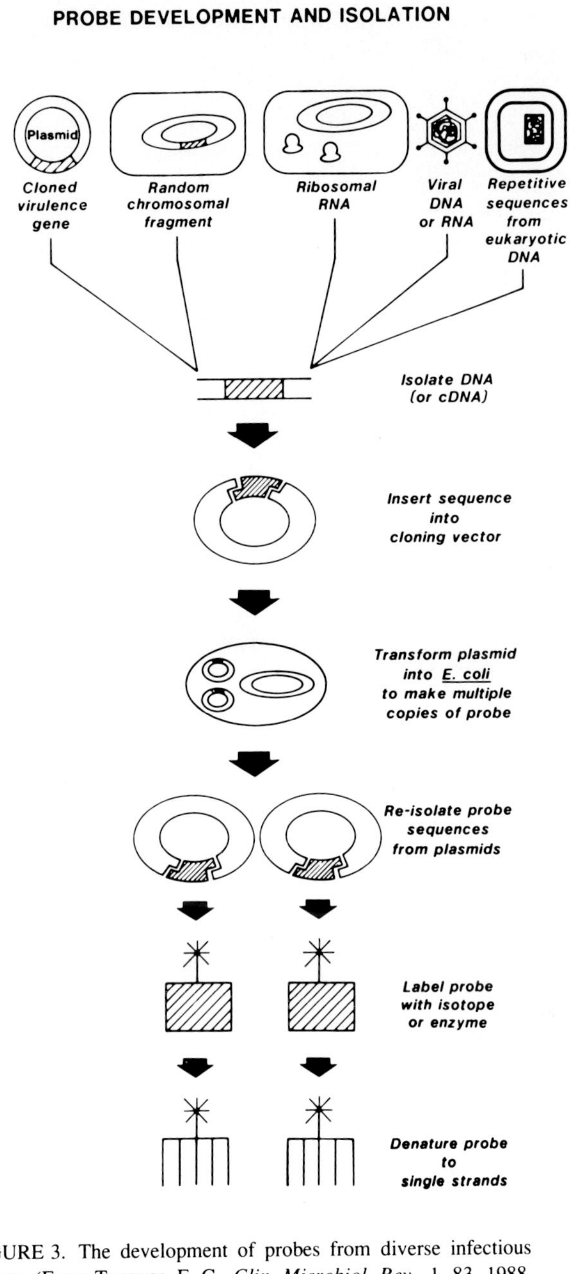

FIGURE 3. The development of probes from diverse infectious agents. (From Tenover, F. C., *Clin. Microbiol. Rev.*, 1, 83, 1988. With permission.)

lowered (from a maximum of 50% to 20 to 30%), more mismatches among base pairs can be tolerated before the DNA hybrid denatures (i.e., comes apart into single strands). This is referred to as "low" stringency.

Moseley and Falkow[7] present an excellent example of how changes in the stringency conditions of a DNA hybridization reaction between two molecules can provide important information. Briefly, if the genes encoding cholera toxin (CT) and the heat labile enterotoxin from *E. coli* (LT) are isolated and allowed to hybridize under stringent conditions, no homology is seen, i.e., the two strands do not remain together after the washing step. If, however, the stringency of the reaction is lowered by reducing the concentration of formamide in the buffer from 50 to 20%, some degree of hybridization takes place, indicating that portions, but not all, of these two DNA molecules are identical.

5. *Effects of Altering Stringency Conditions*

The concept of stringency is important to understand if probes are to be used in the clinical laboratory. By altering the conditions of hybridization, a probe that is highly specific for *Salmonella typhi* at high stringency may well hybridize to the DNA of other species of *Salmonella* at lower stringency, but still not hybridize to DNA from other enteric bacteria. While altering stringency conditions is one way of developing a probe that is genus-specific instead of species-specific, there is an increased risk of detecting homology with other commensal bacteria that may be present in the sample. It is better, in the long run, to identify a sequence of DNA that is present in all *Salmonella* spp. and that will react with the DNA of those species under stringent conditions than to use lower stringency with the risk of reduced specificity. Thus, stringency reactions should be defined and clearly understood by all who will be using DNA probes (or should be defined by the kit manufacturer) and not altered.

B. Principles of Probe Development

The fundamental principle of probe development is that every organism in nature, whether prokaryotic, eukaryotic, or viral, possesses some unique nucleotide sequences that distinguish it from every other organism. These unique sequences may be found in the chromosome of the organism or on plasmids indigenous to that species. They may be part of the coding sequences of a virulence gene, or a stretch of highly repetitive sequences that are interspersed between the genes of a eukaryotic organism. The goal of probe technology is to identify and isolate those unique sequences in such a way that they can be easily reproduced thousands of times without degeneracy. Once isolated (usually by recombinant DNA methods), these unique stretches of nucleotides must then be labeled with a signal generator (a "reporter molecule") that will indicate the binding of the probe to the target nucleic acid. The signal should be easily detectable by scintillation or spectrophotometric techniques if it is to be used in a clinical laboratory. Finally, the labeled probe is incorporated into a diagnostic kit format that can be used by clinical laboratories without an in-depth understanding of recombinant DNA technology. The diagnostic kit, then, consists of a labeled probe, a specimen-processing solution to make the clinical sample amenable to analysis, positive and negative controls, and a means of measuring the signal generated from the probe. Specialized spectrophotometers and scintillation counters have been developed to facilitate use of these kits.

C. The Hybridization Reaction

Hybridization reactions can be carried out with the target DNA either in solution or fixed to a solid support, such as a nitrocellulose or nylon filter. The latter technique is often referred to as "colony hybridization" (for bacteria) or as "spot blots", "dot blots", or "slot blots" for various other types of specimens. Probes can also be attached to plastic surfaces such as microtiter plates or tubes which facilitate the reading of probe reactions using colorimetric labels. The probe, once mixed with sample, either in solution, on the membrane filter, or in a tube or well, seeks out and binds to its complementary nucleotide sequences among the DNA or RNA found in the specimen. The double-stranded nucleic acid is then separated from the rest of the labeled probe DNA and the sample. When the hybridization reaction is carried out in solution, separation of the labeled duplex DNA is accomplished by the addition of hydroxyapatite, which preferentially binds double-stranded nucleic acid (DNA:DNA or DNA:RNA), making the insoluble complexes easy to isolate by centrifugation. Alternately, the probe can be bound to metal spheres and the nascent duplexes removed magnetically from solution at the completion of the reaction. If the reaction is carried out on a solid phase, such as a nitrocellulose filter, the excess probe is washed away. The amount of probe bound to the sample is recorded, and the degree of binding

Table 1
METHODS OF CONSTRUCTING DNA PROBES TO INFECTIOUS AGENTS

Type	Example	Ref.
Cloned virulence determinant	ETEC	20, 21
Random cloned fragments	*Salmonella*	23, 24
Whole-cell chromosomal DNA	*Campylobacter*	25
Ribosomal RNA	*Legionella*	29
Viral genome	Enteroviruses	31
Repetitive eukaryotic DNA	*Brugia* spp.	34, 35
Oligonucleotide probes	Herpesvirus	42

is compared to that found in positive and negative controls to determine whether the sample contained the target sequences in question, and thus, the infectious agent in question.

In the sections that follow, each of the various methods of probe development will be reviewed. Several examples of probe construction are presented in Table 1.

II. PROBES TO SPECIFIC VIRULENCE DETERMINANTS

The first approach to developing a probe to a pathogenic microorganism is to select a key virulence determinant that is unique to that microorganism and that can be cloned using recombinant DNA techniques. Once the gene is isolated from the parent organism and inserted in a cloning vector, a portion of the nucleotide sequence of that cloned gene can be selected as the probe. The sequence chosen is normally one bounded by distinct restriction endonuclease sites that facilitate the isolation of that fragment of DNA away from the remaining DNA of the chimeric plasmid. The probe sequence is labeled and used to determine the presence of that virulence gene among the DNA from organisms present in a patient's specimen by hybridization. This approach has several advantages. First, it obviates the need to isolate and grow the pathogenic microorganism in pure culture, which may be time consuming and technically difficult. Second, organisms that have lost the virulence determinant and are no longer pathogenic, and therefore not clinically relevant, will not be detected (e.g., an enterotoxigenic *E. coli* that has lost a plasmid-encoded toxin gene[8]). Third, if the virulence gene responsible for disease spreads to another microorganism by a genetic event such as conjugation, transduction, or transposition, that novel pathogen will still be recognized by the probe. The major drawback to this approach is that the determinant in question must be cloned and characterized before the probe can be constructed. This can represent a significant amount of time and effort even before a potential diagnostic tool is field tested.

The characterization of a cloned gene consists of (1) locating a series of restriction endonuclease sites within the cloned fragment, (2) identifying the beginning and the end of the coding region of the virulence gene located on the cloned fragment, and (3) identifying a restriction fragment within the cloned gene that can be easily isolated from the rest of the cloning vector.

It is vitally important that the probe be selected from within the open reading frame of the gene and not contain any flanking sequences. In other words, the piece of DNA that will be used to search out the virulence determinant must come from the sequences between the start and stop codons of the gene and not from the DNA on either side of the gene. Flanking sequences may be present by chance in many microorganisms, thus compromising the specificity of the probe. Several investigators have noted problems with false positive results when the entire plasmid, cloning vector plus cloned DNA, was labeled and used as a probe.[9,10] In these instances, bacteria present in the clinical specimen undergoing probe analysis cross-reacted with the cloning vector DNA.

The probes developed to the enterotoxins of *E. coli* are good examples of virulence gene-specific probes. There are five types of enterotoxins that can be found in enterotoxigenic *E. coli*. Two are cholera-like toxins that are heat labile and are thus referred to as heat labile toxins or LTs,[11-13] and three are heat stable, referred to as STaH or ST_{Ib}, which is primarily seen in human infections; STaP or ST_{Ia}, which is the porcine form; and STb which is also an animal-specific toxin.[14-16] The identification of enterotoxigenic *E. coli*, or ETEC, from the stools of patients with diarrhea has always been problematic for clinical microbiologists. Modified Elek tests,[17] cell culture techniques using either Y-1 adrenal cells or Chinese hamster ovary cells,[18] ELISA tests,[19] and suckling mouse assays,[20] all have been employed to distinguish ETEC from their otherwise phenotypically identical cousins in fecal flora. Most, however, were found lacking in either sensitivity or specificity, or were too costly for routine laboratory use.

The detection of ETEC in stool was revolutionized by the introduction of DNA probe tests for these organisms by Moseley et al.[21,22] In this technique, isolated *E. coli* colonies or stool samples from patients were spotted on nitrocellulose filters resting on MacConkey agar plates and grown overnight. The following day, the filters were removed from the agar plates, and the bacterial growth lysed by treatment with alkali. The alkali treatment released the DNA present in the cell, simultaneously denaturing it to single strands. After fixing the DNA released from the cells onto the filters with heat, the DNA was hybridized with radiolabeled probes to LT and STaH. The technique proved to be both highly sensitive and specific when compared to the Chinese hamster overy cell assay for LT and the suckling mouse assay for STs and has proven of value in field studies.[23] It also provides a vital example of the effectiveness of probes for specific virulence determinants.

III. PROBES DEVELOPED FROM RANDOMLY CLONED CHROMOSOMAL FRAGMENTS

The second way of developing a DNA probe for a microorganism is to isolate and purify its chromosomal DNA, cleave the DNA randomly into small fragments with restriction endonucleases, and insert these restriction fragments into a cloning vector. This produces a large number of different cloned fragments, some of which will contain nucleotide sequences unique to the target organism. Each cloned fragment must then be isolated from the vector and tested for its ability to hybridize only to the DNA of the chosen target. It is important to realize that it is not necessary to know the function encoded by the nucleic acid probe. In fact, many probes generated in this fashion may represent only portions of genes, if they code for any protein at all.

This approach to probe development was taken by Fitts in the production of probes for detection of *Salmonella* spp. in food samples.[24] To begin the process of probe development, chromosomal DNA was purified from a strain of *Salmonella typhimurium,* an organism commonly isolated from both food and humans with diarrhea. The DNA was then cleaved with the restriction endonuclease *Bam*HI, and the fragments ligated into the single *Bam*HI site in plasmid YEp13. The chimeric plasmids were transformed into *E. coli.* Two hundred clones containing inserts of *S. typhimurium* DNA were picked and then probed with labeled *E. coli* DNA to determine whether any of these clones contained DNA that did not hybridize to *E. coli*. Fifty-four clones showed no homology to *E. coli* DNA. The DNA inserts from the plasmids contained within these 54 clones were isolated, labeled with ^{32}P, and hybridized to the DNA of 23 isolates of *Salmonella* spp. and 32 non-*Salmonella* spp. These inserts hybridized only to *Salmonella* DNA. The same process was repeated with over 330 isolates of *Salmonella* representing a broad spectrum of serotypes. Among the probes tested, only two reacted with all species of *Salmonella*. These two cloned fragments were then tested and found to be sensitive and specific indicators of the presence of *Salmonella* spp. in food.[25]

IV. LABELED CHROMOSOMAL DNA AS A DIAGNOSTIC PROBE

Some microorganisms have such a unique configuration of the common nucleotide bases (adenine, guanine, cytosine, thymine) in their DNA complement that simple extracts of chromosomal DNA labeled with a radioisotope or biotin, can serve as highly specific diagnostic probes. This has been particularly true of organisms in the genus *Campylobacter* and those of the genus *Mobiluncus*. For instance, Totten and co-workers used labeled whole-cell chromosomal DNA to differentiate among several species of *Campylobacter* as well as several *Campylobacter*-like organisms (CLOs).[26] This technique, referred to as the taxonomic spot blot, demonstrated that four different groups of CLOs could be distinguished among organisms isolated from homosexual males suffering from symptoms of proctitis and proctocolitis. Similarly, *Mobiluncus mulieris*, a pathogen of the female genital tract, can be accurately identified in vaginal specimens by hybridizing labeled chromosomal DNA to vaginal fluid spotted and hybridized on nitrocellulose filters.[27,28] This technique has proven to be a valuable asset to the diagnostic laboratory interested in the pathogenesis of bacterial vaginosis.

While labeling of whole-cell chromosomal DNA has proved effective, it has two drawbacks. First, the technique requires the isolation and purification of chromosomal DNA from the target organisms, which can be a tedious and time-consuming task, especially if the organisms are fastidious and grow slowly, as with the two examples above. Second, the reproducibility of the test may vary with the batch of DNA. Although no problems were noted in either of the above examples, different batches of DNA may give altered sensitivity or specificity, depending on the purity of the chromosomal preparation. The use of cloned chromosomal fragments would greatly facilitate the preparation of probe and would likely aid in reproducibility of results should this technique be applied to other bacterial pathogens.

V. PROBES THAT RECOGNIZE RIBOSOMAL RNA

There is concern that probes chosen to bind to segments of the bacterial chromosome may have limited sensitivity since each bacterial cell may contain only two or three copies of that target sequence. When considering probes to chromosomal determinants, there is also some concern that such probes may only be species specific. To avoid these possibilities, some probe developers have turned to the bacterial ribosome as a source of potential probes. The genes encoding ribosomal RNA are the most highly conserved genetic material in the bacterial kingdom. Ribosomal RNA, for its part, is often present in upwards of 10,000 copies per cell.[29] This makes it an excellent target for a nucleic acid probe, since its high copy number serves as a sort of in vivo amplification system. This concept has been exploited in the development of diagnostic probes for *Legionella* spp., *Mycoplasma* spp., and several mycobacterial species. Because ribosomal RNA is so highly conserved, it is theoretically easier to make a genus-specific probe rather than one that hybridizes only to a single species. This is due to the fact that those same nucleotide sequences will be present in the ribosomes of all members of that genus.[30] These types of probes are described in detail in Chapter 11.

VI. PROBES TO VIRAL PATHOGENS

Viral pathogens are logical targets for diagnostic probes. As a whole, viruses are both difficult and expensive to cultivate in vitro. The cytopathic effect produced by many viruses in cell culture is often subtle, and its recognition and identification require a trained technologist. Many viruses cannot be cultivated outside the human host, but are recognized through serologic conversion, a technique which often detects the presence of a pathogen too late to be of help to the patient.

Probes produced to RNA viruses are usually made to the DNA complement (cDNA) of the viral RNA strand by using the retroviral enzyme reverse transcriptase (see above). The cDNA is then made double stranded by synthesizing a complementary DNA strand using the cDNA as a template. The nascent duplex DNA is inserted into a plasmid vector, and the probe is chosen from one of the open reading frames of the virus. Examples of probes developed to RNA viruses are reviewed by Van Der Werf et al.[31] and Estes et al.[32] Probes to DNA viruses are often made by cloning fragments of the viral genome directly into a vector.[33] The use of probes developed to viral pathogens is reviewed in several chapters of this volume.

VII. PROBES TO EUKARYOTIC ORGANISMS

Probes to eukaryotic organisms are more difficult to develop than those for bacteria and viruses. This is due, in part, to the complexity of the genetic material. The most promising approach to developing probes to eukaryotic pathogens has been to utilize the repetitive sequences often present between genes of higher organisms.[34] These sequences are often unique to specific genera, and their high copy number and ease of isolation make them desirable as probe targets. The use of repetitive sequences as targets for probes directed against the malarial parasite *Plasmodium falciparum* is described in detail in Chapter 18. Repetitive sequences have also been exploited to develop probes to much more complex organisms, such as helminthic worms. For example, Sim et al. have developed probes to *Brugia malayi*, a filarial worm.[35] To construct a probe, DNA was purified from microfilariae and partially digested with the restriction endonuclease *Sau*3A and size-fractionated on sucrose gradients by centrifugation. Fragments ranging in size from 300 bp to 8 kb were inserted into the *Bam*HI site of the cloning vector pBR322 and ligated. Since *Sau*3A and *Bam*HI leave the same single-stranded sticky ends after digestion, DNA cleaved with these two enzymes can still be ligated together. After ligation, the chimeric plasmids were transformed into *E. coli*. A total of 6000 transformants were selected. These clones represent a *B. malayi* gene bank, where, theoretically, every gene present in that organism is represented at least once in one of those transformed organisms. Total genomic DNA from several strains of *Brugia* was labeled with radioactive ^{32}P and allowed to hybridize to DNA from each of the 6000 clones. Four clones demonstrated strong radioactive signals after hybridization with *B. malayi* DNA, but not with DNA from other *Brugia* spp., suggesting that these chimeric plasmids contained cloned sequences that were unique to *B. malayi*. Sim et al. then prepared restriction endonuclease maps of the plasmids containing the *Brugia* DNA inserts. One chimeric plasmid was found to carry three nearly identical sequences which were thought to represent highly repetitive sequences unique to *B. malayi*. This plasmid was chosen as a probe. Further studies documented the effectiveness of this probe in identifying the presence of *B. malayi* microfilariae directly in the mosquito vector by probing blots of aspirates of mosquitoes prepared on nitrocellulose filters.[36] Thus, probes can be developed even to organisms as complex as microfilarial worms.

VIII. OLIGONUCLEOTIDE PROBES

The preceding descriptions of probes have all focused on what have become known as "long probes", i.e., probes of 100 bp or more. A second approach to probe development has focused on chemically synthesizing short pieces of nucleic acid, often only 14 to 40 bp in length, to be used in the identification of infectious agents. These "short probes" or oligonucleotide probes, although a fraction of the size of those described above, show the same high degree of specificity as those probes thousands of base pairs long. In fact, short probes are capable of recognizing even single base pair changes in a given sequence of DNA,[37] when used under stringent conditions. Of course, it is necessary that the DNA or RNA sequence of the target be known before an oligonucleotide can be synthesized.

The major advantage to using oligonucleotide probes is the fact that short probes hybridize

5 to 15 times faster than long probes, often reducing the time of hybridization to less than 30 min.[38,39] Hybridization reactions using long probes may require 6 to 10 hr to come to completion, especially if volume exclusion reagents such as dextran sulfate[40] or polyethylene glycol[41] are not included. And, as cited above, short probes can also be used to detect point mutations in nucleic acid sequences, thus demonstrating remarkable specificity. These probes, once synthesized, are also stable for years, providing a reliable and reproducible source of reagents amenable to use in the clinical laboratory.

The major disadvantage to short probes is the limited number of reporter molecules that can be carried or attached. Although linker arms can be used to label oligonucleotide probes with several enzyme molecules, such as alkaline phosphatase,[42] long probes are still cited as 10 to 100 times more sensitive than short probes.[38] Nonetheless, oligonucleotide probes have been synthesized and shown to be highly specific for several viral and bacterial pathogens. Oligonucleotide probes, for example, have been used to type isolates of herpes simplex virus,[43] to identify the presence of hepatitis B virus in serum,[44] and have shown utility in the diagnosis of enteric stool pathogens.[42] Short probes will definitely become an important diagnostic tool in clinical laboratories in the near future.

IX. CONCLUSIONS

There are several different approaches one can take to develop a probe for an infectious agent. The approach taken depends primarily on the type of organism (bacterial, viral, or eukaryotic) to be identified. Care always must be taken to prove that the probe is highly specific. This is more of a problem with viral and protozoan probes where intervening sequences may prove to be more widespread than originally thought. The probe, if developed from a virulence gene, must be selected from within the open reading frame of the gene and must not include flanking sequences, i.e., nucleotide sequences beyond the initial and terminal codons. Finally, chemically synthesized short probes offer high specificity and decreased hybridizaton times, but may suffer in terms of sensitivity when compared to more traditional long probes.

ACKNOWLEDGMENTS

I wish to thank Steve Moseley and James Kaper for their valuable contributions to this manuscript.

REFERENCES

1. **Engleberg, N. C. and Eisenstein, B. I.,** The impact of new cloning techniques on the diagnosis and treatment of infectious diseases, *N. Engl. J. Med.*, 311, 892, 1984.
2. **Bolivar, F., Rodriguez, R. L., Green, P. J., Betlach, M. C., Heyneker, H. L., Boyer, H. W., Crosa, J. H., and Falkow, S.,** Construction and characterization of new cloning vehicles. II. A multipurpose cloning system, *Gene*, 2, 95, 1977.
3. **Norrander, J., Kempe, T., and Messing, J.,** A new pair of M13 vectors for selecting either strand of a double-digest restriction fragment, *Gene*, 26, 101, 1983.
4. **Arrand, J. E.,** Preparation of nucleic acid probes, in *Nucleic Acid Hybridization, A Practical Approach*, Hames, B. D. and Higgins, S. J., Eds., IRL Press, Oxford, 1985, chap. 2.
5. **Maniatis, T., Fritsch, E., and Sambrook, J.,** *Molecular Cloning: A Laboratory Manual*, Cold Spring Harbor Laboratory Press, Cold Spring Harbor, N.Y., 1982.
6. **Perbal, B.,** *A Practical Guide to Molecular Cloning*, John Wiley & Sons, New York, 1984.
7. **Moseley, S. L. and Falkow, S.,** Nucleotide sequence homology between the heat-labile enterotoxin gene of *Escherichia coli* and *Vibrio cholerae* deoxyribonucleic acid, *J. Bacteriol.*, 144, 444, 1980.

8. **Yamamoto, T., Honda, T., Miwatani, T., and Yokota, T.,** A virulence plasmid in *Escherichia coli* enterotoxigenic for humans: intergenic transfer and expression, *J. Infect. Dis.*, 150, 688, 1984.
9. **Diegutis, P. S., Keirnan, E., Burnett, L., Nightingale, B. N., and Cossart, Y. E.,** False-positive results with Hepatitis B virus DNA dot-hybridization in Hepatitis B surface antigen-negative specimens, *J. Clin. Microbiol.*, 23, 797, 1986.
10. **Ambinder, R. F., Charache, P., Staal, S., Wright, P., Forman, M., Hayward, S. D., and Hayward, G. S.,** The vector homology problem in diagnostic nucleic acid hybridization of clinical specimens, *J. Clin. Microbiol.*, 24, 16, 1986.
11. **So, M., Dallas, W. S., and Falkow, S.,** Characterization of an *Escherichia coli* plasmid encoding for synthesis of heat-labile toxin; molecular cloning of the toxin determinant, *Infect. Immun.*, 21, 405, 1978.
12. **Moseley, S. L. and Falkow, S.,** Nucleotide sequence homology between heat-labile enterotoxin gene of *Escherichia coli* and *Vibrio cholerae* deoxyribonucleic acid, *J. Bacteriol.*, 144, 444, 1980.
13. **Pickett, C. L., Twiddy, E. M., Belisle, B. W., and Holmes, R. K.,** Cloning of genes that encode a new heat-labile enterotoxin of *Escherichia coli, J. Bacteriol.*, 165, 348, 1986.
14. **Moseley, S. L., Hardy, J. W., Huq, I. M., Echeverria, P., and Falkow, S.,** Isolation and nucleotide sequence determination of a gene encoding a heat-stable enterotoxin of *Escherichia coli, Infect. Immun.*, 39, 1167, 1983.
15. **Moseley, S. L., Samadpour-Motalebi, M., and Falkow, S.,** Plasmid association and nucleotide sequence relationships of two genes encoding heat-stable enterotoxin production in *Escherichia coli* H-10407, *J. Bacteriol.*, 156, 441, 1983.
16. **Burgess, M. N., Bywater, R. J., Cowley, C. M., Mullan, N. A., and Newsome, P. M.,** Biological evaluation of a methanol-soluble, heat-stable *Escherichia coli* enterotoxin in infant mice, pigs, rabbits, and calves, *Infect. Immun.*, 21, 526, 1978.
17. **Honda, T., Taga, S., Takeda, Y., and Miwatani, T.,** Modified Elek test for detection of heat-labile enterotoxin of enterotoxigenic *Escherichia coli, J. Clin. Microbiol.*, 13, 1, 1981.
18. **Sack, D. A. and Sack, R. B.,** Test for the detection of enterotoxigenic *Escherichia coli* using Y-1 adrenal cells in miniculture, *Infect. Immun.*, 11, 334, 1975.
19. **Thompson, M. R., Brandwein, H., LaBine-Racke, M., and Gianella, R. A.,** Simple and reliable enzyme linked immunosorbent assay with monoclonal antibodies for detection of *Escherichia coli* heat stable enterotoxin, *J. Clin. Microbiol.*, 20, 59, 1984.
20. **Dean, A. G., Ching, Y.-C., Williams, R. C., and Harden, L. B.,** Test for *Escherichia coli* enterotoxin using infant mice: application in a study of diarrhea in children in Honolulu, *J. Infect. Dis.*, 125, 409, 1972.
21. **Moseley, S. L., Huq, I., Alim, A. R. M. A., So, M., Samadpour-Motalebi, M., and Falkow, S.,** Detection of enterotoxigenic *Escherichia coli* by DNA colony hybridization, *J. Infect. Dis.*, 142, 892, 1980.
22. **Moseley, S. L., Echeverria, P., Seriwatana, J., Tirapat, C., Chaicumpa, W., Sakuldaipeara, T., and Falkow, S.,** Identification of enterotoxigenic *Escherichia coli* by colony hyridization using three enterotoxin gene probes, *J. Infect. Dis.*, 145, 863, 1982.
23. **Echeverria, P., Seriwatana, J., Taylor, D. N., Tirapat, C., Chaicumpa, W., and Rowe, B.,** Identification by DNA hybridization of enterotoxigenic *Escherichia coli* in a longitudinal study of villages in Thailand, *J. Infect. Dis.*, 151, 124, 985.
24. **Fitts, R.,** Development of a DNA-DNA hybridization test for the presence of *Salmonella* in foods, *Food Technol.*, 39, 95, 1985.
25. **Fitts, R., Diamond, M., Hamilton, C., and Neri, M.,** DNA-DNA hybridization assay for detection of *Salmonella* spp. in foods, *Appl. Environ. Microbiol.*, 46, 1146, 1983.
26. **Totten, P. A., Fennell, C. L., Tenover, F. C., Wezenberg, J. M., Perine, P. L., Stamm, W. E., and Holmes, K. K.,** *Campylobacter cinaedi* (sp. nov.) and *Campylobacter fennelliae* (sp. nov.): two new *Campylobacter* species associated with enteric disease in homosexual men, *J. Infect. Dis.*, 151, 131, 1985.
27. **Roberts, M. C., Hillier, S. L., Schoenknecht, F. D., and Holmes, K. K.,** Nitrocellulose filter blots for species identification of *Mobiluncus curtisii* and *Mobiluncus mulieris, J. Clin. Microbiol.*, 20, 826, 1984.
28. **Roberts, M. C., Hillier, S. L., Schoenknecht, F. D., and Holmes, K. K.,** Comparison of gram stain, DNA probe, and culture for the identification of species of *Mobiluncus* in female genital specimens, *J. Infect. Dis.*, 152, 74, 1985.
29. **Chambliss, G., Craven, G., Davies, J., Davis, K., Kahan, L., and Nomura, M.,** *Ribosomes, Structure, Function, and Genetics,* University Park Press, Baltimore, 1980.
30. **Kohne, D., Hogan, J., Jonas, V., Dean, E., and Adams, T. H.,** Novel approach for rapid and sensitive detection of microorganisms: DNA probes to rRNA, in *Microbiology-1986,* Schlessinger, D., Ed., American Society for Microbiology, Washington, D.C., 1986, 110.
31. **Van Der Werf, S. V., Bregegere, F., Kopecka, H., Kitamura, N., Rothberg, P. G., Kourilsky, P., Wimmer, E., and Girard, M.,** Molecular cloning of the genome of poliovirus type 1, *Proc. Natl. Acad. Sci. U.S.A.*, 78, 5983, 1981.
32. **Estes, M. K., Mason, B. B., Crawford, S., and Cohen, J.,** Cloning and nucleotide sequence of the simian rotavirus gene 6 that codes for the major capsid protein, *Nucleic Acids Res.*, 12, 1875, 1984.

33. **Tamashiro, J. C., Hock, L. J., and Spector, D. H.,** Construction of a cloned library of *Eco*RI fragments of the human cytomegalovirus genome (strain AD169), *J. Virol.,* 42, 547, 1982.
34. **Lewin, B.,** Eukaryotic genomes: a continuum of sequences, in *Genes II,* John Wiley & Sons, New York, 1985, chap. 18.
35. **Sim, B. K. L., Piessens, W. F., and Wirth, D. F.,** A DNA probe cloned in *Escherichia coli* for the identification of *Brugia malayi, Mol. Biochem. Parasitol.,* 19, 117, 1986.
36. **Sim, B. K. L., Mak, J. W., Cheong, W. H., Sutanto, I., Kurniawan, L., Marwoto, H. A., Franke, E., Campell, J. R., and Wirth, D. F.,** Identification of *Brugia malayi* in vectors with a species-specific DNA probe, *Am. J. Trop. Med. Hyg.,* 35, 559, 1986.
37. **Wallace, R. B., Shaffer, J., Murphy, R. F., Bonner, J., Hirose, T., and Itakura, K.,** Hybridization of synthetic oligonucleotides to ϕX174 DNA; the effect of single base pair mismatch, *Nucleic Acids Res.,* 6, 3543, 1979.
38. **Bryan, R. N., Ruth, J. L., Smith, R. D., and Le Bon, J. M.,** Diagnosis of clinical samples with synthetic oligonucleotide hybridization probes, in *Microbiology-1986,* Schlessinger, D., Ed., American Society for Microbiology, Washington, D.C., 1986, 116.
39. **Wetmur, J. G.,** Acceleration of DNA renaturation rates, *Biopolymers,* 14, 2517, 1975.
40. **Wahl, G. M., Stern, M., and Stark, G. R.,** Efficient transfer of large DNA fragments from agarose gels to diazobenzyloxymethyl-paper and rapid hybridization by using dextran sulfate, *Proc. Natl. Acad. Sci. U.S.A.,* 76, 3683, 1979.
41. **Amasino, R. M.,** Acceleration of nucleic acid hybridization rate by polyethylene glycol, *Anal. Biochem.,* 152, 304, 1986.
42. **Jablonski, E., Moomaw, E. W., Tullis, R. H., and Ruth, J. L.,** Preparation of oligonucleotide-alkaline phosphatase conjugates and their use as hybridization probes, *Nucleic Acids Res.,* 14, 6115, 1986.
43. **Peterson, E. M., Aarnaes, S. L., Bryan, R. N., Ruth, J. L., and de la Maza, L. M.,** Typing of Herpes Simplex Virus with synthetic DNA probes, *J. Infect. Dis.,* 153, 757, 1986.
44. **Bryan, R. N. and Arnold, L. J., Jr.,** Detection of HBV DNA by hybridization with modified synthetic oligonucleotides containing radioactive and non-radioactive reported groups, *DNA,* 3, 123, 1984.

Chapter 2

DETECTION SYSTEMS FOR HYBRIDIZATION REACTIONS

Edward G. Jablonski

TABLE OF CONTENTS

I. INTRODUCTION

DNA probe hybridization reactions are processes for determining the presence of a specific sequence of DNA in a sample. This DNA probe hybridization assay can be broken down into three intimately related components: a sample preparation procedure, a hybridization format, and a detection system. The detection system consists of a label, the mode of association of the label with the probe, and a method for visualizing or quantitating the label. Simply described, a sample containing the target DNA sequence is denatured and hybridized with its specific probe in one of several formats (denatured on filters or in solution, trapped on beads in a sandwich system, or *in situ).* The label bound to the target DNA, through the hybridized probe, is detected, and the intensity of the signal is proportional to the amount of target present. Inherent in any detection system is its compatibility with the entire DNA probe assay. A detection system that functions well in a particular DNA probe assay may be inappropriate when the hybridization or sample preparation formats are altered.

The simplest measurement of nucleic acid reannealing is the use of the hyperchromic effect. Since this method is relatively insensitive, the detection of a DNA probe hybridization event must ultimately be the detection of a label or reporter group. The first concern in analyzing or developing a detection system is, therefore, the label. Desirable characteristics for a DNA probe label are common to any analytical situation. Obviously, the label must be detectable at a level sufficient for the purpose at hand. The label should be reasonably safe and stable under assay and storage conditions and be capable of attachment to the DNA probe without detrimental effects on the hybridization reaction. Constraints for availability, cost, simplicity, and time of visualization, are also of interest. DNA probe technology demands the determination of extremely small quantities of target DNA on the order of copies or molecules. This consideration alone qualifies relatively few materials as labels for DNA probes. These include: radiolabels, most commonly ^{32}P; enzymes with high turnovers and highly detectable products, such as horseradish peroxidase and alkaline phosphatase; and high quantum yield fluorophores, such as fluorescein and Texas Red and, recently, long-lived lanthanides. The concern over radioactive materials in the laboratory has led to a distinction between radiolabeled and nonradiolabeled detection systems.

The mode of association of the label with the DNA probe has lead to a further classification of detection systems, those of "directly labeled" or "indirectly labeled". These classifications usually apply to nonradioactive labels. Radiolabels are usually direct labels, while enzymes and fluorophores can be either. Directly labeled detection systems have the label covalently attached to the probe prior to hybridization with the target sample. This is accomplished by the enzymatic incorporation of radiolabels or by chemical cross-linking procedures for nonradiolabels. In these systems, the cross-linking process must not inactivate the label, must be stable under hybridization conditions, and must not interfere with probe specificity or kinetics of hybridization. The label is visualized after hybridization, usually after several washing steps to remove unhybridized probe. Indirect systems also involve the use of covalently modified probes. The agent, however, is a relatively small compound capable of binding a secondary complex which incorporates enzyme or fluorescent groups. Examples are avidin/biotin or antibody/antigen systems, where biotin or an antigen is the primary reporter group. Indirect systems, although adding steps to the process, allow the use of materials that would otherwise lose activity as direct labels under hybridization conditions or interfere in the hybridization reaction. In addition, the use of secondary complexes attempts to amplify signal by using multiple copies of the label in the complex.

The method of visualization or quantitation of the label or reporter group is dictated by the nature of the signal, the hybridization system, and the degree of instrumental intervention desired. Radiolabels are usually visualized by autoradiography of hybridized target DNA. Scintillation counting is used as a more quantitative method with high target levels. Fluorescent labels require instrumentation ranging from standard fluorometers to sophisticated time-resolved photon-counting equipment. Enzyme labels are capable of catalyzing the formation of a variety of products with unique properties for quantitation, such as colored precipitates, soluble fluorescent compounds, and luminescence. Enzyme labels are, in a sense, primary labels, and the product of catalysis is the detectable compound. Instrumentation adds more elements to a detection system, reducing simplicity, but lending enhanced sensitivity, reproducibility, and quantitation. Instrumentation will be a necessary component of any automated system.

This chapter attempts to outline the various detection systems that have been devised for DNA probe hybridization reactions. Examples and advantages of each are discussed, and, where applicable, the detection levels reported by investigators for their particular systems are summarized.

II. RADIOLABELED SYSTEMS

A. Labels

Radioisotopes are the earliest and most commonly used labels for the detection of DNA probe hybridization reactions. These unstable atoms decay with the emission of a highly energetic and, therefore, unique signal, easily detected with great sensitivity. The replacement of stable atoms with their radioactive counterpart does not affect the chemical characteristics of the probe. The following is a list of the more common radiolabels and their decay characteristics.

Isotope	Emission type	Energy	T1/2	dpm/mol (100% isotopic enrichment)
^{32}P	β	1.71 MeV	14.3d	2.02×10^{19}
^{125}I	γ	0.035 MeV	60d	3.94×10^{18}
^{3}H	β	0.0181 MeV	12.3y	6.39×10^{16}
^{35}S	β	0.167 MeV	87.4d	3.33×10^{18}

Labeling with ^{32}P is used most often for the detection of filter,[1-4] solution/sandwich,[5] and colony-based[6] hybridizations. Also, it is the most favored isotope for labeling oligonucleotides.[7-9] The most energetic of the above group, ^{32}P, allows reasonable detection times by either autoradiography or scintillation counting techniques. Tritium labels have also been used for filter hybridization detection. The specific activities of probes labeled with ^{3}H are lower than those with ^{32}P, which necessitates longer exposure for visualization. However, resolution with ^{3}H-labeled probes is much improved due to the shallower penetration of the radiation into the photographic emulsion. This characteristic has been exploited for detecting *in situ* hybridizations.[10,11] Due to its higher energy radioactivity, ^{125}I has become widely used as a label for *in situ* hybridization where extreme sensitivity is required. It, however, results in a less precise autoradiograph[12] and gives higher backgrounds.[13] Sandwich-type hybridization techniques have likewise employed ^{125}I labels.[14,15] Bases labeled with these isotopes are commercially available.

B. Labeling Reactions

Radiolabels have been incorporated into nucleic acids by in vivo synthesis in the presence of a radioactive substrate such as $^{32}P_i$.[16,17] The material of interest is isolated by classical separation techniques. The quantity of probe material obtained is low, and contaminating sequences may cause hybridization backgrounds to a degree where the interpretation of results is difficult. Radiolabeled probes with a high degree of specificity are typically produced in vitro from cloned probes. One or more labeled deoxyribonucleotides are used in the synthesis of a DNA strand catalyzed by a polymerase reaction. In addition, DNA, RNA, and short synthetic oligonucleotides can be labeled on the 5′ end with [γ–^{32}P] ATP and polynucleotide kinase or on the 3′ end with [α–^{32}P] cordycepin triphosphate and terminal transferase.

1. Nick Translation

Incorporation of labeled bases into long probes by nick translation involves nicking one of the two strands of double-stranded DNA with DNase I and then excising stretches of the single strands at the nicks with the 5′-3′ exonuclease activity of *Escherichia coli* DNA polymerase I.[18,19] In the process, the polymerase enzyme also adds nucleotides to the 3′ OH. The native bases are replaced with labeled bases in the reaction mixture as the nick moves along the DNA strand. When ^{32}P-labeled bases are incorporated, specific activities of 5×10^8 dpm/μg can be obtained with labeled strands 400 to 800 nucleotides in length. Probes synthesized with ^{3}H-labeled bases have specific activities of 0.5×10^8 to 1.5×10^8 dpm/μg. Labeling by nick translation is a fairly simple reaction resulting in uniformly labeled probes with high specific activities.

2. T4 DNA Polymerase

An alternative labeling technique, devised by O'Farrell, utilizes T4 DNA polymerase.[20,21] The enzyme functions as a 3′-5′ exonuclease in the absence of dNTP, with an activity 200 times greater than that of DNA polymerase I. The undigested strand then serves as a template and primer for the regeneration of an intact double-stranded DNA upon the addition of labeled nucleotides. This process yields intact DNA with no nicks and avoids hairpin structures commonly produced by nick translation. Strand-specific probes can be obtained by cleavage with restriction endonucleases. Although the strand is not uniformly labeled, high specific activities of up to 10^9 dpm/μg with a ^{32}P label can be obtained. DNA labeled to a very high specific activity, however, is unstable due to cleavage from radioactive decay.

3. T4 Polynucleotide Kinase

A method for radiolabeling small DNA fragments or synthetic oligonucleotides, as well as RNA, employs T4 polynucleotide kinase.[22,23] In the exchange reaction, the enzyme catalyzes the removal of an existing 5′ phosphate and adds a γ-phosphate of ATP. The reaction is carried out with excess ATP and ADP. Specific activities of 5×10^5 dpm/pmol ($\sim 8 \times 10^7$ dpm/μg for a 20-mer) of ^{32}P-labeled probes can be produced. Since only the 5′ end is labeled, the probe is well defined. The final product depends solely on the activity of the labeled ATP and, unlike polymerization reactions, more label cannot be added by increasing the enzyme activity in the reaction. This labeling method has been used to demonstrate the ability of synthetic oligonucleotides to detect single base pair mismatches.[24-26]

4. Other Methods

Detailed descriptions of the above labeling procedures as well as end-labeling with terminal deoxynucleotidyl transferase and the Klenow fragment of *E. coli* DNA polymerase I have been made by Arrand[27] and Maniatis et al.[28] The use of M13 universal probe primer and the SP6 polymerase system is also described.

C. Detection Systems

1. Autoradiography

The extent of hybridization with radiolabeled probes on filters and *in situ* is monitored with autoradiographic methods.[29] The procedure is sensitive with good resolution and does not destroy the hybridization support matrix. It lacks the exact quantitative ability of liquid scintillation methods, which in most cases is unnecessary, but it can reveal artifacts not seen in counting.

The hybridized support matrix (filter) is exposed to X-ray film (Kodak XAR5 or X-OMAT AR) in a light-tight cassette. Exposures range from hours to days, depending on the level of radioactivity present. In general, an area of a dot 3 mm across with about 100 dpm of ^{32}P will give a good image in 24 h at room temperature. The sensitivity of the method can be increased with the use of intensifying screens (Du Point Cronex Quanta III, Cronex Lightning Plus, or Fugi Mach II), by eight- to tenfold with two screens. The film is placed between the hybridized filter and screens and is exposed at –70°C in order to prolong the fluorescence emitted in response to a decay event. As little as 5 dpm/7 mm^2 area can be visualized overnight. Given a specific activity of 5×10^8 dpm/µg for a long nick-translated probe of 2 kb (mol wt = 6.5×10^5, containing a total of 40 ^{32}P labels), probe bound to a filter with an area of 7 mm^2 can be visualized at a level of about 10^4 copies. A synthetic oligonucleotide can be detected at 10^5 to 10^6 copies with an equivalent exposure. These levels are given for a general comparison. Absolute detection limits are a function of probe complexity, labeling efficiency, specific activity of the label, and area and time of exposure. The detection of the target itself is also a function of the extent of capture by the support matrix and hybridization efficiency. In terms of weight, the detection sensitivities of target DNA can range from 100 pg for genomic DNA to 0.1 pg for a small plasmid target using ^{32}P-labeled nick-translated probes. *In situ* hybridization with autoradiographic detection of ^{3}H-labeled probes and light microscopy can visualize single copy genes because of the extremely small area of the target and long exposure times. *In situ* detection of ^{3}H-labeled probe involves the application of an emulsion (Kodak NTB-2) directly to the slide. A detailed procedure is given by Pardue.[30] Sandwich hybridization systems using ^{125}I labels have reported detection limits in the 10^5 to 10^6 copy range.[14,15]

For autoradiographs, visual quantitation by comparison to a series of known standards is usually sufficient. More accurate measurements can be gained by densitometry. The amount of probe on the support should be relatively low and the autoradiograph should not be overexposed in order to ensure a linear relationship between exposure area density and probe concentration.

2. Scintillation Counting

If the hybridized support is sufficiently radioactive, the area of interest can be cut out and counted in a liquid scintillation counter. The detection limits for counting are a function of the counting efficiency of the instrument and may be assessed with known standards. Immersion in scintillation fluid prevents rehybridization of the target.

D. Advantages/Disadvantages

1. Advantages

a. The availability and wide use of isotopically labeled DNA components are the greatest advantage for using radiolabels in DNA hybridization detection systems.
b. The labeling reactions and hybridization systems are well detailed and the labels do not affect the hybridization process at moderate densities.
c. The signal, due to its unique nature, is detected directly with adequate sensitivity by relatively simple procedures. Background interferences can arise only from the probe itself and not from the environment of the assay system, a decided advantage in the development and testing of a new probe.

d. Autoradiographs provide a permanent and detailed record of probe location.
e. The commonality of the detection system allows for comparisons or evaluations among investigators.

2. Disadvantages

a. Radioisotopes are hazardous and, therefore, undesirable. The instability of the label requires a continuous supply of raw material and constant resynthesis, with a substantial financial effect from supply, waste disposal, and manpower considerations.
b. Probes labeled with ^{32}P maintain sufficient activity for about 1 week. Heavily labeled probes exhibit extensive degradation from radiolysis.
c. Labeling procedures and purifications, though characterized, are somewhat involved and variations in day-to-day preparations are unavoidable.
d. The most sensitive detections require very low temperatures and long exposure times.

Although radiolabels have pioneered the use of DNA probes for analytical purposes, the acceptance of hybridization technology for routine clinical applications will require a replacement. Radiolabels will continue, however, to be the label of choice for the initial stages of probe development in the research laboratory.

III. NONRADIOLABELED INDIRECT SYSTEMS

A. Labels

The problems and hazards associated with radiolabels have led to the development of sensitive detection systems which are nonradioactive. The first of these systems, and at present the most popular, are indirect systems. These tests rely on the detection of small primary labels that have been incorporated into DNA probes. These small labels, sometimes called reporter groups, do not interfere in the hybridization process of the probe and can survive the harsh conditions of filter-based hybridization systems. The derivatized probes are visualized after hybridization by specifically constructed detector complexes. The detector complexes are capable of binding to the incorporated reporter groups with high affinity. The most sensitive of these techniques use complexes of avidin- and biotin-labeled enzymes which recognize biotin labels on both long cloned probes and short synthetic oligomers. Other systems require the detection of hapten-modified probes with fluorescent or enzyme conjugated antibodies.

The first indirect enzymatic detection system to be successfully demonstrated consisted of a macromolecular complex of avidin and biotinylated horseradish peroxidase, termed ABC peroxidase. This complex binds tightly to hybridized biotinylated probes. Upon treatment with the appropriate substrate, the complex is visualized by the precipitation of a colored product of enzymatic catalysis.[31,32] Horseradish peroxidase systems are available from Enzo Biochemicals in New York. Another sensitive detection system is based upon chemically cross-linked, biotinylated alkaline phosphatase bound to avidin, called Apoly (BAP).[31-33] The phosphatase complex has been shown to be capable of nonisotopically detecting single copy genes in Southern blots of human genomic DNA. Closely related phosphatase detector complexes are commercially available.

B. Labeling Reactions

1. Biotin

Ward and his co-workers developed methods of labeling cloned probes with biotin.[34] Briefly, the Ward method involves the enzymatic incorporation of a biotinylated nucleotide into a purified DNA probe by nick translation. The modified nucleotides are covalent moieties in which the biotin group is attached by a 4- to 16-atom "linker arm" to the C-5 position of pyrimidines. The most common modified nucleotide in use is deoxyuridine triphosphate in which the C-5 position is covalently coupled to the carboxyl group of biotin via an 11-atom linker arm. Incorporation into DNA is at the level of 50 biotin labels per kilobase. Other investigators have reported on novel methods of labeling long DNA probes by chemical methods. Forster et al.[35] have synthesized a photoactivatable analog of biotin for the preparation of large amounts of stable biotin-labeled DNA or RNA probes. Single-stranded M13 probes are chemically labeled with one biotin per 100 to 400 residues. Viscidi et al.[36] prepare biotin-labeled nucleic acids via transamination of unpaired cytosine residues with sodium bisulfite and ethylenediamine. The primary amino groups are then derivatized with biotinyl-ε-aminocaproic acid-*N*-hydroxysuccinimide ester. Another group utilizes biotin covalently linked to a basic macromolecule such as histone H1 or polyethyleneimine, which in turn is cross-linked to DNA via glutaraldehyde or suberate.[37]

Polymerase methods of label incorporation are generally not applicable to short synthetic probes, although Marasugi and Wallace[38] have used enzymatic end-labeling techniques to biotinylate oligonucleotides. Modification of oligonucleotides is generally accomplished with chemical end-labeling methods. Chu et al.[39,40] react the terminal 5′ phosphate of an oligonucleotide with a water-soluble carbodiimide in imidazole buffer to obtain a 5′ phosphorimidazolide, which is capable of reacting with amine-containing molecules in aqueous solution. Chollet and Kawashima[41] have employed very similar chemistries. The method devised at Molecular Biosystems, Inc. for labeling synthetic oligonucleotides uses a nucleoside 3′ phosphoramidite modified with a protected linker arm in the chemical synthesis of a defined sequence.[42-44] The modified nucleoside can be inserted at defined positions in the sequence. Once the oligomer has been synthesized and purified, the blocking groups are removed from the linker arm, leaving a chemically reactive group (e.g., primary amine) on the terminus. The linker arm portion of the oligonucleotide can then be reacted with a biotin derivative or any other label under mild conditions. Short probes labeled with biotin in this fashion do not have altered hybridization characteristics when compared with ^{32}P-kinased probes.

2. Haptens

Hybridization probes are also labeled with hapten groups for immunodetection. Vincent et al.[45] synthesize a dinitrophenol derivative of ATP, which can be used as a substrate for either deoxynucleotidyl terminal transferase for end-labeling or DNA polymerase I for incorporation by nick translation. The incorporated DNA groups are recognized by specific antibodies after hybridization and finally detected by anti-antibodies coupled to peroxidase. Probes are also labeled by in vitro chemical modification of guanine residues with *N*-acetoxy-*N*-2-acetylaminofluorene and its 7-iodo derivative. The resulting fluorene group is recognized by a specific antibody.[46] Syvanen et al.[47] use this chemical technique to demonstrate a fluorescence-based detection system. The second antibody is covalently derivatized with a europium label and quantitated with time-resolved fluorometry.

C. Detection Systems

1. Enzymes

Avidin/biotin- and hapten/antibody-based indirect detection systems depend upon the addition of secondary complexes to visualize the primary label or reporter group. The most common complexes contain enzymes where the catalytic turnover of a substrate results in a product capable of visualization. These complexes are constructed using standard protein modification techniques and are usually commercially available. The size of the complexes and the number of labels per complex are highly variable. The enzyme labels of choice are phosphatase, peroxidase, and β-D-galactosidase, the same labels favored for enzyme-immunoassays. The inherent detectability of an enzyme label is a function of its turnover, or the rate at which product molecules are formed, and the threshold detection limits of the product.

Alkaline phosphatase from calf intestine (mol wt = 140,000) has a specific activity of about 2500 U/mg at 37°C and high pH with the substrate *p*-nitrophenyl phosphate. This corresponds to a turnover number of 3.5×10^5 mol of product per minute per mole of enzyme. Under optimal conditions, horseradish peroxidase (mol wt = 40,000) has a specific activity of about 1000 U/mg, using 2,2′ azino-di-[3-ethyl benzthiazoline-6 sulfonic acid] (ABTS) and hydrogen peroxide, corresponding to 4×10^4 mol/min/mol enzyme. β-D-galactosidase (mol wt = 595,000) has a specific activity of 600 U/mg and a turnover of 3.6×10^5 mol *O*-nitrophenyl-β-D-galactopyranoside per minute per mole of enzyme. β-D-galactosidase is the most detectable enzyme, assuming that the products of catalysis have similar extinction coefficients. The *p*-nitrophenylate anion has an extinction coefficient of 18.2 $mM^{-1}cm^{-1}$, while the product of ABTS conversion is about twice that.

The format of most hybridization systems and the need for extreme sensitivity precludes the use of these simple soluble colorimetric assays to detect enzyme labels. For filter-based and *in situ* hybridization systems, the mechanism of visualizaton is by enzymatic formation of an insoluble dye which deposits around the site of catalysis. These dye deposition assays are better suited for enzyme complexes bound to hybridized probe because the signal (dye) is confined to a very small surface area giving much greater sensitivity over the same amount of dye dissipated in solution. The method is instrumentation independent, but can be semiquantitative if appropriate standards are used for comparison. The obvious drawback in these substrate systems is the loss of direct accuracy. It is difficult, if not impossible, to know the specific activity of an enzyme with dye deposition substrates. In addition, modified, insolubilized enzymes may not turn over at maximum rate due to steric hinderance or diffusional considerations. The actual deposition of the dye depends on crystal size needed for insolubilization. The continuous formation of a precipitate contributes to the complications of steric hinderance and the diffusion of substrates to active sites. The nature of the hybridization support also affects the deposition of product molecules. The detectability of an enzyme label in an indirect DNA probe detection system depends upon the ability of the complex to penetrate to the site of the hybridized reporter group, the turnover of the enzyme when it gets there, and the physical characteristics of the deposited product.

For dye deposition assays, alkaline phosphatase can be visualized with 5-bromo-4-chloro-3-indoyl phosphate in the presence of a tetrazolium salt.[48] A corresponding histochemical substrate for galactosidase is also available.[49] The removal of phosphate generates an indoyl, which condenses to indigo (blue). The reducing equivalent generated in the process causes the conversion of tetrazolium to an insoluble diformazan. If nitroblue tetrazolium is used, the result is a deep purple deposit. Horseradish peroxidase catalyzes the formation of insoluble colored material directly from a number of substrates including 3,3 diaminobenzidine, 3-amino-9-ethylcarbazol, 4-chloro-1-naphthol, *o*-dianisidine, and *p*-phenylenediamine/pyrocatechol. All but the last of these peroxidase substrates are considered carcinogenic.

The application of fluorogenic substrates to detect enzyme labels offers extreme sensitivity and makes it possible to quantitate target DNA. Fluorogenic substrates are converted to highly fluorescent, soluble products and are the most sensitive assay system known for some enzymes. All three enzyme labels previously mentioned catalyze the formation of fluorescent products. The substrates, 4-methylumbelliferyl phosphate, the β-D-galactoside analog, and *p*-hydroxyphenylacetic acid, are all commercially available. Using β-D-galactosidase antibody conjugates, enzyme assays much more sensitive than radioimmunoassays have been developed.[50] The detectability of an enzyme using fluorogenic substrates is a function of the difference in relative quantum yield between the substrate and product. For instance, 4-methylumbelliferone is about 2000 times more fluorescent than the phosphorylated substrate. Alkaline phosphatase, with a K_m in the micromolar range and at specific activity of 1000 U/mg (for 4-methylumbelliferyl phosphate), can be detected at a level of 10^5 copies in 10 min in a volume of 10 μl at twice the signal over background. The substrate for galactosidase is about 50-fold less fluorescent, giving the product that much greater detectability over background.

2. Avidin-Based Detector Complexes

a. Cloned Probes

In general, alkaline phosphatase-based indirect assays using long biotinylated probes have compared well with radioisotope-based detection on filter hybridizations. The classic paper from Ward's laboratory reports a detection sensitivity of 1 to 10 pg of a plasmid DNA target using an alkaline phosphatase-avidin detector complex, incubated for 1 h, on dot blots.[32] Other groups have matched these sensitivities of 10^5 to 10^6 copies of target DNA with their own biotinylated probes and phosphatase-catalyzed dye deposition.[35,36,51] Two groups in England have significantly improved the detection, reporting levels on the order of 10 to 100 fg of phage DNA target on nitrocellulose filters.[37,52] Single copy genes in 0.25 μg of human genomic DNA on Southern blots have been visualized.[37] Horseradish peroxidase systems also function in detection of biotinylated probes on filters, but not as well as phosphatase. Due to the smaller character of the protein, it may be able to penetrate to the site of hybridization more efficiently and so be better suited for the detection of *in situ* hybridization.

Other investigators utilize the catalytic ability of peroxidase to generate a chemiluminescent signal. After hybridization with a biotin probe, the hybrid is labeled with a horseradish peroxidase/avidin complex. The complex is visualized by the addition of hydrogen peroxide and luminol in the presence of an enhancer and recorded by exposure of the filter to high speed film in a light-proof cassette. The detection limit of target DNA is approximately 10^5 copies.[53]

The use of fluorogenic substrates for DNA probe detection is demonstrated by Nagata and colleagues who use an avidin and biotinylated β-D-galactosidase complex with the substrate, 4-methylumbelliferyl-β-D-galactoside, to detect DNA immobilized in microtiter wells. The fluorescent signal is monitored in a microtiter plate fluorescence reader. Sensitivities are equivalent to those of ^{32}P-labeled probes (picogram levels of λ-DNA) and this method adds an element of instrumental quantitation to an enzymatic procedure.[54]

An indirect, nonenzymatic, nonradiometric method using an avidin complex and a biotinylated probe has been developed for detection of cytochemical hybrids with the scanning electron microscope. Manning et al.[55] detects biotin-labeled rRNA of *Drosophila melanogaster* with avidinylated polymethacrylate spheres hybridized to a polytene chromosome preparation.

b. Synthetic Probes

Indirect detection systems using synthetic oligonucleotides are being explored. An oligonucleotide (23-mer) containing a single biotinylated deoxyuridine residue at the 3′ end has been prepared by Murasugi and Wallace.[38] The hybridized probe can be detected by avidin and biotinylated alkaline phosphatase down to 0.5 fmol of plasmid DNA. Chollet and Kawashima[41] have synthesized an oligonucleotide labeled with a single biotin on the 5′ termini. They report on difference in the strength or selectivity of hybridization in comparison to unmodified probes. Detection level is 2 fmol. A third team reports that their 5′-labeled biotin probe exhibits similar hybridization characteristics and is tenfold less sensitive than a corresponding ^{32}P-labeled oligomer in detecting filter-bound target DNA. Background binding of avidin/enzyme complex limits detection to about 0.4 fmol of target.[40]

3. Antibody Detector Complexes

a. Enzyme Labels

The use of antibody/enzyme conjugates as detector complexes for hapten-labeled DNA probes is demonstrated by Tchen et al.[46] Modified probes are hybridized overnight to λ-DNA on nitrocellulose and detected by a peroxidase- or phosphatase-conjugated second antibody at the 2- to 4-pg level.

b. Fluorescent Labels

In addition to enzyme-based systems, hybrids are indirectly detected with fluorescent-labeled macromolecular complexes.[56,57] Fluorescent molecules are the most detectable of materials with the ability to emit thousands of quanta per second under full excitation. On the other hand, ^{32}P emits a signal at the rate of 0.05 β/day. The energetics of the systems, however, partially compensate for the large difference in emission rates. Fluorescent properties of some labels have been exploited by several investigators for DNA probe hybridization detection. The group at Orion in Finland are pioneering the use of fluorescent labels coupled to antibodies as the detector complex.[47] A hapten-labeled probe is complexed with an antibody after hybridization. This antibody is then recognized by a europium-labeled second antibody. Using time-resolved fluorometry to eliminate background fluorescent signal from the environment, detection limits of 2×10^5 to 2×10^8 copies of plasmid pKTH 1202 (6.0 kb) target can be detected.

A different approach to indirect immunofluorescence detection of hybrids is taken by Rudkin and Stollar.[58] Antibodies raised against poly(rA) · poly(dT) in rabbits are shown to recognize RNA:DNA hybrids with high specificity. These antibodies are detected with rhodamine-labeled antibodies directed against rabbit immunoglobins. The bound conjugates are visualized with a fluorescence microscope. An application of this method has been reported.[59]

c. Electron Microscope Labels

A novel *in situ* hybridization protocol incorporates the use of biotinylated probes to hybridize a sample on electron micrograph grids. Detection is via immunogold staining.[60] The method is not restricted to the low resolution of the light microscope, allowing the location of sequences for fine mapping.

D. Advantages/Disadvantages

1. Advantages

a. Biotin or hapten reporter groups are relatively small and do not adversely affect hybridization. This is true for both long nick-translated probes and short synthetic oligonucleotide probes.
b. Biotin and hapten labels are stable under hybridization and storage conditions. Methods have been developed for their incorporation into both cloned and synthetic probes.
c. Biotin and biotin probe conjugates bind avidin both specifically and with high avidity. The dissociation constant for biotin avidin interaction is 10^{-15} *M*.[61] Such a dissociation constant is essentially irreversible. Antibody/antigen associations are also extremely specific, although binding constants are not as strong.
d. Enzyme- and fluorescent-based detector complexes are stable, nonhazardous, and can be visualized with good sensitivity in a relatively short time (1 to 4 h for enzymes, seconds for fluorophores).
e. Enzyme labels, in general, offer amplification of signal and diversity through a variety of substrates.
f. Indirect systems allow the application of highly detectable labels that would otherwise deteriorate under hybridization conditions.
g. Commercial detector complexes are available.
h. Detection sensitivities can be equivalent to radiolabeled detection systems for cloned probes.

2. Disadvantages

a. The detection process is more complex and time-consuming due to the additional binding, washing, and (with enzymes) substrate addition steps.

b. Macromolecular complexes bind nonspecifically to support matrices, giving increased background signal.
c. Substrates convert to product nonenzymatically over long incubation times; thus, the threshold of detection is set by the enzyme complex.
d. Dye precipitates can bleach with time.

IV. NONRADIOLABELED DIRECT SYSTEMS

Direct detection systems combine the best aspects of both radiolabeled and indirect, nonradiometric techniques. Direct detection systems for hybridizations are based on the covalent modification of probes with labels that can be visualized nonradioactively without the use of a secondary label complex. Both fluorophores and enzymes are used for this application, but relatively little has been done in this area. The major obstacles are the development of modification procedures and hybridization conditions suitable for maintaining the activity of the label without adversely affecting hybridization reactions.

A. Cloned Probes

Renz and Kurz describe a direct detection system based on a covalent attachment of enzyme complexes to long DNA probes.[62] The complex is constructed by coupling alkaline phosphatase or horseradish peroxidase to a short polyethyleneimine core with benzoquinone. The unreacted amines on the complex are coupled to the base arylamines via glutaraldehyde. These conjugates contain protein and DNA in a weight ratio of about 30:1 and are capable of hybridizing to specific DNA target sequences in Southern blots. The detection limits approach ^{32}P-labeled systems at about 5×10^5 copies with a substrate incubation time of a few hours. A method for covalently labeling RNA molecules at the 3′ terminus in high yield with hydrazine derivatives of rhodamine and fluorescein is described by Bauman et al.[63] Using fluorescence microscopy to assess hybridization to DNA sepharose, it was determined that the labels had no effect on binding or duplex formation. The method has been applied to cytochemical hybridization.[64] It was not as sensitive as RNA labeled to high specific activity with ^3H and autoradiographic detection, but competitive with in vivo-labeled RNA.

B. Synthetic Probes

Due to the difficulty of conjugating labels to large DNA molecules without polymerase-catalyzed incorporation of modified bases, most of the work on direct systems is being performed on short synthetic probes. Oligonucleotide probes can be synthesized with functional groups at specific locations tailor made for the attachment of enzymes or fluorophores. A number of investigators have described methods to label oligonucleotides with fluorophores by chemical[65-67] and enzymatic reactions,[68] but the use of these as hybridization probes was not examined. This technology is being utilized for the investigation of directly labeled oligonucleotide systems. Heller et al.[69] label a synthetic 22-mer complementary to a portion of the thymidine kinase gene of herpes simplex virus with one Texas Red fluorophore per probe. Special microformatting techniques allow for focusing of the probe into small volumes for subsequent detection by microfluorometry. The system does not depend on amplification, and detection is almost instantaneous. Probes so labeled are quantifiable at the 10^{-20} mol level, but under actual hybridization assay, background emissions limit detection to 10^{-17} mol of target. Extending this concept, probes that hybridize in proximity to one another on the target sequence have been labeled with fluorophores that exhibit fluorescence energy transfer properties. Such systems have enhanced sensitivity due to spectral isolation of the large Stokes-shifted emission wavelength.[70] Jablonski and Ruth[71] and Jablonski et al.[72] directly label short probes with enzymes. Oligonucleotides of 20 to 26 bases in length, containing a single amine-terminated linker arm, are activated with a bifunctional cross-linking reagent such as disuccidimidyl suberate. The modified linker arm probe is conjugated to alkaline phosphatase. The resulting conjugate is composed of one enzyme per probe with full activity and is capable of hybridizing to target DNA on filter supports in 15 min. There is no apparent effect on hybridization kinetics or specificity. Background binding of the conjugate is nonexistent under appropriate stringency. Detection levels of approximately 10^6 copies of target DNA are visualized within a 4-h incubation with substrates. These conjugates have been evaluated by other groups and found to be useful in the diagnosis of falciparum malaria.[73]

Recently, directly labeled enzyme oligonucleotide probes have been applied to *in situ* hybridization systems. Kerschner et al.[74] use these probes to detect and type herpes simplex virus (HSV) in tissue culture and in formalin-fixed tissue sections. Hybridization is detected by depositing a formazan dye which forms an insoluble purple precipitate. This technique offers greater sensitivity than conventional dot-blot hybridization. In addition, it appears to be competitive, both by sensitivity and by rapidity, with other commercially available secondary detection systems. Time elapsed from start to finish is as little as 3 h, depending upon the degree of HSV infection.

C. Advantages/Disadvantages

1. Advantages

a. Directly labeled conjugates are extremely stable (at least 1 year), safe, and convenient.
b. Detection times are fast, due to minimal steps in the development. The detection process is started immediately after hybridization and washing.
c. For short probes, the reagents are chemically well defined, pure, and specific.
d. Background binding of directly labeled probes is reduced compared with that observed with indirect systems.
e. Directly labeled oligonucleotide probes are most suited for clinical diagnosis, in which a variety of applications are desired. Suitably labeled oligonucleotides can be produced easily and relatively inexpensively by synthetic methods.
f. Fluorescent-based direct systems offer the greatest potential for sensitivity with instantaneous detection.

2. Disadvantages

a. It is difficult to label cloned probes directly.
b. Absolute sensitivity is, in general, lower than that of indirect or radiolabeled systems that use cloned probes.
c. Fluorescence-based direct detection systems are, at present, insensitive, and the development of useful clinical systems appears to be a longer-term prospect.

V. DISCUSSION

The detection of minute quantities of hybridized DNA has involved much effort in trying to amplify the intensity of signal from DNA probe labels. The difficulty in achieving more sensitive levels of detection is a serious consideration in the use of DNA probe technology for routine clinical applications. It should be realized, however, that the limitations encountered in target DNA detection levels are not always due to insufficient signal intensity. Most of the labels used for DNA probe assays are inherently detectable at very low levels under appropriate conditions. The inability of a functional label to be adequately detected in a compatible detection system is due to interfering background signal. The design and analysis of any detection system must address this issue. The identification of the sources of interfering background signal allows for an intelligent modification of the system and may offer the most expedient route to the achievement of sufficient detection limits.

Background signal arises from either the label itself or from the environment of the DNA probe assay system. Background from the environment (extrinsic background) is a signal that mimics or interferes with the signal generated by the label and is independent of the label. It is a signal that is generated in the absence of the label. The higher the background from the environment, the lower the detectability of the label in the system. This type of background interference can be avoided by using labels that produce a unique signal in the system. The more unique the signal, the greater is the relative detectability of the label. This is one reason radiolabels have played a major role in analytical biochemistry as well as DNA probe technology. Achieving optimal detection of a label against extrinsic background is not always a matter of increasing signal intensity. Reducing or eliminating the interfering signal will enhance the awareness of the label signal. For example, fluorescence-based detection systems, limited by short-lived blue fluorescence background from the system or simple scattered light, have been improved by using red-emitting fluorophore labels and spectral isolation or long-lived fluorophores with time-resolved detection. Extrinsic background problems most commonly occur in fluorescent-based detection systems where almost everything appears to be fluorescent at very low levels. Enzymatic-based assays can also be restricted by nonenzymatic conversion of substrates or contaminating enzyme activity that survives sample preparation or fixing.

Background signal caused by the label (intrinsic background) is a more general phenomenon. It occurs when the label associates nonspecifically with any of the assay components, including supports, and generates a false result in the absence of target DNA. The sensitivity of the system cannot be increased by boosting signal intensity. Adding more labels to a probe or using enhanced visualization methods will increase specific and nonspecific signal proportionally, resulting in no net increase in signal to noise. This background may be reduced by optimizing the hybridization system to prevent nonspecific binding of the probe to the DNA sample. Support materials and hybridization conditions may also contribute to the adsorption of label or probe. The more complex a system, the greater the possibility of nonspecific binding in the absence of a true target. Nonradiometric, direct detection of synthetic oligonucleotide probes is the least complex and most appealing as a general clinical diagnostic system. The concern that short probes are insensitive because they do not allow for the amplification of signal with multiple reporter groups and multiple labels in the secondary detector complex, as in cloned probe systems, is genuine, but may be misleading. Short probes containing a single enzyme label cannot detect target DNA at the same level of sensitivity as long biotinylated nick-translated probes. The difference is about an order of magnitude, even though long probes may contain as many as 50 biotin per kilobase, each being ligated with multiple labels in the detector complex. The lack of corresponding increase in sensitivity when using nonradiometric indirect long probe systems may be due to a combination of factors, such as differences in hybridization efficiency, diminished enzyme activity in a large molecular weight aggregate, and background binding of the detector complex. Under appropriate conditions, very few copies of an enzyme or fluorophore can be visualized in a reasonable time with present technology. Hybridization systems that can accommodate these conditions are being developed with short probes.

REFERENCES

1. **Ben-Porat, T., Veach, R. A., and Ihara, S.,** Localization of the regions of homology between the genomes of herpes simplex virus, type 1 and pseudorabies virus, *Virology*, 127, 194, 1983.
2. **Berninger, M., Hammer, M., Hoyer, B., and Gorin, J. L.,** An assay for the detection of the DNA genome of hepatitis B virus in serum, *J. Med. Virol.*, 9, 57, 1982.
3. **Bornkamm, G. W., Desgranges, C., and Grissmann, L.,** Nucleic acid hybridization for the detection of viral genomes, *Curr. Top. Microbiol. Immunol.*, 104, 287, 1983.

4. **Maule, A. J., Hull, R., and Donson, J.,** The application of spot hybridization to the detection of DNA and RNA viruses in plant tissues, *J. Virol. Methods,* 6, 215, 1983.
5. **Polsky-Cynkin, R., Parsons, G. H., Allerdt, L., Landes, G., Davis, G., and Rashtchian, A.,** Use of DNA immobilized on plastic and agarose supports to detect DNA by sandwich hybridization, *Clin. Chem.,* 31, 1438, 1985.
6. **Moseley, S. L., Huq, I., Alim, A. R. M. A., So, M., Samadpor-Motalebi, M., and Falkow, S.,** Detection of enterotoxigenic *Escherichia coli* by DNA colony hybridization, *J. Infect. Dis.,* 142, 892, 1980.
7. **Agarwal, K. L., Brunstedt, J., and Noyes, B. E.,** A general method for detection and characterization of an mRNA using an oligonucleotide probe, *J. Biol. Chem.,* 256, 1023, 1981.
8. **Noyes, B. E., Mevarech, M., Stein, R., and Agarwal, K. L.,** Detection and partial sequence analysis of gastrin mRNA by using an oligonucleotide probe, *Proc. Natl. Acad. Sci. U.S.A.,* 76, 1770, 1979.
9. **Peterson, E. M., Aarnaes, S. L., Bryan, R. N., Ruth, J. L., and de la Maza, L. M.,** Typing of herpes simplex virus with synthetic DNA probes, *J. Infect. Dis.,* 153, 757, 1986.
10. **Szabo, P. and Ward, D. C.,** What's new with hybridization *in situ?*, *Tech. Biol. Sci.,* December, 425, 1982.
11. **Pardue, M. L.,** *In situ* hybridization, in *Nucleic Acid Hybridization, A Practical Approach,* Hames, B. D. and Higgins, S. J., Eds., IRL Press, Oxford, 1985, 179.
12. **Prensky, W., Steffensen, D. M., and Hughes, W. L.,** The use of iodinated RNA for gene localization, *Proc. Natl. Acad. Sci. U.S.A.,* 70, 1860, 1973.
13. **Szabo, P., Elder, R., Steffensen, D. M., and Uhlenbeck, O. C.,** Quantitative *in situ* hybridization of ribosomal RNA species to polytene chromosomes of *Drosophila melanogaster*, *J. Mol. Biol.,* 115, 539, 1977.
14. **Ranki, M., Palva, A., Virtanen, M., Laaksonen, M., and Soderlund, H.,** Sandwich hybridization as a convenient method for the detection of nucleic acids in crude samples, *Gene,* 21, 77, 1983.
15. **Syvanen, A.-C., Laaksonen, M., and Soderlund, H.,** Fast quantification of nucleic acid hybrids by affinity-based hybrid collection, *Nucleic Acids Res.,* 14, 5037, 1986.
16. **Southern, E. M.,** Detection of specific sequences among DNA fragments separated by gel electrophoresis, *J. Mol. Biol.,* 98, 503, 1975.
17. **Gall, J. G. and Pardue, M. L.,** Formation and detection of RNA-DNA hybrid molecules in cytological preparations, *Proc. Natl. Acad. Sci. U.S.A.,* 63, 378, 1969.
18. **Rigby, P. W. J., Dieckmann, M., Rhodes, C., and Berg, P.,** Labeling deoxyribonucleic acid to high specific activity *in vitro* by nick-translation with DNA polymerase I, *J. Mol. Biol.,* 113, 273, 1977.
19. **Maniatis, T., Jeffery, A., and Kleid, D. G.,** Nucleotide sequence of the rightward operator of phage lambda, *Proc. Natl. Acad. Sci. U.S.A.,* 72, 1184, 1975.
20. **O'Farrell, P.,** Replacement synthesis method of labeling DNA fragments, *BRL Focus,* 3.3, 1, 1981.
21. **Morris, C. F., Hama-Inaba, H., Mace, D., Sinha, N. K., and Alberts, B.,** Purification of the gene 43, 44, 45 and 62 proteins of the bacteriophage T4 DNA replication apparatus, *J. Biol. Chem.,* 254, 6787, 1979.
22. **Richardson, C. C.,** Phosphorylation of nucleic acid by an enzyme from T4 bacteriophage-infected *Escherichia coli, Proc. Natl. Acad. Sci. U.S.A.,* 54, 158, 1965.
23. **Maxam, A. M. and Gilbert, W.,** Sequencing end-labeled DNA with base specific chemical cleavages, in *Methods in Enzymology,* Vol. 65, Grossman, L. and Muldave, K., Eds., Academic Press, New York, 1980, 499.
24. **Conner, B. J., Reyes, A. A., Morin, C., Itakura, K., Teplitz, R. L., and Wallace, R. B.,.** Detection of sickle cell β^s-globin allele by hybridization with synthetic oligonucleotides, *Proc. Natl. Acad. Sci. U.S.A.,* 80, 278, 1983.
25. **Pirastu, M., Kan, Y. W., Cao, A., Conner, B. J., Teplitz, R. L., and Wallace, R. B.,** Prenatal diagnosis of β-thalassemia. Detection of a single nucleotide mutation in DNA, *N. Engl. J. Med.,* 309, 284, 1983.
26. **Wallace, R. B., Shaffer, J., Murphy, R. F., Bonner, J., Hirose, T., and Itakura, K.,** Hybridization of synthetic oligonucleotides to $\phi\chi$174 DNA: the effect of single base pair mismatch, *Nucleic Acids Res.,* 6, 3543, 1979.
27. **Arrand, J. E.,** Preparation of nucleic acid probes, in *Nucleic Acid Hybridization, A Practical Approach,* Hames, B. D. and Higgins, S. J., Eds., IRL Press, Oxford, 1985, 17.
28. **Maniatis, T., Fritsch, E. F., and Sambrook, J.,** *Molecular Cloning, A Laboratory Manual,* Cold Spring Harbor Laboratory, Cold Spring Harbor, N.Y., 1982, 107.
29. **Anderson, M. L. M. and Young, B. D.,** Quantitative filter hybridization, in *Nucleic Acid Hybridization, A Practical Approach,* Hames, B. D. and Higgins, S. J., Eds., IRL Press, Oxford, 1985, 47.
30. **Pardue, M. L.,** *In situ* hybridization, in *Nucleic Acid Hybridization, A Practical Approach,* Hames, B. D. and Higgins, S. J., Eds., IRL Press, Oxford, 1985, 179.
31. **Gardner, L.,** Non-radioactive DNA labeling: detection of specific DNA and RNA sequences on nitrocellulose and *in situ* hybridization, *BioTechniques,* 1, 38, 1983.
32. **Leary, J. J., Brigati, D. J., and Ward, D. C.,** Rapid and sensitive colorimetric method for visualizing biotin-labeled DNA probes hybridized to DNA or RNA immobilized on nitrocellulose: bio-blots, *Proc. Natl. Acad. Sci. U.S.A.,* 80, 4045, 1983.

33. **Brigati, D. J., Myerson, D., Leary, J. L., Spalholz, B., Travis, S. Z., Fong, C. K. Y., Hsiung, G. D., and Ward, D. C.**, Detection of viral genomes in cultured cells and parafin embedded tissue sections using biotin labeled hybridization probes, *Virology*, 126, 32, 1983.
34. **Langer, P. R., Waldrop, A. A., and Ward, D. C.** Enzymatic synthesis of biotin-labeled polynucleotides: novel nucleic acid affinity probes, *Proc. Natl. Acad. Sci. U.S.A.*, 78, 6633, 1981.
35. **Forster, A. C., McInnes, J. L., Skingle, D. C., and Symons, R. H.**, Non-radioactive hybridization probes prepared by the chemical labeling of DNA and RNA with a novel reagent, photobiotin, *Nucleic Acids Res.*, 13, 745, 1985.
36. **Viscidi, R. P., Connelly, C. J., and Yolken, R. H.**, Novel chemical method for the preparation of nucleic acids for nonisotopic hybridization, *J. Clin. Microbiol.*, 23, 311, 1986.
37. **Al-Hakim, H. H. and Hull, R.**, Studies towards the development of chemically synthesized nonradioactive biotinylated nucleic acid hybridization probes, *Nucleic Acids Res.*, 14, 9965, 1986.
38. **Murasugi, A. and Wallace, R. B.**, Biotin-labeled oligonucleotides: enzymatic synthesis and use as hybridization probes, *DNA*, 3, 269, 1984.
39. **Chu, B. C. F., Wahl, G. M., and Orgel, L. E.**, Derivatization of unprotected polynucleotides, *Nucleic Acids Res.*, 11, 6513, 1983.
40. **Chu, B. C. F. and Orgel, L. E.**, Detection of specific DNA sequences with short biotin-labeled probes, *DNA*, 4, 327, 1985.
41. **Chollet, A. and Kawashima, E. H.**, Biotin-labeled synthetic oligodeoxyribonucleotides: chemical synthesis and use as hybridization probes, *Nucleic Acids Res.*, 13, 1529, 1985.
42. **Ruth, J. L.**, Chemical synthesis of non-radioactively-labeled DNA hybridization probes, *DNA*, 3, 123, 1984.
43. **Ruth, J. L. and Bryan, R. N.**, Chemical synthesis of modified oligonucleotides and their utility as non-radioactive hybridization probes, *Fed. Proc.*, 43, 2048, 1984.
44. **Bryan, R. N. and Arnold, L. J.**, Detection of HBV-DNA with modified synthetic oligonucleotides containing radioactive and non-radioactive reporter groups, *DNA*, 3, 124, 1984.
45. **Vincent, C., Tchen, P., Cohen-Solal, M., and Kourilsky, P.**, Synthesis of 8-(2-4 dinitrophenyl 2-6 aminohexyl) amino-adenosine 5′ triphosphate: biological properties and potential uses, *Nucleic Acids Res.*, 10, 6787, 1982.
46. **Tchen, P., Fuchs, R. P. P., Sage, E., and Leng, M.**, Chemically modified nucleic acids as immunodetectable probes in hybridization experiments, *Proc. Natl. Acad. Sci. U.S.A.*, 81, 3466, 1984.
47. **Syvanen, A.-C., Tchen, P., Ranki, M., and Soderlund, H.**, Time resolved fluorometry: a sensitive method to quantify DNA-hybrids, *Nucleic Acids Res.*, 14, 1017, 1986.
48. **McGady, J.**, A tetrazolium method for non-specific alkaline phosphatase, *Histochemie*, 23, 180, 1970.
49. **Horwitz, J. P., Chua, J., Curby, R. J., Tomson, A. J., DaRooge, M. A., Fisher, B. E., Mauricio, J., and Klundt, I.**, Substrates for cytochemical demonstration of enzyme activity. I. Some substituted 3-indoyl-β-D-galactopyranosides, *J. Med. Chem.*, 7, 574, 1964.
50. **Ishikawa, E. and Kato, K.**, Ultrasensitive enzyme immunoassay, in *Quantitative Enzyme Immunoassay*, Engvall, E. and Pesce, A. J., Eds., Blackwell Scientific Publications, Oxford, 1978, 43.
51. **Syvanen, A.-C., Alanen, M., and Soderlund, H.**, A complex of single-strand binding protein and M13 DNA as hybridization probe, *Nucleic Acids Res.*, 13, 2789, 1985.
52. **Chan, V. T.-W., Fleming, K. A., and McGee, J. O'D.**, Detection of subpicogram quantities of specific DNA sequences on blot hybridization with biotinylated probes, *Nucleic Acids Res.*, 13, 8083, 1985.
53. **Matthews, J. A., Batki, A., Hynds, C., and Kricka, L. J.**, Enhanced chemiluminescent method for the detection of DNA dot-hybridization assays, *Anal. Biochem.*, 151, 205, 1985.
54. **Nagata, Y., Yokota, H., Kosuda, O., Yokoo, K., Takemura, K., and Kikuchi, T.**, Quantification of picogram levels of specific DNA immobilized in microtiter wells, *FEBS Lett.*, 183, 379, 1985.
55. **Manning, J. E., Hershey, N. D., Broker, T. R., Pellegrini, M., Mitchell, H. K., and Davidson, N.**, A new method of *in situ* hybridization, *Chromosoma*, 53, 107, 1975.
56. **Langer-Safer, P. R., Levine, M., and Ward, D. C.**, Immunological method for mapping genes on Drosophila polytene chromosomes, *Proc. Natl. Acad. Sci. U.S.A.*, 79, 4381, 1982.
57. **Manuelidis, P. R., Langer-Safer, P. R., and Ward, D. C.**, High-resolution mapping of satellite DNA using biotin-labeled DNA probes, *J. Cell. Biol.*, 95, 619, 1982.
58. **Rudkin, G. T. and Stollar, B. D.**, High resolution detection of DNA-RNA hybrids *in situ* by indirect immunofluorescence, *Nature (London)*, 265, 472, 1977.
59. **Stuart, W. D. and Porter, D. L.**, An improved *in situ* hybridization method, *Exp. Cell. Res.*, 113, 219, 1978.
60. **Narayanswami, S. and Hamkalo, B. A.**, Electron microscopic *in situ* hybridization using biotinylated probes, *Focus*, 8, 3, 1986.
61. **Green, M. N.**, Avidin, *Adv. Protein Chem.*, 29, 85, 1975.
62. **Renz, M. and Kurz, C.**, A colorimetric method for DNA hybridization, *Nucleic Acids Res.*, 12, 3435, 1984.

63. **Bauman, J. G. J., Weigant, J., and Van Duijn, P.,** Cytochemical hybridization with fluorochrome-labeled RNA. I. Development of a method using nucleic acids bound to agarose beads as a model, *J. Histochem. Cytochem.*, 29, 227, 1981.
64. **Bauman, J. G. J., Weigant, J., and Van Duijn, P.,** Cytochemical hybridization with fluorochrome-labeled RNA. II. Applications, *J. Histochem. Cytochem.*, 29, 238, 1981.
65. **Inoue, H., Imura, A., and Ohtsuka, E.,** Synthesis and hybridization of dodecadeoxyribonucleotides containing a fluorescent pyridopyrimidine deoxynucleoside, *Nucleic Acids Res.*, 13, 7119, 1985.
66. **Smith, L. M., Fung, S., Hunkapiller, M. W., Hunkapiller, T. J., and Hood, L. E.,** The synthesis of oligonucleotides containing an aliphatic amino group at the 5′ terminus: synthesis of fluorescent DNA primers for use in DNA sequences analysis, *Nucleic Acids Res.*, 13, 2399, 1985.
67. **Connolly, B. A.,** Chemical synthesis of oligonucleotides containing a free sulphydryl group and subsequent attachment of thiol specific probes, *Nucleic Acids Res.*, 13, 4485, 1985.
68. **Cosstick, R., McLaughlin, L. W., and Eckstein, F.,** Fluorescent labeling of tRNA and oligodeoxynucleotides using T4 RNA ligase, *Nucleic Acids Res.*, 12, 1791, 1984.
69. **Heller, M. J., Jablonski, E. G., Ruth, J. L., and Hennessy, E. C.,** Microfluorimetric DNA hybridization assay system, *Fed. Proc.*, 45, 1516, 1986.
70. **Heller, M. J., Hennessy, E., Ruth, J. L., and Jablonski, E.,** Fluorescent energy transfer oligonucleotide probes, *Fed. Proc.*, 46, 1968, 1987.
71. **Jablonski, E. and Ruth, J. L.,** Synthesis of oligonucleotide — enzyme conjugates and their use as hybridization probes, *DNA*, 5, 89, 1986.
72. **Jablonski, E., Moomaw, E. W., Tullis, R. H., and Ruth, J. L.,** Preparation of oligodeoxynucleotide-alkaline phosphatase conjugates and their use as hybridization probes, *Nucleic Acids Res.*, 14, 6115, 1986.
73. **McLaughlin, G. L., Ruth, J. L., Jablonski, E., Steketee, R., and Campbell, G. H.,** Use of enzyme-linked synthetic DNA in diagnosis of falciparum malaria, *Lancet*, 714, 1987.
74. **Kerschner, J. H., Cross, B. D., Jablonski, E. G., and Heller, M. J.,** Detection of herpes simplex virus by *in situ* hybridization using alkaline phosphatase labeled synthetic deoxyoligonucleotide probes, *Am. Soc. Microbiol.*, 56(Abstr.), Washington, D.C., 1987, 309.

Part II: Impact of Probe Technology on Diagnostic Laboratories

Chapter 3

IMPACT OF DNA PROBES ON CLINICAL MICROBIOLOGY

Fred C. Tenover

TABLE OF CONTENTS

I. INTRODUCTION

During the last several years, there has been a lot of excitement generated in the clinical microbiology community over the prospect of using DNA probes in the clinical laboratory to detect and identify infectious microorganisms. DNA probes should enable microbiologists to identify even fastidious pathogens rapidly and with high specificity directly in clinical specimens. The use of hybridization technology in diagnostic laboratories, however, is not yet a proven modality; additional studies are needed with commercially available products in order to document the sensitivity, specificity, and cost-effectiveness of this diagnostic method.

Because of the potential impact of DNA probes on the clinical laboratory, in terms of replacing existing methodologies and broadening the scope of infectious agents identified by the laboratory, it is important that laboratory directors be aware of this new technology.

In this chapter, I will review the key issues involved in implementing nucleic acid hybridization techniques in the clinical laboratory and comment on the promises and pitfalls of this methodology.

II. THE CURRENT CHALLENGES OF CLINICAL MICROBIOLOGY

The practice of clinical microbiology, like many subspecialties of laboratory medicine, is undergoing some interesting changes and challenges.[1] The prospective payment system and diagnosis related groups (DRGs) have changed many laboratories from revenue-generating centers to cost centers.[2,3] Changes in supply and personnel funds available to clinical microbiology laboratories have led to abandonment of many of the more technologist-intensive tests. In some cases, it has been the technologist and not the test that has been abandoned. Directors of clinical microbiology laboratories are faced with several alternatives for the future. One alternative is to cut back on the types of services offered and find alternate methods of analysis that are less time consuming and less expensive. The opposite approach would be to attempt to generate additional income for the laboratory by accelerating the turn-around times of current procedures, to allow a greater volume of tests to be performed, and broadening the spectrum of infectious agents that are identified by the laboratory. In this way, the laboratory can attract new markets, such as clinics and smaller hospitals, and increase its revenue-generating potential. In either case, a new technology that is rapid, highly sensitive and specific, and not significantly more expensive than currently utilized methods, could be of considerable value. What role will DNA probes play in these situations? To answer, we need to examine how the clinical microbiology laboratory assists the clinician in establishing the infectious etiology of an illness.

III. THE MISSION OF THE CLINICAL MICROBIOLOGY LABORATORY

The clinical microbiology laboratory assists the clinician in establishing a diagnosis of an infectious disease in several ways. First, the laboratory provides information regarding specimen collection techniques. Often, laboratory personnel will assist in the collection of specimens, particularly for dark-field analysis or blood films. Second, the laboratory detects and identifies pathogens in clinical specimens. This is currently accomplished through the use of microscopy, antigen and antibody testing, and culture techniques. Third, the laboratory provides clinicians with antimicrobial susceptibility data for the organisms recovered from patients' specimens. Finally, the laboratory aids infection control practitioners and public health officers in the identification and tracking of strains causing outbreaks of nosocomial or community-acquired infections.[4-6]

DNA probes could be utilized in lieu of some microscopy techniques for the direct detection of pathogens in specimens, in place of several current methods of identification for bacteria and viruses, as a means of identifying specific strains from outbreaks of nosocomial infections, and even for susceptibility testing of isolates.[7-9] Thus, there is considerable potential for utilization of DNA probes in the diagnostic laboratory in the future.

IV. THE IDEAL DNA PROBE KIT

The ideal DNA probe kit that would be purchased for use in a diagnostic laboratory should have a total processing time of less than 3 h and should demonstrate high degrees of sensitivity (≥95%) and specificity (≥99%) compared to culture. The kit should include all the necessary buffers and control reagents, and directions for use should be simple and concise. The cost should be competitive with existing methods. The test should use a nonradioisotopic detection system but should not require expensive equipment. End points should be clearly defined, and the procedure must be simple enough to allow troubleshooting by existing laboratory personnel.

Perhaps the greatest concern of microbiologists about the new probe-based kits is the cost of this new technology. Although the notion of direct detection of organisms in clinical specimens is attractive, many microbiologists, when faced with their current restricted budgets, will not be willing to sacrifice traditional methods for more costly techniques if some cost savings cannot be demonstrated. At the present time, there is not enough information available to predict whether this technology will be cost-effective for the laboratory. In this era of DRGs, however, laboratory directors should remember to consider the potential savings to the hospital of providing a rapid diagnosis of an infectious illness. A rapid diagnosis of a nontuberculosis mycobacteriosis, for example, could potentially allow the pharmacy to alter antimicrobial therapy to less expensive agents and allow the patient to be transferred out of a costly isolation room and perhaps be discharged at an earlier time, resulting in a considerable cost savings to the institution. These sorts of discussions regarding a more "global" approach to budgeting (i.e., including pharmacy, administration, and laboratory) will certainly be a new, but important, undertaking for laboratory directors who are considering the potential benefits of introducing DNA probe technology into their laboratories.

V. A REVIEW OF SEVERAL COMMERCIALLY AVAILABLE DNA PROBES

A number of the currently available diagnostic kits for identification of pathogenic microorganisms are listed in Table 1. Both filter hybridization and solution hybridization formats have been introduced. The solution hybridization kits can be completed in less than 4 h from the time that the specimen arrives in the laboratory. These kits, for the identification of *Legionella pneumophila*,[10-12] several *Mycobacterium* spp.,[13,14] and *Mycoplasma pneumoniae*, employ a DNA probe to ribosomal RNA and are labeled with ^{125}I. A further review of this technology is presented in Chapter 8.

The second type of kits available are those that use a filter hybridization format and oligonucleotide probes labeled with alkaline phosphatase. These hybridization reactions produce a colorimetric end point which can be detected visually. Eight prelabeled, single-stranded probes are available, including probes to detect *Campylobacter jejuni* and the genes encoding the heat-labile (LT) and heat-stable (ST) enterotoxins of *Escherichia coli*.[15,16] Probes that detect both herpes simplex virus (HSV) type I and type II, together or individually, are available,[17] as are probes to detect the genes encoding hepatitis B surface antigen and core antigen. The stool probes are designed to be used for culture confirmation after the suspected organisms are grown overnight on filters. This technique currently requires a minimum of 30 h to complete from the time the specimen arrives in the laboratory (including 18 to 24 h for growth of the organism on an agar medium). Additional probes for rotavirus type A, an agent of viral gastroenteritis, and *Plasmodium falciparum*, a causative agent of malaria, soon should be available. Other probes, including ^{35}S-labeled DNA and RNA probes to human immunodeficiency virus (HIV), also are available.

Table 1
COMMERCIALLY AVAILABLE DNA PROBE REAGENTS

Organism	Format[a]	Manufacturer[b]
Legionella pneumophila	S	Gen-Probe
Mycoplasma pneumoniae	S	Gen-Probe
Mycobacterium tuberculosis	S	Gen-Probe
M. avium	S	Gen-Probe
M. intracellulare	S	Gen-Probe
Escherichia coli		
LT toxin	F	Dupont/MBI
ST toxin	F	Dupont/MBI
Campylobacter jejuni	F	Dupont/MBI
Herpes simplex virus I and II	F	Dupont/MBI
Hepatitis B virus	F	Dupont/MBI
Plasmodium falciparum	F	Dupont/MBI
Rotavirus type A	F	Dupont/MBI
Human immunodeficiency virus	F/I	Dupont/MBI
Herpes simplex virus	T	Ortho
Campylobacter jejuni	M	Gene Trak
Hepatitis B virus	F	Abbott

[a] S — solution hybridization; F — filter hybridization; T — tissue culture confirmation; M — microtiter/tube-based test; I — *in situ* hybridization.
[b] Gen-Probe, Inc., San Diego, Calif.; E. I. du Pont de Nemours & Co., Wilmington, Del.; Molecular Biosystems, Inc., San Diego, Calif.; Ortho Diagnostic Systems, Raritan, N.J.; Gene Trak, Inc, Framingham, Mass.; Abbott Diagnostics, North Chicago, Ill.

A third DNA probe assay kit for identifying *C. jejuni* will soon be available. This kit uses a tube coated with an antibody which captures DNA:RNA hybrids forming during hybridization assays.

The last of the currently available kits is a HSV culture confirmation test for use on tissue culture specimens.

VI. NEEDS OF CLINICAL MICROBIOLOGISTS

Microbiologists are looking to DNA probes with two goals in mind. First, there is a desire to shorten turn-around times for identification of fastidious pathogens. Second, microbiologists are looking to broaden the range of microorganisms that they can detect in clinical specimens in their own laboratories, thus reducing the costs incurred from sending those specimens to reference laboratories.

A. Rapid Turn-Around Times

Current DNA probe tests require from 1 to 30 h to complete. Those tests requiring 30 h can still be considered rapid when compared to traditional isolation and identification schemes that normally require 4 to 6 weeks to complete. DNA probe tests offer identification of *M. pneumoniae* and *Legionella* species within 4 h of specimen receipt. Compared to the period of time it would take to isolate and identify these organisms using traditional methods, the DNA probe test is indeed a rapid test and is comparable in "hands-on" time to many commercially available direct fluorescent antibody kits.[18]

Those probe tests that can be completed in as little as 1 h certainly could be utilized on a STAT basis. However, probe tests, much like ELISA tests, will still be most cost-effective when run in batches, not individually. The potential for same-day reporting of test results, however, is feasible with current solution hybridization assays.

B. Broadening the Spectrum of Organisms Identified by Smaller Laboratories

Many smaller laboratories send a large number of specimens to reference laboratories for processing. This requires a considerable amount of time and expense. It is possible that DNA probes will allow many laboratories, especially smaller laboratories, to reduce their send-out costs by enabling them to identify some fastidious bacterial and several common viral pathogens in house. At the Seattle Veterans Administration Hospital, Wash., the number of specimens sent to other laboratories increased 31% in 1986 over 1985 and has continued to increase in 1987. Of the increase, 85% was attributable exclusively to viral cultures. The use of DNA probes for HSV alone would reduce the number of specimens sent to a reference laboratory by at least 50%. Similar reductions in send-out costs could surely be realized by other laboratories that do not routinely perform viral cultures. Other nonroutine cultures, such as for *Legionella* species, also could be replaced with a DNA probe assay.

C. Disadvantages of DNA Probes

There are several drawbacks to using probes in lieu of cultures, however, that should be recognized. First, only a limited number of organisms can be detected by using DNA probes. For example, by using a DNA probe to detect HSV as the sole diagnostic test on a specimen from a skin lesion, the ability to detect the presence of cytomegalovirus, varicella-zoster virus, or other viral agents in that lesion is lost. Second, as noted by Fung et al.[19] in their comparison of DFA and probe techniques to culture, not all specimens are amenable to analysis with nonculture-based techniques. In their study, only 77% of culture-positive specimens contained adequate numbers of cells for probe or DFA analysis. Thus, false negative results can be obtained with probe assays unless proper specimen collection techniques and effective screening procedures for determining specimen adequacy are in place. For specimens containing low numbers of cells, culture should be available.

Corey[20] also expressed concerns about the quality control of nonculture-based tests and about the costs, both economic and psychological, of false positive and false negative tests (e.g., the impact of a false negative test when screening a neonate suspected of having disseminated herpes simplex infection in light of possible acyclovir therapy, or the psychological trauma of a false positive test to a patient reported to harbor HSV or HIV). Such new tests will require a very high degree of sensitivity and specificity to be used with confidence in lieu of culture, even though culture itself is known not to be 100% sensitive. Since few laboratory directors have been faced before with the choice of traditional culture methods vs. alternate nonculture methods, these sorts of "costs" may be difficult to assess. However, this impact of DNA probe technology must be weighted carefully when considering the replacement of culture with probes.

VII. INCORPORATING DNA PROBES INTO LABORATORY WORK FLOW

Probes can be used in the laboratory for confirming the identity of organisms grown in vitro or for direct detection of pathogenic microorganisms in patients' specimens. *M. tuberculosis, C. jejuni*, and enterotoxigenic *E. coli* (ETEC), for instance, can all be identified, once grown on solid media, by using commercially available probe kits. This approach, which can be completed in less than 3 h, may be more costly than traditional biochemical methods for organisms such as *Legionella* spp., but will likely be cost-effective for the identification of organisms such as *M. tuberculosis.*

The second use for DNA probes is the direct detection of infectious agents in patients' specimens. Examples of such tests would include screening of throat swabs for *M. pneumoniae* or lung biopsy material for *Legionella* spp. These tests could be run on the day they were received or batch processed several times per week and still produce results faster than traditional culture methods. The question of whether probes will be helpful in an outpatient clinic setting, where rapid response is important, is difficult to judge at this point because there are no published reports with the probe kits now available that address this issue. Since the majority of laboratories are not currently configured to run rapid-response tests, outside of microscopic-based tests, the initiation of probe-based tests in the laboratory for this purpose may require some changes in the way specimens are processed, including the number of personnel assigned to this task. The traditional set up area may require a second technologist to process STAT probe- and monoclonal antibody-based tests. The expansion of rapid-response sections with microbiology laboratories may be a direct consequence of the availability of new reagents, such as DNA probes and monoclonal antibody-based tests, that make rapid, accurate results available to clinicians in reasonable time frames.

VIII. FATE OF ROUTINE CULTURING TECHNIQUES

There is speculation that between probes and monoclonal antibody-based tests, agar-based bacterial cultures will soon disappear from the clinical laboratory. In some smaller laboratories, including outpatient clinics and physicians' offices, where culture techniques are traditionally limited, this may well occur. Cultures of throats for *Streptococcus pyogenes* may soon be replaced by latex agglutination or DNA probe tests that decrease the need for technical competence and are amenable to either STAT or batch testing. In these situations, where technical competence in the processing of traditional cultures may be questionable, probes offer a reasonable alternative, providing there is adequate training of the technical personnel, sufficient quality control to insure accuracy of results, and periodic monitoring of test results and quality control records by a qualified consultant. While most microbiologists would recommend culturing those specimens that are negative by rapid techniques, few clinics currently using such rapid methods appear to be following that advice, since it adds additional expense.

The probability of probes totally replacing culture in larger laboratories in the next several years, however, is low for two reasons. First, to obtain information, such as the antimicrobial susceptibility or serotype of an organism, it must first be isolated in pure culture. Since such data cannot be replaced at present by other means, culture of probe-positive specimens is still required. Second, microbiologists and infectious disease specialists alike know that they are occasionally surprised by the organisms that appear in culture from clinical specimens. Reference laboratories and larger referral centers need to be able to detect the unusual or unexpected organism, such as the *Pseudomonas aeruginosa* in cerebral spinal fluid or the *Streptobacillus moniliformans* in blood. If culture techniques were eliminated and replaced by a limited set of probes, our ability to find new or unusual organisms would be greatly diminished, as would be our capacity to find more common organisms in unusual body sites. On the other hand, even a select group of probes extend the ability of the laboratory to detect and identify unusual organisms, such as the ETEC in the stool of a traveler to a foreign country.[21]

It is clear that our current culture methodologies are not always optimal. Even in larger laboratories, the attention to detail in identifying isolates to species may well be unnecessary in certain situations. Bartlett and others have argued that identification of microorganisms completed 7 days after culture submission is often of little clinical value.[22-24] Would it not follow that an identification of a bacterium to genus in 4 h with a DNA probe, with antimicrobial susceptibility results provided the following day (from isolates obtained in pure culture), would be preferable to a genus and species identification and susceptibility tests in 72 h? Although some exceptions could certainly be cited where some members of a genus may possess a greater pathogenic potential than others (i.e., *Salmonella typhi* vs. *S. typhimurium*), as more probe tests are used routinely in clinical laboratories, this issue should receive greater attention.

The other issue raised by the use of DNA probes is the need for antimicrobial susceptibility data. If probes are used to indicate those specimens that should be cultured for a specific pathogen, i.e., stools positive for *Shigella* spp., then probes will be a very expensive culture enhancement. If, on the other hand, probes replace culture, then antibiograms and serotyping data for analysis will be lost. While the use of statistical methods predicting the antimicrobial susceptibility of an organism recognized in a clinical specimen by a DNA probe holds some promise, such methods cannot be used to determine strain-specific data, such as serotyping provides. All laboratorians should consider this option with care, since its impact on public health may be significant. Probes may not be suitable alternatives to culture until antimicrobial susceptibility tests and reasonable alternatives to serotype results are available through additional DNA probe tests, monoclonal antibodies, or statistical models.

IX. THE UTILITY OF USING BATTERIES OF PROBE TESTS

Batteries of culture media and enrichment broths are commonly used in the clinical laboratory to aid in the detection of those pathogens most commonly encountered in a given body site. For instance, stool specimens are cultured on a variety of differential and selective media to aid in the isolation of *Salmonella*, *Shigella*, and *Campylobacter* species, the most common causes of bacterial enteritis. Other media are included, depending on the geographic location of the hospital and the patient population, to aid in the detection of *Clostridium difficile*, *Vibrio parahaemolyticus*, and *Yersinia enterocolitica*. The key point is that, depending on the media used, some pathogens will not be recovered by the laboratory, even if present in the specimen. Stools represent the least valuable and most inefficient of the specimen types submitted to the clinical laboratory, with the cost of identifying cultures often exceeding $900 per positive specimen.[25,26] Rosenblatt has suggested that the physician should take responsibility for ordering laboratory tests only for those pathogens most likely to cause diarrheal disease on the basis of history and physical exam.[27] At least one controlled trial aimed at educating physicians in cost-containment strategies[28] and the anecdotal experiences of several clinical microbiology laboratory directors, however, suggest that the burden of what pathogens to screen still falls on the clinical laboratory the majority of the time. In this regard, a set of probes to common stool pathogens would appear to be of value. A stool probe battery, consisting of campylobacter, salmonella, and shigella probes, could be used routinely for outpatient cultures. Additional probes for *Y. enterocolitica*, enterohemorrhagic *E. coli*, rotavirus, and, perhaps, even enteric adenoviruses, could be added, depending on the patient's history and the incidence of those pathogens in the community. The laboratory director, then, must decide whether to culture probe-positive stools in order to obtain susceptibility and serotype data, whether to culture probe-negative specimens for pathogens other than those for which probes were available (e.g., *C. difficile* or *Aeromonas* spp.), or whether to culture at all. These decisions must be made by each laboratory depending on the patient population, the extent of laboratory personnel available, and associated costs.

Similarly, a battery of probes to common bacterial and viral respiratory pathogens could broaden the range of organisms that could be reported by a small laboratory in a timely fashion. Rapid probe results could be of significant value in a pediatric hospital during an epidemic of respiratory disease which could be related to respiratory syncytial virus, *Bordetella pertussis*, or a variety of other pediatric respiratory pathogens. Probe data could be used by clinicians and infection control personnel to guide antimicrobial or antiviral therapy as indicated. The rapid response time provided by the probe data would clearly have clinical impact in this setting.

Another area that could benefit from a battery of probes is the area of sexually transmitted diseases (STDs). Although rapid fluorescent antibody tests are available for HSV, and ELISA tests are available for *Neisseria gonorrhoeae* and *Chlamydia trachomatis*, an STD probe kit would have the advantage of utilizing a single technology where multiple analyses could be run by a single technologist simultaneously without the confusion of multiple types of assays. Even newer pathogens such as *Mobiluncus* spp. have proven amenable to direct probe analysis.[29]

X. TYPES OF PROBE ASSAYS

DNA probe tests can take on three forms: solution hybridization, hybridization on solid supports, and *in situ* hybridization. It is possible that all three modes of hybridization will be used in the clinical laboratory in the future.

The ribosomal RNA-based probes are solution hybridization reactions that can be completed in ≤4 h. Currently, probes to several *Mycobacterium* species, *Legionella* species, and *Mycoplasma pneumoniae* are available. This type of hybridization is the easiest for microbiologists to perform and lends itself well to kit form.

Other probes currently available are designed to confirm the identity of *C. jejuni* and ETEC in stool, either blotted on nitrocellulose or nylon filters and grown overnight, or in isolated colonies picked from stool cultures and inoculated onto filters. This is a more traditional probe assay, one that is technically more demanding, and one unlikely to find a home in many clinical microbiology laboratories for those reasons. However, novel methods for the extraction of DNA from stool samples to be used directly in solution hybridization reactions or for filter hybridizations have shown promising results. Thus, direct detection of diarrheal pathogens without an initial growth enrichment phase may soon be feasible. Filter assays for the confirmation of HSV types I and II and for detection of hepatitis B virus also are available, although they have yet to find their niche in the clinical laboratory.

Soon to be available in kit format is another hybridization technique that may find widespread use in clinical laboratories. This technique, called sandwich hybridization, utilizes a solid support, such as a test tube or microtiter plate, and a two-probe system for capturing and labeling target DNA in a sample. The technique, described in greater detail in Chapter 17, attaches the first probe to the solid support, such as the bottom of a test tube. The target DNA is captured by this probe that binds to only a portion of the target, leaving a stretch of exposed single-stranded target DNA still available for binding. A second probe, this one carrying a reporter molecule, binds to the target, creating a sort of sandwich.[30] Although similar to the solution hybridization reaction, this approach uses a washing step to remove excess probe rather than a centrifugation step to separate excess label from the nascent duplexes. A similar capture technique using monoclonal antibodies to DNA:RNA hybrids has been described.[31] This method will soon be available for detection of *C. jejuni* in stool samples, after overnight enrichment.

The last approach to hybridization is *in situ* hybridization, where the specimen is fixed on microscope slides, hybridized *in situ*, and examined under light microscopy to discern the results. While this method would seem to be best suited for cytology and histology laboratories, it is simple enough to be performed by any larger laboratory. Studies by Myerson et al.[32] suggest that this technique is as sensitive as monoclonal antibody tests for the detection of cytomegalovirus in lung biopsies and much more rapid than culture.

It should be noted that not all specimens will be suitable for probe analysis without some sort of enrichment phase. Specimens, such as blood and cerebral spinal fluid, may contain too few organisms for direct detection with DNA probes. Saiki et al.,[33] however, have developed a method for amplifying the presence of particular DNA or RNA species in solution by using a pair of DNA primers that are complementary to each end of the desired sequence, and DNA polymerase to make the new strands of nucleic acid. After allowing the primers to seek out and bind to their complementary sequences at either end of the region to be amplified, DNA polymerase is added which makes a complementary copy of the flanked region, first off the nucleic acid strand in the sample, then off of the nascent nucleic acid strand. As reaction times are increased and additional polymerase is added, the number of copies of the flanked sequences increases substantially. After a given period, even sequences present in low copy number in a cell (such as HIV) can be readily detected with a DNA probe. Although this procedure has not yet been exploited for detection of the nucleic acid of infectious agents, its potential is indeed impressive.

The ultimate goal of those developing hybridization tests is to automate the entire process of staining, reading, and reporting. Currently, microprocessor-driven, automatic slide-staining machines are being adapted for this purpose. Specimens can be placed on multiple slides or similar membranes, hybridized with a variety of DNA oligonucleotide probes (see Chapter 2), washed, and reacted with reporter molecules to determine the presence of hybrids in a totally automated fashion. The slides can then be read visually or by using a spectrophotometer to determine the extent of hybridization between the target sample and the various probes. At present, short probes that can be labeled with alkaline phosphatase or other enzymes via linker arm technology pose the highest probability of success for this method, since they are easy to manufacture in large quantities and have well-defined hybridization kinetics.

XI. DNA PROBES VS. MONOCLONAL ANTIBODY-BASED TESTS

Monoclonal antibody- and DNA probe-based kits are now in direct competition for the microbiology diagnostics market. Monoclonal antibodies will soon replace many of the polyclonal reagents currently on the market. Owing to their higher specificity (albeit sometimes at the expense of avidity) and reproducibility, these reagents are expected to occupy a $1 billion market by 1990.[34] Probes, on the other hand, are just now coming to market, but are predicted to be in the same range by 1995. These two reagent types will compete for the direct detection market where a microorganism is sought directly in a patient's specimen. For some agents, such as latent viral infections, probes hold the advantage in sensitivity. Probes also will be preferable to culturing for infectious agents in complex specimens, such as stool. Monoclonal antibody-based tests, however, are still easier to set up and will likely be less expensive than their DNA probe counterparts. Monoclonal-based tests, much like probes, do require extra equipment, and those that employ fluorescent labels require expertise in microscopy. Studies directly comparing the two techniques are rare, but in one published recently, monoclonal antibodies to HSV were superior.[19] Probe kits, however, are in their infancy. Many problems are yet to be discovered, due to the limited availability of the test kits. Given a year of market exposure, this will no doubt change and the "ease of use" advantage of monoclonals may well disappear. The choice of a monoclonal antibody over a DNA probe kit will then depend on the traditional standards of sensitivity, specificity, accuracy, and precision.[35,36]

XII. ECONOMICS

Now, more than ever, laboratory directors need to be concerned with the cost of specimen processing and analysis. Because only a few laboratories have been using probes routinely, cost analysis of these tests is difficult.

It is of interest that probes may well open up new lines of reference testing for many laboratories, including smaller laboratories. A recent article in the journal *Medical Laboratory Observer* noted that of the hospital laboratories surveyed (not just microbiology laboratories), 75% had increased their outpatient workload in 1985 while inpatient tests declined in 4% of those same institutions during that same period.[37] Of laboratory directors interviewed, 84% thought that DRGs were either closely (36%) or somewhat (48%) related to these changes. The surprising data showed that nearly two thirds of those laboratories actively marketed their outpatient services. Because the turn-around time for DNA probe tests is relatively short, and because hybridization tests lend themselves well to batch testing, probes may well prove to be a relatively simple way for laboratories to increase the spectrum of tests offered for outpatient testing without significantly increasing the number of technologists required to perform those tests.

A final word regarding probes and the prevalence of a given disease in the community is in order. Probes will be too expensive to be used for the diagnosis of diseases that occur rarely in the community served by the hospital or clinic. To be economical, probe tests should be reserved for high volume tests that can be batch processed, or for those cases where rapid diagnosis has a significant impact on the care of the patient. There will soon be probes available for a wide variety of infectious agents, many of which will be rarely seen by a majority of microbiologists in this country — ETEC and *P. falciparum* are but two examples. Such kits should not be used routinely by laboratories without a clear indication that they will affect clinical care and that comparable data cannot be obtained at lesser costs from reliable sources.

XIII. CONCLUSIONS

DNA probes are just now appearing in clinical microbiology laboratories. They have the potential for benefiting the laboratory, both by reducing the time necessary to identify infectious agents with fastidious growth requirements and by extending the spectrum of infectious agents that can be recognized by the laboratory. Probes will be used both for culture confirmation and for direct detection of pathogens in patient specimens. The cost-effectiveness of these tests is still uncertain. Probes offer smaller laboratories the opportunity to reduce fee-basis testing and the opportunity to increase revenue from outpatient testing. When deciding to use probes, one must carefully weigh the cost of false positive and false negative results, as well as the need for follow-up tests such as antimicrobial susceptibility tests and serological identification of isolates. The potential of probes to alter the course of clinical microbiology is indeed great.

ACKNOWLEDGMENTS

I gratefully thank Drs. Richard Tilton, Carol Spiegel, and Marie Coyle for their valuable input and comments on this manuscript.

REFERENCES

1. **Tilton, R. C.,** Where do we go from here?, *Clin. Microbiol. Newsl.,* 8, 80, 1986.
2. **Bartlett, R. C.,** Effect of DRGs on utilization of the microbiology laboratory, *Infect. Control,* 6, 323, 1985.
3. **Ferraro, M. J.,** Rapidity, completeness, and accuracy: their roles in cost effective patient care-laboratory perspectives, in *The Role of Clinical Microbiology in Cost-Effective Health Care,* Smith J. W., Ed., College of American Pathologists, Skokie, Ill., 1985, 99.
4. **Isenberg, H. D., Washington, J. A., II, Balows, A., and Sonnenwirth, A. C.,** Collection, handling and processing of specimens, in *Manual of Clinical Microbiology, 4th Ed.,* Lennette, E. H., Balows, A., Hausler, W. J., Jr., and Shadomy, H. J., Eds., American Society for Microbiology, Washington, D.C., 1985, 73.
5. **Barry, A. L.,** *The Antimicrobial Susceptibility Test: Principles and Practices,* Lea & Febiger, Philadelphia, 1976.
6. **Washington, J. A., II,** The clinical microbiology laboratory: utilization and cost-effectiveness, *Am. J. Med.,* 78(6B), 8, 1985.
7. **Perine, O. L., Totten, P. A., Holmes, K. K., Sng, E. H., Ratnam, A. V., Widy-Wersky, R., Nsanze, H., Habet-Gabr, E., and Westbrook, W. G.,** Evaluation of a DNA-hybridization method for detection of African and Asian strains of *Neisseria gonorrhoeae* in men with urethritis, *J. Infect. Dis.,* 152, 59, 1985.
8. **Ogle, J. W., Janda, J. M., Woods, D. E., and Vasil, M. L.,** Characterization and use of a DNA probe as an epidemiological marker for *Pseudomonas aeruginosa, J. Infect. Dis.,* 155, 119, 1987.
9. **Tenover, F. C., Gootz, T. D., Gordon, K. P., Tompkins, L. S., Young, S. A., and Plorde, J. J.,** Development of a DNA probe for the structural gene of the 2″-*O*-adenyltransferase aminoglycoside modifying enzyme, *J. Infect. Dis.,* 150, 678, 1984.
10. **Edelstein, P. H.,** Evaluation of the Gen-Probe DNA probe for detection of *Legionellae* in culture, *J. Clin. Microbiol.,* 23, 481, 1986.
11. **Wilkinson, H. W., Sampson, J. S., and Plikaytis, B. B.,** Evaluation of a commercial gene probe for identification of *Legionella* cultures, *J. Clin. Microbiol.,* 23, 217, 1986.

12. **Edelstein, P. H., Bryan, R. N., Enns, R. K., Kohne, D. E., and Kacian, D. L.,** Retrospective study of Gen-Probe rapid diagnostic system for detection of Legionellae in frozen clinical respiratory tract samples, *J. Clin. Microbiol.*, 25, 1022, 1987.
13. **Drake, T. A., Hindler, J. A., Berling, O. G. W., and Bruckner, D. A.,** Rapid identification of *Mycobacterium avium* complex in culture using DNA probes, *J. Clin. Microbiol.*, 25, 1442, 1987.
14. **Kiehn, T. E. and Edwards, F. F.,** Rapid identification using a specific DNA probe of *Mycobacterium avium* complex from patients with acquired immunodeficiency syndrome, *J. Clin. Microbiol.*, 25, 1551, 1987.
15. **Echeverria, P., Taylor, D. N., Seriwatana, J., and Moe, C.,** Comparative study of synthetic oligonucleotide and cloned polynucleotide enterotoxin gene probes to identify enterotoxigenic *Escherichia coli*, *J. Clin. Microbiol.*, 25, 106, 1987.
16. **Seriwatana, J. P., Echeverria, P., Taylor, D. N., Sakuldaipeara, T., Changchawalit, S., and Chivoratanond, O.,** Identification of enterotoxigenic *Escherichia coli* with synthetic alkaline phosphatase-conjugated oligonucleotide probes, *J. Clin. Microbiol.*, 25, 1438, 1987.
17. **Peterson, E. M., Aarnaes, S. L., Bryan, R. N., Ruth, J. L., and de la Maza, L. M.,** Typing of Herpes Simplex Virus with DNA probes, *J. Infect. Dis.*, 153, 757, 1986.
18. **Edelstein, P. H., Beer, K. B., Sturge, J. C., Watson, A. J., and Goldstein, L. C.,** Clinical utility of a monoclonal direct fluorescent reagent specific for *Legionella pneumophila*: comparative study with other reagents, *J. Clin. Microbiol.*, 22, 419, 1985.
19. **Fung, J. C., Shanley, J., and Tilton, R. C.,** Comparison of the detection of Herpes Simplex Virus in direct clinical specimens with Herpes Simplex virus-specific DNA probes and monoclonal antibodies, *J. Clin. Microbiol.*, 22, 748, 1985.
20. **Corey, L.,** Laboratory diagnosis of Herpes Simplex Virus infections. Principles guiding the development of rapid diagnostic tests, *Diagn. Microbiol. Infect. Dis.*, 4, 111S, 1986.
21. **Kean, B. H.,** Travelers' diarrhea: an overview, *Rev. Infect. Dis.*, 8, S111, 1986.
22. **Bartlett, R. C.,** *Medical Microbiology: Quality, Cost, and Clinical Relevance*, John Wiley & Sons, New York, 1974.
23. **Ryan, K. J.,** Laboratory use under DRGs: laboratory perspective, in *The Role of Clinical Microbiology in Cost-Effective Health Care*, Smith, J.W., Ed., College of American Pathologists, Skokie, Ill., 1985, 86.
24. **Westerman, E. L.,** Medical perspective–rapidity, completeness, and accuracy: their roles in cost-effective patient care, in *The Role of Clinical Microbiology in Cost-Effective Health Care*, Smith, J. W., Ed., College of American Pathologists, Skokie, Ill., 1985, 91.
25. **Koplan, J. P., Fineberg, H. V., Ferraro, M. J. B., and Rosenberg, M. L.,** Value of stool cultures, *Lancet*, 2, 413, 1980.
26. **Guerrant, R. L., Shields, D. S., Thorson, S. M., Schorling, J. B., and Groschel, D. H. M.,** Evaluation and diagnosis of acute infectious diarrhea, *Am. J. Med.*, 78(6B), 91, 1985.
27. **Rosenblatt, J. E.,** Laboratory diagnosis of infectious diarrhea: a clinician's viewpoint, in *The Role of Clinical Microbiology in Cost-Effective Health Care*, Smith, J. W., Ed., College of American Pathologists, Skokie, Ill., 1985, 105.
28. **Schroeder, S. A., Myers, L. P., McPhee, S. J., Showstack, J. A., Simborg, D. W., Chapman, S. A., and Leong, J. K.,** The failure of physician education as a cost containment strategy: report of a prospective controlled trial at a university hospital, *JAMA*, 252, 225, 1984.
29. **Roberts, M. C., Hillier, S. L., Schoenknecht, F. D., and Holmes, K. K.,** Comparison of gram stain, DNA probe, and culture for the identification of species of *Mobiluncus* in female genital specimens, *J. Infect. Dis.*, 152, 74, 1985.
30. **Polsky-Cynkin, R., Parsons, G. H., Allerdt, L., Landes, G., Davis, G., and Rashtchian, A.,** Use of DNA immobilized on plastic and agarose supports to detect DNA by sandwich hybridization, *Clin. Chem.*, 31, 1438, 1985.
31. **Boguslawski, S. J., Smith, D. E., Michalak, M. A., Mickelson, K. E., Yehle, C. O., Patterson, W. L., and Carrico, R. J.,** Characterization of monoclonal antibody to DNA·RNA and its application to immunodetection of hybrids, *J. Immunol. Methods*, 89, 123, 1986.
32. **Myerson, D., Hackman, R. C., and Meyers, J. D.,** Diagnosis of cytomegalovirus pneumonia by *in situ* hybridization, *J. Infect. Dis.*, 150, 272, 1984.
33. **Saiki, R. K., Scharf, S., Faloona, F., Mullis, K. B., Horn, G. T., Erlich, H. A., and Arnheim, N.,** Enzymatic amplification of beta-globin genomic sequences and restriction site analysis for diagnosis of sickle cell anemia, *Science*, 230, 1350, 1985.
34. **Noel, J. K.,** DNA probes vs monoclonal antibodies, *Am. Clin. Prod. Rev.*, 5, 10, 1986.
35. **Schachter, J.,** Rapid diagnosis of sexually transmitted diseases-speed has its price, *Diagn. Microbiol. Infect. Dis.*, 4, 185, 1986.
36. **Galen, R. S. and Gambino, S. R.,** *Beyond Normality: The Predictive Value and Efficiency of Medical Diagnoses*. John Wiley & Sons, New York, 1975.
37. **Hallam, K.,** The rising trend in outpatient testing, *Med. Lab. Obs.*, 18, 24, 1986.

Chapter 4

DETECTION OF BACTERIA IN FOODS USING DNA HYBRIDIZATION

Walter E. Hill

TABLE OF CONTENTS

I. INTRODUCTION

Traditionally, a major concern of food microbiologists has been the rapid identification of particular bacterial species in foods. The identification was usually accomplished through the use of selective enrichments followed by biochemical classification or serological confirmation of the resulting pure cultures. Considerable importance was placed on developing methods to enumerate viable microbial cells in contaminated foodstuffs. Effort was also expended to solve the special problems arising from the physical or chemical properties of natural or processed foods and from indigenous microbial populations, which caused interference in the examination of some foods.

Recently, research emphasis has been shifted toward developing assays for the identification of pathogenic strains of certain species, although schemes for determining precise taxonomic assignments are still of considerable interest. The development of methods to identify virulence determinants and to quantify microbial loads is an outgrowth of the increasing need to evaluate the microbiological health hazards posed by contaminated foods. Nevertheless, the overall microbiological quality of food, which is in part reflected by the numbers and types of microbes present, is of primary importance to food microbiologists and public health officials.

The recent emphasis on the assessment of pathogenicity is reflected in some of the current compilations of microbiological methods for the examination of foods,[1,2] most of which deal with virulence testing using immunological techniques and animal models. Only recently have gene probes been considered for the detection of foodborne pathogens. The advantages and disadvantages of these probes as compared with other methods have been discussed.[3]

Techniques to determine the presence of foodborne viruses in foods have not yet proved efficacious, and the problems associated with them will be discussed here only briefly. To date, no reports have described methods for the detection of foodborne fungi, parasitic protozoa, or worms.

II. PROBLEMS WITH TESTING FOODS

Two major problems must be overcome to successfully apply DNA probe tests to detect bacteria in foods: the many types of laboratory samples to be prepared and low cell numbers. Many laboratory samples are handled by blending a tenfold dilution in buffer and then plating various dilutions. However, this procedure may reduce the sensitivity below that required for clinically significant levels of bacterial contamination. Recent research has been focusing on enrichments of homogenized foods so that bacteria can be collected on filters suitable for use in DNA hybridization or so that liquid hybridizations can be performed.

Although some methods used in the hospital laboratory can be applied to the microbiological analysis of foods, differences between clinical specimens and food samples often require quite separate procedures. An important difference is the heterogeneous and variable nature, in both composition and microflora, of foods as compared with clinical specimens. There are many more classes of foods than the dozen or so types of patient specimens usually encountered in a clinical microbiology laboratory. The diversity of bacterial, fungal, and protozoan species and the numbers of such organisms that are considered to be significant may also be much greater in foods than in clinical specimens.

The viable cell titers in foods may be quite low in contrast to the numbers found in the human body, which often serves as an excellent growth and enrichment medium. Thus, although the bacteria responsible for a disease may be abundant in clinical specimens, they may be present in very low numbers in foods suspected as being the vehicles for the etiologic agent of an outbreak. It is not surprising, then, that food microbiologists have expended a considerable amount of effort to devise a variety of selective enrichments to increase the proportion of bacteria that they are seeking to isolate and identify.

Because bacterial cell numbers per gram of food may be quite low, cells capable of reproduction must be present to allow amplification of the bacterial DNA sequences to detectable levels. Therefore, colony hybridization methods may not be useful for the analysis of foods where preformed toxin is present, such as in outbreaks caused by toxins produced by *Staphylococcus aureus* or *Clostridium botulinum.* The correct term for these food poisoning diseases should be intoxications rather than infections. Viable cells of these genera may not be present in the implicated foods; therefore, other methods of analysis must be used to search for the presence of the gene products themselves, in this case, the toxins. In fact, because of food processing steps such as pasteurization, large quantities of dead cells of certain species may be present in sufficient numbers to yield positive results with a sensitive genetic probe. The ingestion of bacterial DNA, notwithstanding, it may be more reliable to search only for viable cells using growth as a selective criterion when assessing biological health hazards.

Competition with indigenous microflora may prevent the proportion or the cell numbers of a pathogenic strain from increasing to a level required for detection. When dealing with bacteria with low infectious doses (ID) (for example, *Salmonella* and *Campylobacter* species, where the ID_{50} for humans may be less than 10^3 cells),[4,5] proper selective enrichment procedures are a critical part of the method. If a sufficient number of *Salmonella* cells can be obtained (about 10^6 to 10^7 cells per ml of a selective enrichment), false negative results decrease, but rare false positives can still occur.[6]

The considerable amount of food and bacterial colony debris encountered after homogenization of foods by blending has so far discouraged the use of biotin-labeled probes because of problems with high background; most reported work has been with ^{32}P probes. Also, because bacterial contamination of foods may be at levels considerably below those found in clinical specimens, the 10- to 100-fold greater sensitivity of ^{32}P is often critical for analysis.[7] However, much proprietary research is being conducted, and a wide range of substrates are being considered as candidates for probe labels. Even the sensitivity of biotin-labeled probes has been improved.[8]

Physiologically demanding cultural conditions, such as elevated or lowered temperatures, and the addition of selective agents, such as detergents, toxic inhibitors, or antibiotics, can have serious consequences for the recovery of pathogenic strains of bacteria from foods. Although the method used to detect *Escherichia coli* in foods may be well suited for the identification and enumeration of "environmental" isolates or so-called "normal" flora, these procedures may fail to recover up to 50% of the strains pathogenic for humans.[9-11]

Bacterial cells that have been cured of plasmids often have a faster growth rate[12] under conditions of physiological stress during selective enrichment. Since many of the genes that encode virulence determinants are found on plasmids,[13] these cells may be isolated from foods, but cannot subsequently be identified as pathogenic strains in the laboratory, even though they may be pathogenic when recovered from patient specimens that have been analyzed in a clinical setting by quite different procedures.

Bacterial cells in foods may be injured by the treatment of food during preparation or because of exposure to components of the food product such as preservatives. A parallel situation may occur in the clinical laboratory when recovery of bacteria from patients undergoing antibiotic therapy is attempted. Schemes for recovering microorganisms from foods may include a resuscitation step, that is, a period of nonselective growth before harsher selective environments are imposed.[15,16]

III. ADVANTAGES OF PROBE TESTS

Compared with many of the current methods in widespread use for the microbiological analysis of foods, genetic probes have two major advantages: they are faster, and they obviate the need for pure cultures. In the traditional scheme, after cultures have been isolated and the genus and (perhaps) the species are determined, virulence tests are conducted to demonstrate the presence of pathogenic determinants such as toxins, hemolysins, and/or colonization factors. Currently, such tests may require special facilities or may be time consuming and quite expensive to perform. Results can often be quite variable, especially when compared with those of other laboratories. In addition, these methods may not be well suited for rapid screening of the large numbers of samples collected during an outbreak of foodborne disease or during surveys of food processing establishments, raw commodities, and retail products.

To be valid, tests for virulence that rely on gene expression (so-called phenotypic tests) must often be carried out by using pure bacterial cultures. Such a requirement may delay the reporting of results for a considerable period. For example, it may take more than a week to isolate and confirm the presence of *E. coli*,[17] and a month may be necessary to detect *Listeria monocytogenes*.[18] Such lags in obtaining data can seriously hamper the investigation of an outbreak, particularly when further action depends on initial results. The response required to pinpoint the source of disease and to institute a fecal of contaminated foodstuffs can be seriously delayed. Furthermore, long testing periods are not suitable for the analysis of a food with a short shelf life. If foods must be held until testing is complete, distribution is delayed and storage costs can add considerably to the cost of processing.

A last consideration of procedures to be adopted for widespread use is reliability, especially when the food samples and the results of tests on them may become evidence in a legal proceeding. Confidence in the results of any test accepted for use as an industry or regulatory standard is essential.

IV. DETECTION OF VIRULENT BACTERIA

An important question for both public health officials and food producers who are concerned about the microbial quality of foods is, "Does this food contain microorganisms that pose a health hazard to consumers?" The answer to this question is not easily determined for two major reasons. First, only some of the bacterial species in a food may have pathogenic potential. Because not all strains of a particular species are pathogenic to humans, it is necessary to develop methods for testing the virulence of foodborne bacteria, particularly among enteric bacteria. *E. coli*, for example, is often considered to be normal flora, but is known to cause human disease by several different mechanisms;[19-22] however, much progress has been made in elucidating the genetic bases of diseases caused by this organism.[23] Second, it is not always a straightforward matter to assess the potential virulence for humans of a strain in the laboratory. The advantages and disadvantages of genetic vs. phenotypic tests has been discussed.[3]

Many aspects of DNA hybridization tests make them attractive candidates as methods for the examination of food to identify and enumerate specific organisms or particular virulence determinants. However, only a few studies have reported data on their reliability (especially between-laboratory repeatability) compared with other methods.[24,25] DNA hybridization procedures have proven their utility as tools for bacterial identification and for elucidating taxonomic relationships.[26,27] When ^{32}P-labeled DNA probes are used to test the DNA of bacterial cells lysed on filters (colony hybridization), an accuracy of about 95% is observed for pure cultures tested with a particular natural (cloned) probe.[25,28] When the nucleotide sequence of the gene being sought is known (the target sequence), recent advances in oligodeoxyribonucleotide synthesis have allowed the construction, in the laboratory, of short sequences of DNA usually 15 to 25 bases in length. These "synthetic" DNA probes are as accurate in colony hybridization tests as natural probes.[29]

The first reports of the use of DNA colony hybridization to detect pathogenic strains involved clinical specimens.[30,31] Shortly thereafter, this technique was shown to be efficient in detecting and enumerating strains of enterotoxigenic *E. coli*[32] and virulent *Yersinia enterocolitica*[33] in seeded foods such as dairy products, seafoods, vegetables, and meats. DNA hybridization has also been used as a qualitative method for determining the presence or absence of many species of *Salmonella*.[34,35]

The sensitivity of DNA colony hybridization methods makes it possible to detect ten enterotoxigenic *E. coli* colonies in a background of one million nonenterotoxigenic colonies on a single agar plate.[32] However, this level of sensitivity may be severely reduced by heavy loads of indigenous microflora that will overgrow the strains being sought. For example, foods contaminated with *Y. enterocolitica* may have high microbial titers of other species, making it difficult for *Y. enterocolitica* to grow sufficiently to increase the number of target gene copies to detectable levels.[36] However, the use of alkaline enrichment[37] allowed detection by colony hybridization of 66 to 100% of virulent *Y. enterocolitica* cells seeded into nine food types, even when background flora reached 3×10^8 per gram.[36] After enterotoxigenic *E. coli* were seeded into ten food types, 37 to 100% of the seeded cells were enumerated by DNA colony hybridization, using a cloned heat-labile enterotoxin gene probe.[38] No significant difference was observed between plate count and eosin methylene blue agars, demonstrating that a selective agar does not appreciably decrease the efficiency of the colony hybridization method.

The food types examined so far do not inhibit the detection of *E. coli* and *Y. enterocolitica*. The most serious obstacles appear to be the number and nature of the indigenous microflora that may grow faster than the particular species being sought. Intelligent use of selective enrichments may do much to alleviate this problem.

V. OTHER FOODBORNE PATHOGENS

In addition to bacteria, other classes of foodborne organisms posing a biological health hazard to humans include viruses, fungi, protozoa, and worms. Because of the scarcity of gene probe work on the latter three groups, this section will concentrate on viruses.

Many viruses, including hepatitis A and other enteroviruses, Norwalk-like viruses, rotaviruses, enteric adenoviruses, caliciviruses, and parvoviruses may be transmitted through foods.[39,40] The analysis of foods for the presence of these viruses using hybridization methods is still in its infancy. Hybridization methods have been described for determining the presence of enteroviruses,[41] rotaviruses,[42] enteric adenoviruses,[43] and parvoviruses.[44,44a] For most other possible food-associated gastroenteritis viruses, difficulties in obtaining enough virus particles have hampered the cloning and/or sequencing efforts necessary for the preparation of nucleic acid hybridization probes. Naturally contaminated foods usually contain very few virions, and many of the viruses are refractory to efforts aimed at stimulating them to replicate in the laboratory.

There have been no reports describing the use of hybridization methods for the detection of food-associated virions. Basically, the two most serious problems preventing implementation of plaque hybridization methods are sensitivity and amplification of target DNA to detectable levels. Viruses cannot replicate in processed foods. When they are present, it is usually in very low numbers, and they are difficult to separate quantitatively from food materials.[40,45,45] The cited methods usually do not give total recovery of some viruses from foods, and it is not known whether such procedures will work with other foodborne viral pathogens.

Since low particle numbers prevent the direct detection of viruses by hybridization procedures, the number of particles must be increased before analysis. The amount of viral material is usually increased by growing viruses that have been recovered from foods on sensitive tissue culture cell systems; however, these systems have proved effective for only a few viruses. Tissue culture cells which are able to replicate hepatitis A virus[47,48] and rotaviruses[49,50] require that the virus first be adapted to the culture system. Therefore, such tests are not suitable for the routine or rapid amplification of viral isolates directly from fresh foods. At present, no such system can be used with the Norwalk group of viruses.[51] However, once these obstacles are overcome and sufficient numbers of viral nucleic acid target molecules can be readily obtained, the develop-

Table 1
DNA PROBES FOR FOODBORNE PATHOGENIC BACTERIA (IN USE OR UNDER DEVELOPMENT)[a]

Organism	Gene(s)/virulence determinant	Ref.
Aeromonas hydrophila	Cytotonic toxin[b]	60
Campylobacter jejuni	Species specific	61, 62
Clostridium botulinum	Neurotoxin A[b,c]	63
	Neurotoxins B[b,c] and E[b,c]	64, 65
Escherichia coli	Heat-labile enterotoxins (LT_h,[b] LT_p[b])	32, 66
	Heat-stable enterotoxins (ST-H,[b] ST-P[b])	57, 66, 67
	"Invasiveness" (also *Shigella*)	68
	Shiga-like (Vero) toxin	69, 70
Listeria monocytogenes	β-Hemolysin	83
Salmonella spp.	Genus specific	34, 35
Shigella	"Invasiveness" (also *E. coli*)	71
Staphylococcus aureus	Enterotoxin A	72
	Enterotoxins B[b,c] and C_1[b,c]	73, 74
Vibrio cholerae	O1 and non-O1 enterotoxin[b]	75—77
	Cytotoxin-hemolysin	78
V. parahaemolyticus	Thermostable direct hemolysin[b]	79—81
V. vulnificus	Cytotoxin-hemolysin	82
Yersinia enterocolitica	"Invasiveness"	33

[a] If a report on the use of a synthetic probe has not yet been published, citation is made to the sequence data.
[b] Synthetic probe constructed or under development.
[c] Oligonucleotide probes based on amino acid sequence of gene product.

ment of methods for analyzing foods for viruses will probably parallel that for foodborne bacteria. DNA probes are used with clinical specimens to quantitate viral genomes in dot blots[43,52-54] or to characterize particular isolates by hybridization of Southern blots of restriction endonuclease fragment patterns obtained after gel electrophoresis.[55]

Further tests to detect viruses in foods will probably rely heavily on methods developed for viruses in clinical specimens.[43,52] Clearly, improved procedures are needed to isolate viruses from foods, to amplify them, and to detect them by hybridization. Current tests are generally not satisfactory for the routine use of DNA hybridization for determining the presence of viruses in foods; however, with further development of probe signal enhancement systems,[56] hybridization may be sensitive enough to detect viruses without amplification.

VI. SUMMARY

DNA probes for the detection of bacteria in foods using hybridization has almost come of age. Because of the extraordinary amount of current research and development of probes, labeling methods, and sampling systems, we will soon see a dramatic increase in the frequency of their use and a rapid expansion of the scope of their application, at least for the detection of bacteria. Much research needs to be applied to developing adequate methods for the detection of fungi, protozoa, and worms.

Because of the large amount of debris in homogenated foods, some nonisotopic labels that work well with purified DNAs may develop unsatisfactorily high background levels. Although some methods overcome the problems of debris, radioactive ^{125}I and ^{32}P are often used as probe labels.[32,57-59] Wider acceptance of DNA hybridization techniques for use with foods depends on the development of sensitive and rapid assay systems that do not require radioactive components. The high level of research activity in these areas will soon bring about methods that will enjoy broad implementation.

The exquisite sensitivity of the techniques of analytical chemistry, such as nuclear magnetic resonance and electron spin resonance, may yield exceedingly rapid results. Attomole quantities of target DNA sequences might be determined in a few seconds as the course of hybridization is followed in real time. With such detection systems, sensitive probes capable of detecting fewer than ten cells may be developed.

The use of synthetic probes will speed assay time, and with the ready availability of laboratory-assembled oligodeoxyribonucleotides, higher probe concentrations can be used to drive reactions faster; hybridizations may be conducted in 1 h or less by conventional detection systems. These rapid assay systems may provide results in less than one laboratory shift. Such protocols would be a boon to the microbiological testing of foods with short shelf lives or which require relatively expensive storage facilities.

Finally, as scientists, technicians, and production laboratory supervisors become more familiar with DNA hybridization methods, they will appreciate the comparative simplicity, reliability, and rapidity of the tests. A number of natural and synthetic probes, for species and for particular genes, are already in use or under development (Table 1). Furthermore, because of the common methodology in all such tests, personnel training costs will be greatly reduced. For those involved in the microbiological analysis of foods, these new techniques hold the promise of a radical departure from traditional procedures, taking full advantage of powerful screening methods based on genetic assays.

ACKNOWLEDGMENT

The author thanks Shay Fout, Food and Drug Administration, Cincinnati, Ohio, for assistance in preparing the section on viruses.

REFERENCES

1. **Speck, M. L., Ed.,** *Compendium of Methods for the Microbiological Examination of Foods.,* 2nd ed., American Public Health Association, Washington, D.C., 1984.
2. Food and Drug Administration, *Bacteriological Analytical Manual,* 6th ed., Association of Official Analytical Chemists, Arlington, Va., 1984.
3. **Hill, W. E.,** DNA hybridization method for detecting enterotoxigenic *Escherichia coli* in human isolates and its possible application to food samples, *J. Food Saf.,* 3, 233, 1981.
4. **Blaser, M. J. and Newman, L. S.,** A review of human salmonellosis; I. Infective dose, *Rev. Infect. Dis.,* 4, 1096, 1982.
5. **Robinson, D. A.,** Infective dose of *Campylobacter jejuni* in milk, *Br. Med. J.,* 282, 1584, 1981.
6. **Flowers, R. S.,** Comparison of rapid *Salmonella* screening methods and the conventational culture method, *Food Technol. (Chicago),* 39, 103, 1985.
7. **Zwadyk, P., Cooksey, R. C., and Thornsberry, C.,** Commercial detection methods for biotinylated gene probes: comparison with ^{32}P labeled DNA probes, *Abstr. Annu. Meet. Am. Soc. Microbiol.,* American Society for Microbiology, Washington, D.C., 1985, 355.
8. **Bialkowska-Hobrzanska, H.,** Evaluation of the sensitivity and specificity of biotinylated DNA probe hybridization (BDPH) assay in the diagnosis of enterotoxigenic *Escherichia coli* (ETEC), *Abstr. Annu. Meet. Am. Soc. Microbiol.,* American Society for Microbiology, Washington, D.C., 1986, 355.
9. **Ferreira, J. L., Hamday, M. K., White, W. D., McCay, S., and Zapatka, F.,** Effect of isolation procedures on survival and plasmid loss of enterotoxigenic *Escherichia coli, Ga. J. Sci.,* 43, 71, 1985.
10. **Mehlman, I. J. and Romero, A.,** Enteropathogenic *Escherichia coli*: methods for recovery from foods, *Food Technol. (Chicago),* 36, 73, 1982.

11. **Hill, W. E., Ferreira, J. L., Payne, W. L., and Jones, V. M.,** Probability of recovering pathogenic *Escherichia coli* from food, *Appl. Environ. Microbiol.*, 49, 1374, 1985.
12. **Zund, P. and Lebek, G.,** Generation time-prolonging R plasmids: correlation between increases in the generation time of *Escherichia coli* caused by R plasmids and their molecular size, *Plasmid*, 3, 65, 1980.
13. **Elwell, L. P. and Shipley, P. L.,** Plasmid-mediated factors associated with virulence of bacteria to animals, *Annu. Rev. Microbiol.*, 34, 465, 1980.
14. **Hill, W. E. and Carlisle, C. L.,** Loss of plasmids during enrichment for *Escherichia coli*, *Appl. Environ. Microbiol.*, 41, 1046, 1981.
15. **Egan, A. F.,** Enumeration of stressed cells of *Escherichia coli*, *Can. J. Microbiol.*, 25, 116, 1979.
16. **Andrews, W. H.,** Resuscitation of injured *Salmonella* spp. and coliforms from foods, *J. Food Prot.*, 49, 62, 1986.
17. **Mehlman, I. J., Andrews, W. H., and Wentz, B. A.,** Coliform bacteria, in *Bacteriological Analytical Manual*, 6th ed., Association of Official Analytical Chemists, Arlington, Va., 1984, chap. 5.
18. **Hayes, P. S., Feeley, J. C., Graves, L. M., Ajello, G. W., and Fleming, D. W.,** Isolation of *Listeria monocytogenes* from raw milk, *Appl. Environ. Microbiol.*, 51, 438, 1986.
19. **Sack, R. B.,** Human diarrheal disease caused by enterotoxigenic *Escherichia coli*, *Annu. Rev. Microbiol.*, 29, 333, 1975.
20. **Konowalchuk, J., Speirs, J. I., and Stavric, S.,** Vero response to a cytotoxin of *Escherichia coli*, *Infect. Immun.*, 18, 775, 1977.
21. **Nandadasa, H. G., Sargent, G. F., Brown, M. G., McNeish, A. S., and Williams, P. H.,** The role of plasmids in adherence of invasive *Escherichia coli* to mammalian cells, *J. Infect. Dis.*, 143, 286, 1981.
22. **Gross, R. J.,** *Escherichia coli* diarrhoea, *J. Infect.*, 7, 177, 1983.
23. **Smith, H. R., Scotland, S. M., and Rowe, B.,** Genetics of *Escherichia coli* virulence, in *The Virulence of Escherichia coli*, Sussman, M., Ed., Academic Press, New York, 1985.
24. **Edberg, S. C.,** Principles of nucleic acid hybridization and comparison with monoclonal antibody technology for the diagnosis of infectious diseases, *Yale J. Biol. Med.*, 58, 425, 1985.
25. **Echeverria, P., Taylor, D. N., Seriwatana, J., Chatkaeomorakot, A., Khungvalert, V., Sakuldaipeara, T., and Smith, R. D.,** A comparative study of enterotoxin gene probes and tests for toxin production to detect enterotoxigenic *Escherichia coli*, *J. Infect. Dis.*, 153, 255, 1986.
26. **Brenner, D. J., Fanning, G. R., Rake, A. V., and Johnson, K. E.,** Batch procedure for thermal elution of DNA from hydroxyapatite, *Anal. Biochem.*, 28, 447, 1969.
27. **Brenner, D. J.,** Impact of modern taxonomy on clinical mcrobiology, *Am. Soc. Microbiol. News*, 49, 58, 1983.
28. **Hill, W. E. and Payne, W. L.,** Genetic methods for the detection of microbial pathogens. Identification of enterotoxigenic *Escherichia coli* by DNA colony hybridization: collaborative study, *J. Assoc. Off. Anal. Chem.*, 67, 801, 1984.
29. **Hill, W. E., Wentz, B. A., Payne, W. L., Jagow, J. A., and Zon, G.,** DNA colony hybridization method using synthetic oligonucleotides to detect enterotoxigenic *Escherichia coli*: collaborative study, *J. Assoc. Off. Anal. Chem.*, 69, 531, 1986.
30. **Moseley, S. L., Huq, I., Alim, A. R. M. A., So, M., Samadpour-Motalebi, M., and Falkow, S.,** Detection of enterotoxigenic *Escherichia coli* by DNA colony hybridization, *J. Infect. Dis.*, 142, 892, 1980.
31. **Moseley, S. L., Echeverria, P., Seriwatana, J., Tirapat, C., Chaicumpa, W., Sakuldaipeara, T., and Falkow, S.,** Identification of enterotoxigenic *Escherichia coli* by colony hybridization using three enterotoxin gene probes, *J. Infect. Dis.*, 145, 863, 1982.
32. **Hill, W. E., Madden, J. M., McCardell, B. A., Shah, D. B., Jagow, J. A., Payne, W. L., and Boutin, B. K.,** Foodborne enterotoxigenic *Escherichia coli*: detection and enumeration by DNA colony hybridization, *Appl. Environ. Microbiol.*, 45, 1324, 1983.
33. **Hill, W. E., Payne, W. L., and Aulisio, C. C. G.,** Detection and enumeration of virulent *Yersinia enterocolitica* in food by DNA colony hybridization, *Appl. Environ. Microbiol.*, 46, 636, 1983.
34. **Fitts, R., Diamond, M., Hamilton, C., and Neri, M.,** DNA-DNA hybridization assay for detection of *Salmonella* spp. in foods, *Appl. Environ. Microbiol.*, 46, 1146, 1983.
35. **Fitts, R.,** Development of a DNA-DNA hybridization test for the presence of *Salmonella* in foods, *Food Technol. (Chicago)*, 39, 95, 1985.
36. **Jagow, J. and Hill, W. E.,** Enumeration by DNA colony hybridization of virulent *Yersinia enterocolitica* colonies in artificially contaminated food, *Appl. Environ. Microbiol.*, 51, 441, 1986.
37. **Aulisio, C. C. G., Mehlman, I. J., and Sanders, A. C.,** Alkali method for rapid recovery of *Yersinia enterocolitica* and *Yersinia pseudotuberculosis* from foods, *Appl. Environ. Microbiol.*, 39, 135, 1980.
38. **Ferreira, J. L., Hill, W. E., Hamdy, M. K., Zapatka, F. A., and McCay, S. G.,** Detection of enterotoxigenic *Escherichia coli* in foods by DNA colony hybridization, *J. Food Sci.*, 51, 665, 1986.
39. **Larkin, E. P.,** Food contaminants — viruses, *J. Food Prot.*, 44, 320, 1981.
40. **Larkin, E. P.,** Detection, quantitation and public health significance of foodborne viruses, in *Foodborne Microorganisms and Their Toxins*, Pierson, M. D. and Stern, N. J., Eds., Marcel Dekker, New York, 1986, 439.

41. **Rotbart, H. A., Levin, M. J., and Villarreal, L. P.,** Use of subgenomic poliovirus DNA hybridization probes to detect the major subgroups of enteroviruses, *J. Clin. Microbiol.*, 20, 1105, 1984.
42. **Flores, J., Perez-Schael, I., Boeggeman, E., White, L., Perez, M., Purcell, R., Hoshino, Y., Midthun, K., Chanock, R. M., and Kapikian, A. Z.,** Genetic relatedness among human rotaviruses, *J. Med. Virol.*, 17, 135, 1985.
43. **Takiff, H. E., Seidlin, M., Krause, P., Rooney, J., Brandt, C., Rodriguez, W., Yolken, R., and Straus, S. E.,** Detection of enteric adenoviruses by dot-blot hybridization using a molecularly cloned viral DNA probe, *J. Med. Virol.*, 16, 107, 1985.
44. **Cotmore, S. F. and Tattersall, P.,** Characterization and molecular cloning of a human parvovirus genome, *Science*, 226, 1161, 1984.
44a. **Fout, G. S.,** Food and Drug Administration, unpublished data.
45. **Larkin, E. P.,** Detection of viruses in foods, in *Methods in Environmental Virology*, Gerba, C. P. and Goyal, S. M., Eds., Marcel Dekker, New York, 1982, 221.
46. **Sullivan, R., Peeler, J. T., Tierney, J. T., and Larkin, E. P.,** Evaluation of a method for recovering poliovirus 1 from 100-gram oyster samples, *J. Food Prot.*, 47, 108, 1984.
47. **Siegl, G., de Chastonay, J., and Kronauer, G.,** Propagation and assay of hepatitis A virus in vitro, *J. Virol.*, 9, 53, 1984.
48. **Venuti, A., Di Russo, C., del Grosso, N., Patti, A.-M., Ruggeri, F., De Stasio, P. R., Martiniello, M. G., Pagnotti, P., Degener, M., Midulla, M., Pana, A., and Perez-Bercoff, R.,** Isolation and molecular cloning of a fast-growing strain of human hepatitis A virus from its double-stranded replicative form, *J. Virol.*, 56, 579, 1985.
49. **Agliano, A. M., Rossi, A., and Sanna, A.,** Isolation from faecal specimens of new strains of human rotavirus primarily cytopathic for stationary cell cultures without trypsin, *Arch. Virol.*, 84, 119, 1985.
50. **Offit, P. A., Clark, H. F., Stroop, W. G., Twist, E. M., and Plotkin, S. A.,** The cultivation of human rotavirus strain 'Wa' to high titer in cell culture and characterization of the viral structural polypeptides, *J. Virol. Methods*, 7, 29, 1983
51. **Herrmann, J. E., Nowak, N. A., and Blacklow, N. R.,** Detection of Norwalk virus in stools by enzyme immunoassay, *J. Med. Virol.*, 17, 127, 1985.
52. **Spector, S. A., Rua, J. A., Spector, D. H., and McMillan, R.,** Detection of human cytomegalovirus in clinical specimens by DNA-DNA hybridization, *J. Infect. Dis.*, 150, 121, 1984.
53. **Brandsma, J. and Miller, G.,** Nucleic acid spot hybridization: rapid quantitative screening of lymphoid lines for Epstein-Barr viral DNA, *Proc. Natl. Acad. Sci. U.S.A.*, 77, 6851, 1980.
54. **Virtanen, M., Syvanen, A.-C., Oram, J., Soderlund, H., and Ranki, M.,** Cytomegalovirus in urine: detection of viral DNA by sandwich hybridization, *J. Clin. Microbiol.*, 20, 1083, 1984.
55. **Blum, H. E., Figus, A., Haase, A. T., and Vyas, G. N.,** Laboratory diagnosis of hepatitis B virus infection by nucleic acid hybridization analyses and immunohistologic detection of gene products, in *Developments in Biological Standardization, Standardization and Control of Biologicals Produced by Recombinant DNA Technology*, Vol. 59, Perkins, F. T. and Hennessen, W., Eds., S. Karger, Basel, 1983, 125.
56. **Leary, J. J., Brigati, D. J., and Ward, D. C.,** Rapid and sensitive colorimetric method for visualizing biotin-labeled DNA probes hybridized to DNA or RNA immobilized on nitrocellulose: bio-blots, *Proc. Natl. Acad. Sci. U.S.A.*, 80, 4045, 1983.
57. **Hill, W. E., Payne, W. L., Zon, G., and Moseley, S. L.,** Synthetic oligodeoxyribonucleotide probes for detecting heat-stable enterotoxin-producing *Escherichia coli* by DNA colony hybridization, *Appl. Environ. Microbiol.*, 50, 1187, 1985.
58. **Palva, A. M.,** *omp*A gene in the detection of *Escherichia coli* and other Enterobacteriaceae by nucleic acid sandwich hybridization, *J. Clin. Microbiol.*, 18, 92, 1983.
59. **Ranki, M., Palva, A., Virtanen, M., Laaksonen, M., and Soderlund, H.,** Sandwich hybridization as a convenient method for the detection of nucleic acids in crude samples, *Gene*, 21, 77, 1983.
60. **Schultz, A. J., Hill, W. E., and McCardell, B. A.,** DNA homology and immunological cross reactivity between *Aeromonas hydrophila* cytotonic toxin and cholera toxin, *Abstr. Annu. Meet. Am. Soc. Microbiol.*, American Society for Microbiology, Washington, D.C., 1986, 64.
61. **Rashtchian, A. and Abbott, M.,** Utilization of DNA probes for detection of *Campylobacter* species, *Abstr. Annu. Meet. Am. Soc. Microbiol.*, American Society for Microbiology, Washington, D.C., 1985, 356.
62. **Rashtchian, A., Abbott, M., Lovern, D. H., Mock, G. A., and Shaffer, M. M.,** Use of synthetic oligonucleotides complementary to rRNA for detection of *Campylobacter*, *Abstr. Annu. Meet. Am. Soc. Microbiol.*, American Society for Microbiology, Washington, D.C., 1986, 343.
63. **Schmidt, J. J., Sathyamoorthy, V., and DasGupta, B. R.,** Partial amino acid sequence of the heavy and light chains of botulinum neurotoxin type A, *Biochem. Biophys. Res. Commun.*, 119, 900, 1984.
64. **Sathyamoorthy, V. and DasGupta, B. R.,** Partial amino acid sequences of the heavy and light chains of botulinum neurotoxin type E, *Biochem. Biophys. Res. Commun.*, 127, 768, 1985.

65. **Schmidt, J. J., Sathyamoorthy, V., and DasGupta, B. R.,** Partial amino acid sequences of botulinum neurotoxins types B and E, *Arch. Biochem. Biophys.*, 238, 544, 1985.
66. **Smith, R. D., Trepod, C. M., Tu, E.,** Synthetic oligonucleotide hybridization probes for enterotoxigenic *Escherichia coli, Fed. Proc.*, 44, 527, 1985.
67. **Moseley, S. L., Hardy, J. W., Huq, M. I., Echeverria, P., and Falkow, S.,** Isolation and nucleotide sequence determination of a gene encoding a heat-stable enterotoxin of *Escherichia coli, Infect. Immun.*, 39, 1167, 1983.
68. **Boileau, C. R., d'Hauteville, H. M., and Sansonetti, P. J.,** DNA hybridization technique to detect *Shigella* species and enteroinvasive *Escherichia coli, J. Clin. Microbiol.*, 20, 959, 1984.
69. **Willshaw, G. A., Smith, H. R., Scotland, S. M., and Rowe, B.,** Cloning of genes determining the production of Vero cytotoxin by *Escherichia coli, J. Gen. Microbiol.*, 131, 3047, 1985.
70. **Newland, J. W., Strockbine, N. A., Miller, S. F., O'Brien, A. D., and Holmes, R. K.,** Cloning of Shiga-like toxin structural genes from a toxin converting phage of *Escherichia coli, Science*, 230, 179, 1985.
71. **Maurelli, A. T., Baudry, B., d'Hauteville, H., Hale, T. L., and Sansonetti, P. J.,** Cloning of plasmid DNA sequences involved in invasion of HeLa cell by *Shigella flexneri, Infect. Immun.*, 49, 164, 1985.
72. **Betley, M. J. and Mekalanos, J. J.,** The complete nucleotide and derived amino acid sequences of staphylococcal enterotoxin A, *Abstr. Annu. Meet. Am. Soc. Microbiol.*, American Society for Microbiology, Washington, D.C., 1985, 257.
73. **Huang, I.-Y. and Bergdoll, M. S.,** The primary structure of staphylococcal enterotoxin B. III. The cyanogen bromide peptides of reduced and amino-ethylated enterotoxin B and the complete amino acid sequence, *J. Biol. Chem.*, 245, 3518, 1970.
74. **Schmidt, J. J. and Spero, L.,** The complete amino acid sequence of staphylococcal enterotoxin C_1, *J. Biol. Chem.*, 258, 6300, 1983.
75. **Mekalanos, J. J., Swartz, D. J., Pearson, G. D. N., Harford, N., Groyne, F., and de Wilde, M.,** Cholera toxin genes: nucleotide sequence, deletion analysis and vaccine development, *Nature (London)*, 306, 551, 1983.
76. **Spicer, E. K., Kavanaugh, W. M., Dallas, W. S., Falkow, S., Konigsberg, W. H., and Schafer, D. E.,** Sequence homologies between A subunits of *Escherichia coli* and *Vibrio cholerae* enterotoxins, *Proc. Natl. Acad. Sci. U.S.A.*, 78, 50, 1981.
77. **Hanchalay, S., Seriwatana, J., Echeverria, P., Holmgren, J., Tirapat, C., Moseley, S. L., and Taylor, D. N.,** Non-O1 *Vibrio cholerae* in Thailand: homology with cloned cholera toxin genes, *J. Clin. Microbiol.*, 21, 288, 1985.
78. **Goldberg, S. L. and Murphy, J. R.,** Molecular cloning of the hemolysin determinant from *Vibrio cholerae* El Tor, *J. Bacteriol.*, 160, 239, 1984.
79. **Kaper, J. B., Campen, R. K., Seidler, R. J., Baldini, M. M., and Falkow, S.,** Cloning of the thermostable direct or Kanagawa phenomenon-associated hemolysin of *Vibrio parahaemolyticus, Infect. Immun.*, 45, 290, 1984.
80. **Nishibuchi, M. and Kaper, J. B.,** Nucleotide sequence of the thermostable direct hemolysin gene of *Vibrio parahaemolyticus, J. Bacteriol.*, 162, 558, 1985.
81. **Nishibuchi, M., Hill, W. E., Zon, G., Payne, W. L., and Kaper, J. B.,** Synthetic oligodeoxyribonucleotide probes to detect Kanagawa phenomenon-positive *Vibrio parahaemolyticus, J. Clin. Microbiol.*, 23, 1091, 1986.
82. **Wright, A. C., Morris, J. G., Jr., Maneval, D. R., Jr., Richardson, K. and Kaper, J. B.,** Cloning of the cytotoxin-hemolysin gene of *Vibrio vulnificus, Infect. Immun.*, 50, 922, 1985.
83. **Datta, A. R., Wentz, B. A., and Hill, W. E.,** Detection of hemolytic *Listeria monocytogenes* by using DNA colony hybridization, *Appl. Environ. Microbiol.*, 53, 2256, 1987.

Part III: Nucleic Acid Probes for Diarrheal Pathogens

Chapter 5

DNA PROBES FOR BACTERIAL STOOL PATHOGENS

Fred C. Tenover

TABLE OF CONTENTS

I. INTRODUCTION

Processing stool for the detection of bacterial pathogens is a time-consuming and costly undertaking in a clinical microbiology laboratory. The number of potentially pathogenic microorganisms is large, and many require selective media or special culture conditions for growth. Recognition and identification of pathogens, even after growth on agar media, is often imprecise and involves the use of combinations of biochemical and serological tests. Thus, there is an acute need for more rapid and cost effective methods of detecting and identifying pathogens in stool.

While DNA probes have been used in research situations to detect and identify enteric pathogens, these studies have not been without problems, the most critical of which has been the lower sensitivity of some probes compared to traditional methods of detection. Because of this, probes for enteric pathogens have not yet found their way to the clinical laboratory.

Herein, I will review the use of probes to detect enteric pathogens and explore some of the problems relating to their use in the clinical laboratory. A summary of the probes described is presented in Table 1.

II. REVIEW OF BACTERIAL ENTERIC PATHOGENS

There is a growing number of bacterial pathogens that are recognized as causing diarrhea in humans. *Salmonella* species, *Shigella* species, and *Campylobacter* species are the common pathogens that most laboratories in the U.S. attempt to isolate from stool submitted for culture. In addition, many laboratories include selective media to aid in the isolation of *Yersinia enterocolitica*, *Clostridium difficile*, and *Vibrio parahemolyticus*. Depending on the geographic location of the laboratory, the list of important diarrheal pathogens also may include several pathovars of *Escherichia coli* (EC) that mediate illness either through the elaboration of toxins or by a variety of other mechanisms. These include the heat-labile (LT) and heat-stable (ST) enterotoxin-producing strains (ETEC), the enteroinvasive strains (EIEC) that mediate a *Shigella*-like illness, enteropathogenic strains (EPEC), and the enterohemorrhagic strains (EHEC), such as those belonging to the 0157:H7 serotype. Yet other laboratories include *Aeromonas* spp. and *Plesiomonas* spp. in the list of organisms that they attempt to recover from stool.

To look for even a few of these organisms is time consuming and expensive. In fact, no other procedure is less cost efficient in the microbiology laboratory than stool culture.[1,2] Yet, the identification of individuals harboring stool-borne pathogens is of significance not only in terms of managing the patient's illness, but because of the public health problem related to spread of the organism within the community. Thus, the detection and identification of enteric pathogens is an important aspect of clinical diagnostic microbiology.

III. ENTEROTOXIGENIC *ESCHERICHIA COLI*

Enterotoxigenic *E. coli* are a common cause of diarrhea in developing countries [3] and in Americans who travel abroad.[4] Prior to the development of probes, laboratories had to rely on Y-1 adrenal cell [5] or Chinese hamster ovary cell assays[6] to identify LT-producing strains of *E. coli*, and the suckling mouse assay[8] for detection of ST-producing strains. Both techniques are expensive and require a great deal of expertise to perform and interpret. Subsequently, there have been a number of alternate tests developed to detect ETEC including ELEK tests,[8] ELISA assays,[9] latex particle agglutination tests,[10] and DNA probes.

The first DNA probes to bacterial pathogens were those developed by Moseley and co-workers for the detection of LT-producing ETEC.[11] The technique, a modified version of the colony hybridization test described by Grunstein and Hogness,[12] proved very successful in

Table 1
DNA PROBES TO HUMAN ENTERIC PATHOGENS

Organism	Probe type[a]	(Label[b])	Ref.
Enterotoxigenic *E. coli*			
LT1	RF	(^{32}P)	11, 14—20, 23, 33
	O	(^{32}P, AP)	15, 33
LT2	RF	(^{32}P)	32
STA1	RF	(^{32}P, AP)	14—20, 23, 31, 33
STA2	RF	(^{32}P, AP)	11, 14—20, 31, 33
	O	(^{32}P, AP)	15, 33
STII	RF	(^{32}P)	29, 30
Enteropathogenic *E. coli*	RF	(^{32}P)	38
Shigella spp.	RF	(^{32}P)	42, 45—47
Enteroinvasive *E. coli*	RF	(AP)	45
	RF	(^{32}P)	43, 47
Enterohemorrhagic *E. coli*	RF	(^{32}P)	48
Salmonella spp.	RF	(^{32}P)	54—56
Salmonella typhi	RF	(^{32}P)	57
Campylobacter jejuni	RF	(^{32}P)	64, 65
	O	(AP)	66, 67
Yersinia enterocolitica	RF	(^{32}P)	68, 69

[a] RF = restriction fragment, O = oligonucleotide probe.
[b] AP = alkaline phosphatase.

detecting those colonies that carried the LT gene. Using a probe isolated from the cloned LT enterotoxin gene, [13] Moseley and co-workers were able to identify the presence of the LT enterotoxin gene in colonies of *E. coli* grown overnight on nitrocellulose filters. This technique was also effective in detecting the presence of ETEC in stool samples that had been inoculated onto nitrocellulose filters resting on MacConkey agar plates and grown overnight. Moseley et al. extended the utility of using DNA probes for detecting ETEC by adding probes for STa (one of the heat-stable enterotoxins) to the screening procedure.[14] These, too, proved successful in detecting ST-producing ETEC in stool, showing a high correlation with the suckling mouse assay method.

The results of Moseley et al. were confirmed by Echeverria et al.,[15-18] Patamaroj et al.,[19] Seriwatana et al.,[20] and Georges et al.[21] in several field studies. Echeverria et al., however, reported difficulty with probe preparations contaminated with vector sequences,[17] underlying the need for strict quality control of probe DNA.

In one longitudinal study of the prevalence of ETEC in Thai villages, Echeverria et al. noted that stool samples processed in the reference laboratory in Bangkok demonstrated 94% sensitivity and 100% specificity when compared to the standard Y-1 adrenal cell and suckling mouse assays for LT and ST, respectively.[17] However, when stool samples were processed in the field for subsequent hybridization analysis in the reference laboratory, the sensitivity dropped to an alarming 18%. These investigators attributed this decline in sensitivity to their inability in a field situation to adequately fix the DNA samples to the nitrocellulose filters. Since DNA probes have often been cited as an excellent means of performing field studies of the distribution of infectious agents, the apparent failure to adequately prepare the filters in the field for future processing is a significant finding and one that deserves more attention. It also has important implications for the clinical laboratory, since this same method of processing[11,22] is recommended for preparing filters for hybridization with commercially available probes.

Several other key issues associated with the use of DNA probes for the identification of diarrheal pathogens were recently addressed by Lanata et al.[23] During a study of ETEC infections in human volunteers, these investigators used DNA proves and traditional Y-1 adrenal cell assays to monitor the presence of ETEC in stool samples collected from the study subjects. Less than 1% of the hybridization reactions obtained with the ^{32}P-labeled DNA probes gave indeterminate signals on X-ray film. Thus, the results of the hybridization tests were easy to interpret. The specificity of the test was also very high. Of the 35 isolates showing discrepancies between probe reaction and Y-1 adrenal cell assay, 34 of the probe-negative, assay-positive isolates were shown to have spontaneously lost the LT-encoding plasmid. In terms of the probes themselves, these investigators noted that the LT-A probe gave more definitive results than the LT-B probe, and that freshly labeled ST probe gave better signals than probes nick translated previously, probably due to degradation of the short fragments. This emphasizes the importance of choosing the appropriate probe for the hybridization reaction and the need to test isolates quickly, since storage can result in the loss of the plasmid-encoded target genes. Other ways of optimizing probe results for ETEC have been explored by Yam et al.[24]

The ST toxins have proven to be a heterogenous group of proteins,[25,26] and DNA hybridization studies have led to an understanding of the significance of the various STs in human disease. There are two STs capable of causing a positive reaction in the suckling mouse assay, termed variously STp and STh, ST_{Ia} and ST_{Ib}, or STal and STa2.[14,27] The genes encoding these toxins are related, but can be distinguished with appropriate hybridization probes. Genes encoding both of these toxins are found in human ETEC isolates.[14,27] An unrelated ST, active only in an infant pig intestinal loop model, has been termed STb or STII.[28] Gene probe studies have demonstrated that this toxin is not significant in human disease.[29,30]

Recently, Thompson et al. reported the results of a study to determine the optimum method of detecting *E. coli* ST. [31] A monoclonal-based ELISA, a radioimmunoassay (RIA) using polyclonal antibodies, the suckling mouse assay, and gene probes were used to examine 100 isolates of *E. coli* in a blinded, two-laboratory study. Total agreement was noted between the suckling mouse assay, RIA, and ELISA. The gene probe, however, detected three additional ST^+ isolates, phenotypically negative for toxin production. The authors concluded that the gene probe, in fact, was detecting the presence of ST genes that were not expressed. Thus, hybridization was thought to be an excellent method for screening large numbers of isolates. The ELISA method, which could be completed within 28 h after colonies were isolated on agar medium, offered the advantages of not requiring specialized instrumentation or training of personnel, while retaining very high sensitivity and specificity,[31] although primary culture was still required.

Pickett and co-workers have recently described a second LT gene in a strain of *E. coli*. This second heat labile toxin, while capable of activating adenyl cyclase, is not neutralized by antisera against either cholera toxin or LT.[32] An 800-bp probe developed from the cloned gene failed to show any homology with the original LT from *E. coli*. The distribution of this gene in nature and its association with human disease have yet to be determined.

At this point in time, there is little reason for the majority of laboratories in the U.S. to screen for ETEC. The incidence of these infections is simply not high enough to warrant pursuing these pathogens in stool, even if a rapid method becomes available. The two possible exceptions would be laboratories in the southwestern U.S., particularly those close to the Mexican border, where the incidence of ETEC is high, and those laboratories associated with travellers' clinics, since ETEC is a common cause of travellers' diarrhea.[4]

Commercially prepared DNA probes for LT1, ST_{Ia}, and ST_{Ib} are now available (E. I. du Pont de Nemours & Co., Wilmington, Del.). The sensitivity and specificity of these probes has been reviewed recently by Seriwatana et al.[33] In this study, these alkaline phosphatase-labeled oligonucleotide probes were 59% sensitive and 86% specific in the detection of ETEC in blots of stool grown overnight on filters resting on MacConkey agar, as compared to the traditional Y-1 adrenal cell and suckling mouse assays. These same probes, when radiolabeled with ^{32}P, proved to be 71% and 91% sensitive and specific, respectively. The alkaline phosphatase probes proved much more sensitive and specific when used for confirmation of ETEC previously isolated from MacConkey agar; however, their relatively poor performance when used against direct stool blots would suggest that a reporter molecule with a more intense signal output is necessary if they are to be used in this context.

IV. ENTEROPATHOGENIC *E. COLI*

Those strains of *E. coli* that do not elaborate enterotoxins and are not invasive, but are epidemiologically linked to infantile diarrhea, are referred to as the enteropathogenic *E. coli* or EPEC.[34,35] The majority of these strains belong to a defined set of serotypes. Studies of the small intestines of both human infants and animal models infected with these organisms demonstrate destruction of the normal enterocyte microvilli and show a close adherence of bacteria to the brush border.[34] EPEC strains have been divided into two classes based on their ability to adhere to Hep-2 tissue culture cells.[36] This adherence phenomenon apparently is coded on a 60-Mdalton plasmid, and mapping studies of this plasmid have indicated the regions involved in the adherence phenomenon.[37] Nataro and colleagues have developed a DNA probe for the EPEC adherence factor (EAF) and used it to screen the stools of Peruvian infants with and without diarrhea. The probe demonstrated both a high specificity and sensitivity for identifying the presence of *E. coli* carrying the gene for EAF.[38] This probe, however, does not detect Class II EPEC which do not possess the EAF gene. Nonetheless, this would be a valuable probe for investigations of diarrheal illness in children.

V. *SHIGELLA* AND ENTEROINVASIVE *E. COLI*

Sansonetti and co-workers[39] and others[40,41] have shown that several key determinants involved in the pathogenicity of *Shigella* species and enteroinvasive *E. coli* (EIEC) are plasmid mediated. These large molecular weight plasmids appear in both types of organisms and are very similar in size and structure. Two groups of investigators have independently developed DNA probes to different regions of these plasmids and shown them to be successful in differentiating *Shigella* and EIEC from other nonpathogenic stool flora.[42,43]

Boileau and co-workers were the first to describe a probe for the detection of *Shigella* and EIEC.[42] The probe consisted of a 15-kb *Eco*RI fragment from the virulence plasmid of *Shigella flexneri* serotype 5. This probe was found to be highly specific for *Shigella* spp. and EIEC, but did not hybridize to a wide variety of other enteric flora. Only one *Shigella boydii* isolate was not detected with the probe. That strain proved to be noninvasive in the Sereny test[44] and did not contain the large virulence plasmid. The ability of the probe to detect these organisms directly in stool was not reported.

Sethabutr et al.[45] used the probe described by Boileau and colleagues to examine stools from children in Thailand for the presence of EIEC and *Shigella*. They also compared the sensitivity and specificity of probes that were biotinylated to probes labeled with the radioactive isotope ^{32}P. Both the radiolabeled and biotinylated probes detected the same number of EIEC and *Shigella* in stool samples that had been spotted on MacConkey agar, grown overnight, lifted from the agar with Whatman 541 paper, and hybridized. While all EIEC, which had been previously isolated and characterized by the Sereny test, were positive with both probes, two stool samples that were culture-positive for *Shigella* were not detected by hybridization analysis.

A similar result was reported by Taylor et al. who also used the probe described by Boileau et al. in their study of the agents of dysentary in Thai children.[46] In this study, there was complete agreement between the results of hybridization assays carried out on isolated *E. coli* colonies, Sereny tests, and ELISA assays for detecting EIEC. Only nine of ten stool samples that yielded EIEC by culture, however, were positive by hybridization when the stool sample was spotted on filters and grown overnight before analysis. Using this same probe, only 31% (27 of 88) of stool samples yielding *Shigella* spp. by culture were positive by direct hybridization. Two stool samples which were negative for both EIEC and *Shigella* by culture were positive with the probe. Therefore, the probe was felt to be a reliable means of identifying EIEC isolated by culture, an acceptable means of detecting EIEC directly in stool, but a poor method for detecting *Shigella* spp. directly in stool. The authors suggest that EIEC are present in much higher concentrations and are more evenly distributed in stool than are *Shigella*. Interestingly, all EIEC in this study fermented lactose, an unusual feature for most EIEC.

Small and Falkow also developed a probe for *Shigella* and EIEC.[43] Their probe, a 2.5-kb *Hin*dIII fragment from the 210-kb virulence plasmid pSF201, was shown to be specific for these two groups of organisms by colony hybridization studies. No clinical samples were examined.

Using the probes developed by Boileau et al.[42] and Small and Falkow,[43] Wood et al. examined 20 isolates of *Shigella* spp. and 19 isolates of EIEC that were shown to be invasive in the Sereny test.[47] While both probes detected all of the EIEC by colony hybridization, two *Shigella* isolates, *S. sonnei*, and *S. flexneri,* that were negative by the Sereny test, were positive by the hybridization assay. One of these isolates had previously been shown to be positive in the Sereny test by another laboratory. Both isolates contained a large virulence plasmid, suggesting that deletions or other changes in virulence determinants had rendered them noninvasive.

Thus, the two probes that have been developed for *Shigella* and EIEC have proven to be excellent for culture confirmation of these pathogens, moderately effective for the detection of EIEC in stool, but poor for the detection of *Shigella* in stool, even after enrichment on MacConkey agar. This suggests that if probes are to be used for the identification of stool pathogens, enrichment will be necessary. The likelihood of detecting *Shigella* species by this method, however, is currently low.

VI. ENTEROHEMORRHAGIC *E. COLI*

Levine et al. have recently described a DNA probe for the enterohemorrhagic *E. coli* (EHEC) that cause hemorrhagic colitis and are associated with hemolytic uremic syndrome.[48] The pathogenicity of these organisms is related to the production of a Vero cell toxin.[49] Both the very young[50] and the elderly seem particularly susceptible to these infections.[51] The laboratory diagnosis of these infections has recently been reviewed by Karmali.[52] Although studies by Harris et al.[53] suggest that EHEC were not widespread within the greater Chicago area, they concluded that the distribution of these organisms may vary in different geographic regions. Thus, certain communities may well benefit from a DNA probe, such as that described by Levine et al., to screen stools from suspected patients.

VII. *SALMONELLA* SPECIES

The first probes to *Salmonella* species were those developed by Fitts for the detection of salmonellae in food.[54] Starting with a library of *Salmonella* chromosomal DNA cloned into a plasmid vector, the inserts were systematically tested for their ability to hybridize only to the DNA from other *Salmonella*. Two probes were eventually found that hybridized to DNA from all 330 *Salmonella* spp., but not to DNA from other enteric organisms. These probes have been successfully used to screen food products for the presence of salmonellae[55] and have been used to screen stool and food samples in two recent outbreaks of salmonellosis.[56] The probes demonstrated sensitivities in the range of 87 to 100% and specificities ranging from 98 to 100%; however, the numbers of samples in each case were small. Nonetheless, this report is encouraging.

A probe to detect *Salmonella typhi* has been reported by Rubin et al.;[57] however, no clinical samples were analyzed in the study. Thus, the sensitivity and specificity of the probe in a clinical situation have not been evaluated. None of the probes developed for *Salmonella* are commercially available for clinical use.

VIII. *CAMPYLOBACTER* SPECIES

Campylobacters are prevalent human and animal pathogens,[58] and their recognition in clinical samples, while not technically difficult, is costly.[59] Campylobacters are microaerophilic organisms that grow best at 42°C. Our understanding of the pathobiology of these organisms is incomplete. Both heat-labile, cholera-like cytotonic enterotoxins, and Shiga-like Vero cell cytotoxic factors have been noted in several strains of these organisms.[60-62] Other strains appear to be invasive.[63] The relationship of each of these virulence properties to the ability of the organisms to produce human illness is still in question.

Several attempts have been made to develop diagnostic probes for these organisms. The first probe reported for *C. jejuni* was by Claus et al.[64] Against a total of 42 organisms, including *C. jejuni, C. coli, C. fetus, C. faecalis*, and "atypical" campylobacters, the probe, of unknown origin, was 100% sensitive and specific. These investigators also said the probe was capable of detecting campylobacter sequences in stool that had been stored for 1 month in transport media.

Tompkins and Krajden have also attempted the development of a diagnostic probe for campylobacters.[65] However, these investigators encountered problems with high background using radiolabeled probes.

Recently, Gene-Trak, Inc. reported the development of a biotinylated oligonucleotide probe for *Campylobacter*. Compared to a two-culture method on a selective medium, the probe demonstrated a 97% sensitivity and 99% specificity.[66] The probe assay did require an 8- to 16-h enrichment phase before the hybridization step. The assay is said to detect as few as 5×10^4 organisms.[67] These are very encouraging data for the clinical laboratory, although it does not remove the need for special culture conditions.

IX. *YERSINIA ENTEROCOLITICA*

Although DNA probes have been developed that detect *Yersinia enterocolitica*, they have been used primarily to detect this organism in food samples rather than stool.[68,69] A probe to *Y. enterocolitica* could be of value to the clinical laboratory, if the incidence of this pathogen were high in the community.

X. OTHER PATHOGENS

Although a number of other bacterial pathogens are associated with diarrheal illness, including *C. difficile*,[70] *Aeromonas* species,[71,72] and *Plesiomonas* species,[73] no probes to these agents have been developed, as of yet, to aid in their detection and identification.

XI. STOOL PROBES AND THE CLINICAL MICROBIOLOGY LABORATORY

Studies by Echeverria et al.,[16-18] Lanata et al.,[23] Moseley et al.,[11] and Seriwatana et al.[33] suggest that direct detection of ETEC in stool is still less sensitive than probing isolated colonies. *Shigella*, in particular, are very difficult to detect by hybridization assays.[45,46] Although the isolation of colonies from the stool sample adds an additional 18 to 24 h to the procedure, this delay may be acceptable for some reference laboratories because the alternatives, such as the use of Y-1 adrenal cell tissue culture assay[5] or the suckling mouse assay,[7] are simply too costly and technically difficult for routine use.

On the other hand, probes are not yet ready for use in the hospital diagnostic laboratory. Considering the fact that probes are probably going to be expensive, the 18 to 24 h of incubation required to produce significant growth of the organism before probe analysis may be too long to be considered worthy of the additional cost. To be effective in the clinical laboratory, the probes must be able to produce an answer within 4 h of specimen receipt. To accomplish this, it will probably be necessary to extract DNA from the stool sample and perform the hybridization tests on the crude DNA extract. A method for extracting DNA rapidly from clinical specimens, including stool, has been developed. Although not yet available for diagnostic use, preliminary results look promising.[74] This system of probing DNA directly, instead of stool lysed *in situ*, should greatly improve the sensitivity and specificity of the stool assay.

The ideal DNA probe kit for detection of enteric pathogens would have a rapid DNA extraction method and offer reagents for testing *Salmonella, Shigella,* and *Campylobacter*, either individually or together. If packaged individually, it would allow the laboratory to screen only those pathogens they felt were appropriate, and if positive, a report could be generated immediately. However, since the symptoms of the illnesses produced by these three pathogens can be similar, a combination of the three probes would allow rapid reporting of negative results which, in most laboratories, is 90 to 95% of all stool specimens submitted for culture. If the three-pathogen screen were positive, individual probe assays, other agglutination tests, or routine culture could be set up, depending on the need for antibiograms, serotyping, or other strain identification procedures.

XII. CONCLUSIONS

A variety of DNA probes have been described for enteric pathogens. The most widely tested are those for enterotoxigenic *E. coli.* Some commercially prepared probe kits are already available, and others should be on the market soon. Probes, especially if available in kits containing a battery of reagents to detect *Salmonella, Shigella,* and *Campylobacter*, will be a very welcome addition to the clinical microbiology laboratory. However, they ultimately should be capable of detecting these pathogens directly in stool and not after overnight culture. In this regard, the use of a rapid DNA extraction method may be of significant benefit to the clinical laboratory.

ACKNOWLEDGMENTS

I wish to thank Dr. S. Moseley for his help in the preparation of this manuscript.

REFERENCES

1. **Koplan, J. P., Fineburg, H. V., Ferraro, M. J. B., and Rosenberg, M. L.,** The value of stool culture, *Lancet,* 2, 413, 1980.
2. **Rosenblatt, J. E.,** Laboratory diagnosis of infectious diarrhea: a clinician's viewpoint, in *The Role of Clinical Microbiology in Cost-Effective Health Care,* Smith, J. W., Ed., College of American Pathologists, Skokie, Ill., 1985, 105.
3. **Black, R. E., Merson, M. H., Rahman, A. S. M. M., Yunus, M., Alim, A. R. M. A., Huq, I., Yolken, R. H., and Curlin, G. T.,** A two-year study of bacterial, viral, and parasitic agents associated with diarrhea in rural Bangladesh, *J. Infect. Dis.,* 142, 660, 1980.
4. **Kean, B. H.,** Travelers' diarrhea: an overview, *Rev. Infect. Dis.,* 8, S111, 1986.
5. **Sack, D. A. and Sack, R. B.,** Test of the detection of enterotoxigenic *Escherichia coli* using Y1 adrenal cells in miniculture, *Infect. Immun.,* 11, 334, 1975.
6. **Guerrant, R. L., Brunton, L. L., Schnaitman, T. C., Rebhun, L. I., and Gilman, A. G.,** Cyclic adenosine monophosphate and alteration of Chinese hampster ovary cell morphology: a rapid, sensitive in vitro assay for the enterotoxins of *Vibrio cholerae* and *Escherichia coli, Infect. Immun.,* 10, 320, 1974.
7. **Dean, A. G., Ching, Y. C., Williams, R. G., and Harden, L. B.,** Test of *Escherichia coli* using infant mice: application in a study of diarrhea in children in Honolulu, *J. Infect. Dis.,* 125, 407, 1972.
8. **Honda, T., Taga, S., Takeda, Y., and Miwatani, T.,** Modified Elek test for the detection of heat-labile enterotoxin of enterotoxigenic *Escherichia coli, J. Clin. Microbiol.,* 13, 1, 1981.

9. **Thompson, M. R., Brandwein, H., LaBine-Racke, M., and Gianella, R. A.,** Simple and reliable enzyme linked immunosorbent assay with monoclonal antibodies for detection of *Escherichia coli* heat-stable enterotoxin, *J. Clin. Microbiol.*, 20, 59, 1984.
10. **Finkelstein, R. A. and Yang, Z.,** Rapid test for the identification of heat-labile enterotoxin-producing *Escherichia coli* colonies, *J. Clin. Microbiol.*, 18, 23, 1983.
11. **Moseley, S. L., Huq, I., Alim, A. R. M. A., So, M., Samadpour-Motalebi, M., and Falkow, S.,** Detection of enterotoxigenic *Escherichia coli* by DNA colony hybridization, *J. Infect. Dis.*, 142, 892, 1980.
12. **Grunstein, M. and Hogness, D.,** Colony hybridization: a method for the isolation of cloned DNA's that contain a specific gene, *Proc. Natl. Acad. Sci. U.S.A.*, 72, 3961, 1975.
13. **Dallas, W. S., Gill, D. M., and Falkow, S.,** Cistrons encoding *Escherichia coli* heat-labile toxin, *J. Bacteriol.*, 139, 850, 1979.
14. **Moseley, S. L., Echeverria, P., Seriwatana, J., Tirapat, C., Chaicumpa, W., Sakuidalpeara, T., and Falkow, S.,** Identification of enterotoxigenic *Escherichia coli* by colony hybridization using three enterotoxin gene probes, *J. Infect. Dis.*, 145, 863, 1982.
15. **Echeverria, P., Taylor, D. N., Seriwatana, J., Chatkaeomorakot, A., Khungvalert, V., Sakuldaipeara, T., and Smith, R. D.,** A comparative study of enterotoxin gene probes and tests for toxin production to detect enterotoxigenic *Escherichia coli, J. Infect. Dis.*, 153, 255, 1986.
16. **Echeverria, P,. Seriwatana, J., Leksomboon, U., Tirapat, C., Chaicumpa, W., and Rowe, B.,** Identification by DNA hybridization of enterotoxigenic *Escherichia coli* in homes of children with diarrhea, *Lancet*, 1, 63, 1984..
17. **Echeverria, P., Seriwatana, J., Taylor, D. N., Tirapat, C., Chaicumpa, W., and Rowe, B.,** Identification by DNA hybridization of enterotoxigenic *Escherichia coli* in a longitudinal study of villages in Thailand, *J. Infect. Dis.*, 151, 124, 1985.
18. **Echeverria, P., Seriwatana, J., Chityothin, O., Chaicumpa, W., and Tirapat, C.,** Detection of enterotoxigenic *Escherichia coli* in water by filter hybridization with three enterotoxin gene probes, *J. Clin. Microbiol.*, 16, 1086, 1982.
19. **Patamaroj, U., Seriwatana, J., and Echeverria, P.,** Identification of enterotoxigenic *Escherichia coli* isolated from swine with diarrhea in Thailand by colony hybridization using three enterotoxin gene probes, *J. Clin. Microbiol.*, 18, 1429, 1983.
20. **Seriwatana, J., Echeverria, P., Escamilla, J., Glass, R., Huq, I., Rockhill, R., and Stoll, B. J.,** Identification of enterotoxigenic *Escherichia coli* in patients with diarrhea in Asia with three enterotoxin gene probes, *Infect. Immun.*, 42, 152, 1983.
21. **Georges, M. C., Wachsmuth, I. K., Birkness, K. A., Moseley, S. A., and Georges, A. J.,** Genetic probes for enterotoxigenic *Escherichia coli* isolated from childhood diarrhea in the Central Africa Republic, *J. Clin. Microbiol.*, 18, 199, 198.
22. **Maas, R.,** An improved colony hybridization method with significantly increased sensitivity for detection of single genes, *Plasmid*, 10, 296, 1983.
23. **Lanata, C. F., Kaper, J. B., Baldini, M. M., Black, R. E., and Levine, M. M.,** Sensitivity and specificity of DNA probes with the stool blot technique for detection of *E. coli* enterotoxins, *J. Infect. Dis.*, 152, 1087, 1985.
24. **Yam, W. C., Lung, M. L., and Ng, M. H.,** Evaluation and optimization of the DNA filter assay for direct detection of enterotoxigenic *Escherichia coli* in the presence of stool coliforms, *J. Clin. Microbiol.*, 24, 149, 1986.
25. **Moseley, S. L., Samadpour-Motalebi, M., and Falkow, S.,** Plasmid association and nucleotide sequence relationships of two genes encoding heat stable enterotoxin, *J. Bacteriol.*, 156, 441, 1983.
26. **So, M. and McCarthy, B. J.,** Nucleotide sequence of the bacterial transposon Tn*1681* encoding heat-stable (ST) toxin and its identification in enterotoxigenic *Escherichia coli* strains, *Proc. Natl. Acad. Sci. U.S.A.*, 77, 4011, 1980.
27. **Moseley, S. L., Hardy, J. W., Huq, M. I., Echeverria, P., and Falkow, S.,** Isolation and nucleotide sequence of a gene encoding a heat stable enterotoxin of *Escherichia coli, Infect. Immun.*, 39, 1167, 1983.
28. **Burgess, M. N., Bywater, R. J., Cowley, C. M., Mullin, N. A., and Newsome, P. M.,** Biological evaluation of a methanol-soluble, heat-stable *Escherichia coli* enterotoxin in infant mice, pigs, rabbits, and calves, *Infect. Immun.*, 21, 526, 1978.
29. **Echeverria, P., Seriwatana, J., Patamaroj, U., Moseley, S. L., MacFarland, A., Chityothin, O., and Chaicumpa, W.,** Prevalence of heat-stable II enterotoxigenic *Escherichia coli* in pigs, water, and people at farms in Thailand as determined by DNA hybridization, *J. Clin. Microbiol.*, 19, 489, 1984.
30. **Weikel, C. S., Tiemens, K. M., Moseley, S. L., Huq, I. M., and Guerrant, R. L.,** Species specificity and lack of production of STb enterotoxin by *Escherichia coli* strains isolated from humans with diarrheal illness, *Infect. Immun.*, 52, 323, 1986.
31. **Thompson, M. R., Jordan, R. L., Luttrell, M. A., Brandwein, H., Kaper, J. B., Levine, M. M., and Giannella, R. A.,** Blinded, two-laboratory comparative analysis of *Escherichia coli* heat-stabile enterotoxin production by using monoclonal antibody enzyme-linked immunosorbent assay, radioimmunoassay, suckling mouse assay, and gene probes, *J. Clin. Microbiol.*, 24, 753, 1986.

32. **Pickett, C. L., Twiddy, E. M., Belisle, B. W., and Holmes, R. K.,** Cloning of genes that encode a new heat-labile enterotoxin of *Escherichia coli, J. Bacteriol.,* 165, 348, 1986.
33. **Seriwatana, J., Echeverria, P., Taylor, D. N., Sakuldaipeara, T., Changchawalit, S., and Chivoratanond, O.,** Identification of enterotoxigenic *Escherichia coli* with synthetic alkaline phosphatase-conjugated oligonucleotide DNA probes, *J. Clin. Microbiol.,* 25, 1438, 1987.
34. **Edelman, R. and Levine, M. M.,** Summary of a workshop on enteropathogenic *Escherichia coli, J. Infect. Dis.,* 147, 1108, 1983.
35. **Levine, M.,** *Escherichia coli* that cause diarrhea: enterotoxigenic, enteropathogenic, enteroinvasive, enterohemorrhagic, and enteroadherent, *J. Infect. Dis.,* 155, 377, 1987.
36. **Nataro, J. P., Scaletsky, I. C. A., Kaper, J. B., Levine, M. M., and Trabulski, L. R.,** Plasmid-mediated factors conferring diffuse and localized adherence of enteropathogenic *Escherichia coli, Infect. Immun.,* 48, 378, 1985.
37. **Baldini, M. M., Kaper, J. B., Levine, M. M., Candy, D. C. A., and Moon, H. W.,** Plasmid-mediated adhesion in enteropathogenic *Escherichia coli, J. Pediatr. Gastroenterol. Nutr.,* 2, 534, 1983.
38. **Nataro, J. P., Kaper, J. B., Black, R. E., Bravo, N., and Levine, M. M.,** Detection of an adherence factor of enteropathogenic *Escherichia coli* with a DNA probe, *J. Infect. Dis.,* 152, 560, 1985.
39. **Sansonetti, P. J., d'Hauteville, H., Formal, S. B., and Toucas, M.,** Plasmid mediated invasiveness of "Shigella-like" *Escherichia coli, Ann. Microbiol. (Paris),* 133, 351, 1982.
40. **Silva, R. M., Toledo, M. F. R., and Trabulsi, L. R.,** Correlation of invasiveness with plasmid in enteroinvasive strains of *Escherichia coli, J. Infect. Dis.,* 146, 706, 1982.
41. **Harris, J. R., Wachsmuth, I. K., Davis, B. R., and Cohen, M. L.,** High-molecular weight plasmid correlates with *Escherichia coli* enteroinvasiveness, *Infect. Immun.,* 37, 1295, 1982.
42. **Boileau, C. R., d'Hauteville, H. M., and Sansonetti, P. J.,** DNA hybridization technique to detect *Shigella* species and enteroinvasive *Escherichia coli, J. Clin. Microbiol.,* 20, 959, 1984.
43. **Small, P. L. C. and Falkow, S.,** Development of a DNA probe for the virulence plasmid of *Shigella* spp. and enteroinvasive *Escherichia coli,* in *Microbiology—1986,* Schlessinger, D., Ed., American Society for Microbiology, Washington, D.C., 1986, 121.
44. **Sereny, B.,** Experimental keratoconjunctivitis shigellosa, *Acta Microbiol. Acad. Sci Hung.,* 4, 367, 1957.
45. **Sethabutr, O., Echeverria, P., Hanchaly, S., Taylor, D. N., and Leksomboon, U.,** A non-radioactive DNA probe to identify *Shigella* and enteroinvasive *Escherichia coli* in stools of children with diarrhea, *Lancet,* 2, 1095, 1985.
46. **Taylor, D. N., Echeverria, P., Pal, T., Sethabutr, O., Saiborisuth, S., Sricharmorn, S., Rowe, B., and Cross, J.,** The role of *Shigella* spp., enteroinvasive *Escherichia coli,* and other enteropathogens as causes of childhood dysentary in Thailand, *J. Infect. Dis.,* 153, 1132, 1986.
47. **Wood, P. K., Morris, J. G., Jr., Small, P. L. C., Sethabutr, O., Toledo, M. R. F., Trabulsi, L., and Kaper, J. B.,** Comparison of DNA probes and the Sereny test for the identification of invasive *Shigella* and *Escherichia coli* strains, *J. Clin. Microbiol.,* 24, 498, 1986.
48. **Levine, M. M., Jian-gou, X., Kaper, J. B., Lior, H., Prado, V., Tall, B., Nataro, J., Karch, H., and Wachsmuth, I. K.,** A DNA probe to identify enterohemorrhagic *Escherichia coli* of 0157:H7 and other serotypes that cause hemorrhagic colitis and hemolytic uremic syndrome, *J. Infect. Dis.,* 156, 175, 1987.
49. **Riley, L. W., Remis, R. S., Helgerson, S. D., McGee, H. B., Wells, J. G., Davis, B. R., Hebert, R. J., Olcott, E. S., Johnson, L. M., Hargrett, N. T., Blake, P. A., and Cohen, M. L.,** Hemorrhagic colitis associated with a rare *Escherichia coli* serotype, *N. Engl. J. Med.,* 308, 681, 1982.
50. **Neill, M. A., Tarr, P. I., Clausen, C. R., Christie, D. L., and Hickman, R. O.,** *Escherichia coli* 0157:H7 as the predominant pathogen associated with the hemolytic uremic syndrome: a prospective study in the Pacific Northwest, *Pediatrics,* 80, 37, 1987.
51. **Ryan, C. A., Tauxe, R. V., Hosek, G. W., Wells, J. G., Stoesz, P. A., McFadden, H. W., Jr., Smith, P. W., Wright, G. F., and Blake, P. A.,** *Escherichia coli* 0157:H7 diarrhea in a nursing home: clinical, epidemiological, and pathological findings, *J. Infect. Dis.,* 154, 631, 1986.
52. **Karmali, M. A.,** Laboratory diagnosis of verotoxin-producing *Escherichia coli* infections, *Clin. Microbiol. Newsl.,* 9, 65, 1987.
53. **Harris, A. A., Kaplan, R. L., Goodman, L. J., Doyle, M., Landau, W., Segreti, J., Mayer, K., and Levin, S.,** Results of a screening method used in a 12-month stool survey for *Escherichia coli* 0157:H7, *J. Infect. Dis.,* 152, 775, 1985.
54. **Fitts, R.,** Development of a DNA-DNA hybridization test for the presence of *Salmonella* in foods, *Food Technol.,* 39, 95, 1985.
55. **Fitts, R., Diamond, M., Hamilton, C., and Neri, M.,** DNA-DNA hybridization assay for detection of *Salmonella* spp. in foods, *Appl. Environ. Microbiol.,* 46, 1146, 1983.
56. **Harris, J., Maker, S., Bicchiere, R., George, H., Prinz, H., and Klinger, J.,** Evaluation of a nucleic acid hybridization assay during two foodborne *Salmonella* outbreaks, *Abstr. Annu. Meet. Am. Soc. Microbiol.,* 376, 1987, C-319.

57. **Rubin, F. A., Kopecko, D. J., Noon, K. F., and Baron, L. S.,** Development of a DNA probe to detect *Salmonella typhi, J. Clin. Microbiol.,* 22, 600, 1985.
58. **Blaser, M. J. and Reller, L. B.,** Campylobacter enteritis, *N. Engl. J. Med.,* 305, 1444, 1981.
59. **Pennie, R. A., Zunino, J. N., Rose, C. E., Jr., and Guerrant, R. L.,** Economical, simple method for production of the gaseous environment required for cultivation of *Campylobacter jejuni, J. Clin. Microbiol.,* 20, 320, 1984.
60. **Johnson, W. M. and Lior, H.,** Cytotoxic and cytotonic factors produced by *Campylobacter jejuni, Campylobacter coli,* and *Campylobacter laridis, J. Clin. Microbiol.,* 24, 275, 1986.
61. **Goossens, H., Butzler, J. P., and Takeda, Y.,** Demonstration of cholera-like enterotoxin production by *Campylobacter jejuni, FEMS Microbiol. Lett.,* 29, 73, 1985.
62. **Ruiz-Palacios, G. M., Torres, J., Torres, N. I., Escamilla, E., Ruiz-Palacios, B., and Tamayo, J.,** Cholera-like enterotoxin produced by *Campylobacter jejuni;* characterization and clinical significance, *Lancet,* 2, 250, 1983.
63. **Newell, D. G., McBride, H., Saunders, F., Dehele, Y., and Pearson, A. D.,** The virulence of clinical and environmental isolates of *Campylobacter jejuni, J. Hyg.,* 94, 45, 1985.
64. **Claus, P., Moseley, S. L., and Falkow, S.,** The use of gene-specific DNA probes for the identification of enteric pathogens, *Prog. Food Nutr. Sci.,* 7, 139, 1983.
65. **Tompkins, L. S. and Krajden, M.,** Approaches to the detection of enteric pathogens, including *Campylobacter,* using nucleic acid hybridization, *Diagn. Microbiol. Infect. Dis.,* 4, 71S, 1986.
66. **Rashtchian, A., Abbott, M., Ottaviani, M., Eldredge, J., and Klinger, J.,** GENE-TRAK *Campylobacter:* a rapid non-isotopic DNA probe assay for detection of *Campylobacter, Abstr. Annu. Meet. Am. Soc. Microbiol.,* 376, 1987, C-317.
67. **Rashtchian, A., Abbott M., Eldredge, M., Logan, T., Lovern, D., Mock, G., and Ottaiani, M.,** Novel method for detection of DNA:RNA hybridization using non-isotopically labeled synthetic probes, *Abstr. Annu. Meet. Am. Soc. Microbiol.,* 376, 1987, C-318.
68. **Hill, W. E., Payne, W. L., and Aulisio, C. C. G.,** Detection and enumeration of virulent *Yersinia enterocolitica* in foods by DNA colony hybridization, *Appl. Environ. Microbiol.,* 46, 636, 1983.
69. **Jagow, J. and Hill, W. E.,** Enumeration by DNA colony hybridization of virulent *Yersinia enterocolitica* colonies in artificially contaminated food, *Appl. Environ. Microbiol.,* 51, 441, 1986.
70. **Bartlett, J. G., Chang, T. W., Gurwith, M., Gorbach, S. L., and Onderdonk, A. B.,** Antibiotic-associated pseudomembranous colitis due to toxin-producing clostridia, *N. Engl. J. Med.,* 298, 531, 1978.
71. **Asao, T., Kozaki, S., Kato, K., Kinoshita, Y., Otsu, K., Uemura, T., and Sakaguchi, G.,** Purification and characterization of an *Aeromonas hydrophila* hemolysin, *J. Clin. Microbiol.,* 24, 228, 1986.
72. **George, W. L., Nakata, M. M., Thompson, J., and White, M. L.,** *Aeromonas*-related diarrhea in adults, *Arch. Intern. Med.,* 145, 2207, 1985.
73. **Rolston, K. V. and Hopper, R. L.,** Diarrhea due to *Plesiomonas shigelloides* in cancer patients, *J. Clin. Microbiol.,* 20, 597, 1984.
74. **Van Brunt, J. and Klausner, A.,** Pushing probes to market, *Bio/Technology.,* 5, 211, 1987.

Chapter 6

DNA PROBES FOR PATHOGENIC *VIBRIO* SPECIES

James B. Kaper, J. Glenn Morris, Jr., and Mitsuaki Nishibuchi

TABLE OF CONTENTS

I. *VIBRIO* SPECIES IN HUMAN DISEASE

Vibrio species have been associated with a wide range of disease manifestations, including severe, dehydrating diarrhea, wound infections, and septicemia.[1] Six species (*Vibrio cholerae, V. parahaemolyticus, V. mimicus, V. hollisae, V. fluvialis,* and *V. furnissii)* have been associated primarily with diarrheal disease (Table 1). Strains that produce cholera toxin (strains of *V. cholerae* O-group 1, and some strains of non-O1 *V. cholerae* and *V. mimicus*) can cause profuse watery diarrhea, resulting in dehydration and circulatory collapse in an adult (cholera). Diarrhea caused by strains that do not produce cholera toxin tends to be relatively mild and self limited, although severe disease and occasional deaths associated with these organisms have been reported.[2] Five other Vibrio species have been associated primarily with disease manifestations other than diarrhea (Table 1). V. vulnificus appears to be the most virulent; it can cause severe, rapidly progressive wound infections, and, in susceptible hosts, a syndrome of "primary septicemia".[3] Most, if not all, Vibrio species are free living in marine or brackish water. Vibrio-associated illness has tended to occur in coastal areas in the summer and fall (when the water is warmer and vibrio counts are higher) and to be associated with seafood or seawater.

The disease cholera is caused by cholera-toxin-producing strains of *V. cholerae.* Seven cholera pandemics have occurred during the past 2 centuries; the disease is currently endemic in the Indian subcontinent and part of Asia and Africa, with sporadic cases reported from North America, Australia, and Europe. A majority of persons infected with the organism remain asymptomatic, with only 2 to 11% (in infections due to El Tor and classical biotypes, respectively) developing the most severe form of the disease, cholera gravis.[5] Peak stool losses occur within 24 h of the onset of symptoms and, in cholera gravis, may exceed 1 liter of stool per hour, resulting in rapid dehydration and circulatory collapse. Fluid replacement is the key element in therapy; while antibiotics shorten the duration of symptoms and excretion of the organism, they are of secondary importance. In areas with high rates of disease, transmission occurs through contamination of water by fecal material. However, it is becoming increasingly clear that toxigenic strains may be present in the environment as free-living organisms. Such environmental reservoirs appear to be responsible for sporadic cases in Australia[6] and along the U.S. Gulf Coast (where crabs have been implicated as the vehicle of transmission),[7] and may help explain the observed seasonality of cases in areas such as Bangladesh.[8]

V. parahaemolyticus is regarded primarily as a gastrointestinal pathogen, although it can also cause wound and ear infections and septicemia.[1,9] The organism has been implicated as the cause of 24% of reported cases of foodborne disease in Japan and has been associated with major common-source outbreaks in the U.S. Patients with gastroenteritis generally have a self-limited disease, with a median duration of 3 d; a dysentery-like syndrome associated with the organism has been described in India and Bangladesh. *V. parahaemolyticus* is a common free-living organism in the estuarine environment, with reported isolations from water, sediment, suspended particulates, plankton, fish, and shellfish. *V. mimicus* (so named because of its phenotypic similarity to *V. cholerae*) has been implicated as a cause of sporadic cases of diarrheal disease (with occasional reports of associated bloody diarrhea), as well as seawater-associated ear infections.[10] The organism has been isolated from a number of environmental sources, including oysters; illness has been associated with eating raw oysters. *V. hollisae* is another newly described *Vibrio* species that has been implicated as a cause of diarrhea; illness has again been linked to eating raw seafood.[11]

V. vulnificus (also referred to as a "lactose-positive halophilic marine *Vibrio*") has been implicated as a cause of wound infections and septicemia. Wound infections are usually seawater associated, and may range from mild, self-limited infections to severe and rapidly progressive cellulitis and myositis.[1,3,11] Primary septicemia (septicemia without an obvious focus) tends to occur in patients who have cirrhosis, hemochromatosis, or an underlying malignancy. Approximately one third of patients are in shock when first seen clinically, or

Table 1
CLINICAL PRESENTATIONS OF INFECTIONS WITH VIBRIO SPECIES ASSOCIATED WITH HUMAN ILLNESS

	Clinical presentations[a]				
Species	**GI**	**Wound**	**Ear**	**1° Sept[b]**	**2° Sept[c]**
V. cholerae					
O1	++	(+)			
Non-O1	++	+	+	(+)	(+)
V. parahaemolyticus[d]	++	+	(+)		(+)
V. fluvialis	++				
V. mimicus	++		+		
V. hollisae	++			(+)	
V. furnissii	++				
V. vulnificus	(+)	++		++	+
V. alginolyticus		++	+		
V. damsela		++			
V. metschnikovii	?			?	
V. cincinnatiensis[4]				?	

[a] A ++ denotes the most common presentation, + other clinical presentations, and (+) very rare presentation. GI denotes gastrointestinal.
[b] Septicemia occurring without an obvious primary source.
[c] Septicemia with a primary source, usually a wound infection.
[d] In sporadic U.S. cases, extraintestinal infections have been reported more commonly than gastroenteritis.

become hypotensive within 12 h of hospital admission. Over 70% have distinctive bullous skin lesions. Thrombocytopenia is common and there is often evidence of disseminated intravascular coagulation. The overall mortality rate exceeds 50%,[1] even with aggressive supportive therapy and use of appropriate antibiotics. The organism is present in the estuarine environment, and there is a strong association between septicemia and consumption of raw oysters.[12] *V. alginolyticus* and *V. damsela* have also been associated with wound infections, although infections do not appear to be as severe as those seen with *V. vulnificus*. There are single-case reports of *V. metschnikovii* septicemia and *V. cincinnatiensis* meningitis.[1,4]

II. PROBES FOR *VIBRIO CHOLERAE*

V. cholerae is a relatively easy organism to isolate and identify if the laboratory knows to look for it. The existence of readily available isolation media and diagnostic antiserum plus the relatively small role of rapid antibiotic administration for treatment obviates the need for the rapid and highly specific identification that DNA probes can provide. However, DNA probes have yielded much valuable information about this pathogen. The probe most commonly employed for this species detects sequences encoding cholera enterotoxin, but, in addition, probes to detect sequences encoding hemolysin, lysogenic phage, and a toxin regulatory element have been reported. DNA probes have been used for a variety of applications, including identification in large-scale ecological studies, pathogenesis studies, molecular epidemiology, and vaccine development. The various probes and their applications are discussed below.

A. Probes to Detect Genes Encoding Cholera Enterotoxin

The major virulence factor for *V. cholerae* is the cholera enterotoxin (CT), and the first DNA probes to detect cholera toxin genes were cloned fragments encoding the closely related heat-labile enterotoxin (LT) of enterotoxigenic *E. coli*. Moseley and Falkow[13] reported that DNA probes for the A and B subunits of LT did not hybridize to *V. cholerae* under conditions of high stringency, but did hybridize under low stringency conditions. Information gleaned from this probe also suggested that more than one copy of the cholera toxin gene existed in some strains of *V. cholerae*, a phenomenon later confirmed by Mekalonos.[14]

The LT-A probe, specifically, a 1200-bp *Hin*cII fragment from the plasmid pEWD299, was also used by Kaper et al.[15] to answer an important epidemiological question about the potential pathogenicity of environmental strains of *V. cholerae* O1. While nearly all strains of *V. cholerae* O1 isolated from patients produce cholera toxin, there are a number of isolates from environmental sources, such as oysters, brackish water, and sewage from such diverse areas as Bangladesh, Brazil, Guam, Great Britain, and the U.S., that do not produce any toxin detectable by standard phenotypic assays. Prior to this probe study, there was concern that these strains were mutants of toxigenic strains which still retained the potential to revert to toxigenicity and could serve as a source of infection. The concern about these strains was so great that they were tested in volunteers for pathogenicity.[16] When these strains were tested with the LT probe, no homologous sequences were found, thus demonstrating that these strains were not simple nontoxigenic mutants, but rather did not possess any genetic material encoding cholera toxin.[15] In contrast, all *V. cholerae* O1 and non-O1 strains which produced cholera toxin possessed sequences homologous to the LT probe. Thus, the nontoxigenic strains, since they lack any toxin gene sequences, cannot revert and serve as a reservoir of cholera.

The use of toxin probes in molecular epidemiology (or "molecular fingerprinting") was also demonstrated with several isolates of *V. cholerae* O1 from cases of cholera in the U.S. Gulf Coast. Strains from Texas and Louisiana from 1973 to 1981 were analyzed by hybridizing restricted chromosomal DNA with the LT probe.[17] An identical restriction fragment pattern homologous to the LT probe was seen in all Gulf Coast isolates, and this pattern was not seen in other strains isolated elsewhere in the world (see Figures 1 and 2). Thus, the restriction fragment length polymorphisms (RFLP) detected by the toxin gene probe demonstrated that a single endemic strain has persisted in the U.S. since at least 1973 and is responsible for sporadic cases of cholera in this country.

With the cloning and sequencing of genes encoding cholera toxin (*ctx*),[18-20] the LT probe has been replaced by a cholera toxin probe for screening *Vibrio* species. The sequence data shows that the genes for LT and CT share approximately 76% homology. A convenient CT probe is found in the plasmid pCVD27 which contains a 554 bp *Xba*I-*Cla*I fragment encoding 94% of the gene encoding the A subunit ligated to *Eco*RI linkers and cloned into pBR325. Cholera toxin gene probes have confirmed and extended observations made with the LT gene probe. The ability of the DNA probe to rapidly screen thousands of isolates for the presence of specific genes was utilized in a study of *Vibrio* species in the Louisiana Gulf Coast environment.[21] Over 2500 *V. cholerae* isolates were rapidly tested for the presence of toxin genes, thus elucidating the distribution of these genes in the Louisiana environment. In this study, enterotoxigenic *V. cholerae* O1 strains were recovered only from sewage stations or human disease, whereas enterotoxigenic *V. cholerae* non-O1 strains were recovered from environmental sources, in addition to clinical and sewage samples.[21] The molecular epidemiology studies have been extended with a 1983 *V. cholerae* O1 isolate from Mexico and a 1984 isolate from Maryland (crabs shipped from Texas were implicated as the source), both showing the same restriction enzyme pattern as the 1973 through 1981 isolates.[22,23] The large survey of Louisiana *V. cholerae* isolates cited above yielded a number of isolates of *V. cholerae* non-O1 and *V. mimicus* strains which hybridized to the probe. The restriction enzyme pattern of the toxin genes in these strains differed from that seen with the Gulf Coast *V. cholerae* O1 strains and showed greater genetic divergence than the highly conserved O1 strains.[24] Based on the toxin gene RFLP and total chromosomal restriction patterns, at least two separate clusters of *V. cholerae* non-O1 could be distinguished in addition to a distinct cluster of *V. mimicus* strains. These results demonstrate that the appearance of toxigenic *V. cholerae* O1 in the Gulf Coast was not due to a mutation of the somatic O1 antigen, but represents a distinct clone of toxigenic strains in this area.

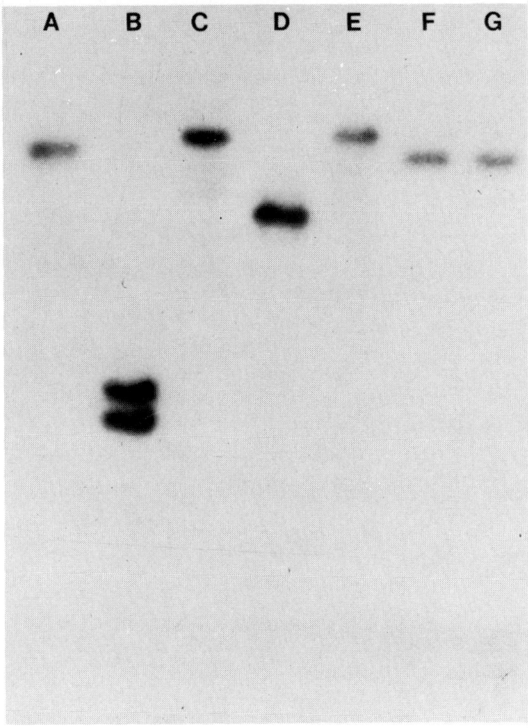

FIGURE 1. Autoradiograph showing hybridization of the *E. coli* LT probe to *Hin*dIII restriction fragments of El Tor *V. cholerae* O1 strains isolated from a variety of countries. Isolates from the U.S. (lane B) gave a unique pattern not seen in other strains. Lane A, Stokes 1 (Australia, 1977); lane B, 4808 (U.S., 1978); lane C, E9120 (indonesia, 1961); lane D, E9950 (Philippines, 1960); lane E, E8439 (Hong Kong, 1961); lane F, 62746 (Bangladesh, 1976); lane G, 30167 (Bangladesh, 1976). (From Kaper, J. B., Bradford, H. B., Roberts, N. C., and Falkow, S., *J. Clin. Microbiol.*, 16, 129, 1982. With permission.)

The study of *V. cholerae* in the U.S. Gulf Coast is an excellent example of the important contributions DNA probes can make in understanding the epidemiology of an infectious disease. DNA probes first demonstrated that environmental nontoxigenic strains of *V. cholerae* O1 cannot revert and serve as a reservoir of cholera, thus resolving an important public health issue. Probes for genes encoding cholera toxin also revealed that a single strain of *V. cholerae* O1 has been resident in the U.S. Gulf Coast since at least 1973 and that cases of cholera in the U.S. unconnected with foreign travel are due to infection with this strain. The probes have also demonstrated the presence of homologous sequences in *V. cholerae* non-O1 and *V. mimicus* strains, both clinical and environmental isolates. RFLP patterns using toxin gene probes have also allowed very sensitive "molecular fingerprinting" of these strains in the absence of useful bio-, sero-, or phage-typing systems. Finally, the DNA probe for the cholera toxin gene has permitted large ecological and epidemiological surveys to be carried out for *V. cholerae* O1, non-O1, and *V. mimicus*, by allowing the rapid, sensitive, and highly specific determination of toxin gene sequences in thousands of isolates.

One final application of a DNA probe for the detection of genes encoding cholera toxin is in vaccine development. Work at the Center for Vaccine Development in Baltimore, Md. has resulted in the development of an attenuated strain of *V. cholerae* that shows promise as an oral vaccine for the prevention of cholera.[25,26] Strains of *V. cholerae* were attenuated by constructing

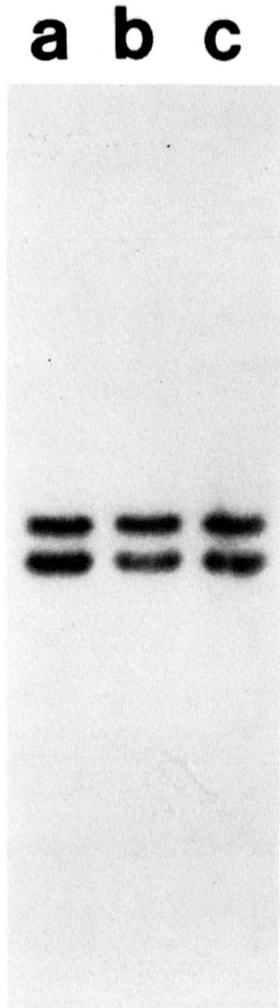

FIGURE 2. Autoradiograph showing hybridization of the *E. coli* LT probe to *Hin*dIII restriction fragments of *V. cholerae* O1 strains isolated from the U.S. All U.S. strains isolated since 1973 have shown the identical pattern. Lane A, E506 (Texas, 1973); lane B, 4808 (Louisiana, 1978); lane C, SGN7277 (Louisiana, 1980). (From Kaper, J. B., Bradford, H. B., Roberts, N. C., and Falkow, S., *J. Clin. Microbiol.*, 16, 129, 1982. With permission.)

specific deletions in the genes encoding cholera toxin. Cholera toxin consists of a single A subunit and five identical B subunits. The A subunit is responsible for the ADP-ribosylating activity characteristics of this toxin, while the B subunit serves to bind the holotoxin to the ganglioside G_{M1} receptor. Antibodies directed against the B subunit will prevent binding of the cholera toxin to intestinal cells. The cloned cholera toxin genes were mutated in vitro by removing a 554-bp *Xba*I-*Cla*I fragment containing sequences encoding the A subunit. The promoter for the *ctx* operon and the complete B subunit sequence were retained, thus allowing the production of the nontoxic B without the A subunit of cholera toxin. This mutation was then introduced into the chromosome of virulent *V. cholerae* strains by homologous recombination, specifically replacing the active toxin genes without affecting other genes encoding critical antigens. The desired mutation was detected using the 554-bp probe to screen colonies for the absence of this sequence. Vaccine candidates derived by this method (i.e., probe-negative strains that no longer produce the A subunit) have proven to be highly immunogenic in volunteers and have conferred excellent protection against challenge with virulent *V. cholerae*.[27]

B. Probes Detecting Genes Encoding Hemolysin

Many strains of *V. cholerae* produce a hemolysin active on sheep and other species of erythrocytes. This characteristic was initially used to distinguish the classical biotype (hemolysin negative) from the El Tor biotype (hemolysin positive) of this species. However, more recent El Tor isolates have been only weakly hemolytic or nonhemolytic. Hemolysin has also been suggested as a possible virulence factor, particularly for non-O1 strains of *V. cholerae* that do not produce cholera toxin.[28] The gene encoding this factor, designated *hly*, has been cloned by several groups,[29-31] and its DNA sequence has recently been determined.[32] The initial reports describing this gene used fairly large probes of a 6.2-kb *Pst*I fragment[30] or a 2.3-kb *Eco*RI-*Nru*I fragment[29] to detect homologous sequences in *V. cholerae* strains. These probe results have shown that nonhemolytic *V. cholerae* strains possess sequences homologous to the probe, thus suggesting mutations in a regulatory rather than structural gene. Using the 6.2-kb *Pst*I probe with Southern blots, Brown and Manning[33] have shown homologous sequences in chromosomal restriction digests of *V. cholerae* non-O1 strains. Differences in sizes of restriction fragments hybridizing to the hemolysin gene probe can distinguish hemolytic from nonhemolytic O1 strains and hemolytic non-O1 strains from nonhemolytic O1 strains.[33]

Work in our laboratories at the Center for Vaccine Development has employed a hemolysin gene probe for taxonomic and pathogenicity studies. Using a 400-bp *Hpa*I fragment that is internal to the *V. cholerae* hemolysin structural gene, we probed ten pathogenic *Vibrio* species and found homologous sequences in all strains of *V. cholerae* O1 and non-O1 tested, but not in any other species.[34] This result was true even when examining a number of sucrose-negative *V. cholerae*, which are easily confused with *V. mimicus* unless several confirmatory tests are performed. The hemolysin probe correctly detected the sucrose-negative *V. cholerae* strains, but did not hybridize with any *V. mimicus* strain tested. Similar to our work in developing an attenuated vaccine strain of *V. cholerae* described above, we used this 400-bp probe to construct a mutant strain lacking this fragment.[31] This mutation was constructed in strains lacking cholera enterotoxin and these nonhemolytic strains were evaluated in volunteers. The results indicated that the hemolysin did not contribute to the pathogenicity of these strains in man.[35]

C. Other Probes

V. cholerae has also been probed using labeled total DNA from lysogenic vibriophage. Goldberg and Murphy[36] employed the lysogenic phage VcA3 as a probe to examine the U.S. Gulf Coast *V. cholerae* strains. Their results confirmed the previous molecular epidemiological studies using the LT and CT probes and showed that all human *V. cholerae* O1 strains isolated from the U.S. Gulf Coast region since 1973 are lysogenic for this temperate phage. Furthermore, Southern blots of restricted genomic DNA show that the prophage is integrated at an identical chromosomal site in each of the human strains, thus further strengthening the clonal relationship among these iolates. The phage probe also revealed an additional relationship when a human isolate from Florida, which was toxin negative (tox^-) and lacked the *ctx* genes, was examined. This tox^- strain contained the prophage in an identical chromosomal site, thus suggesting a clonal relationship between the tox^+ and the tox^- strains from the Gulf Coast that would be impossible to detect using the CT probe alone.

Another epidemiological question was addressed by Cook et al.[37] using a number of molecular techniques, including probing with the CT probe and another vibrio prophage, VcA1. This epidemiological question concerned the fact that the El Tor biotype of *V. cholerae* had completely replaced the classical biotype in Bangladesh between 1973 and 1979. The classical biotype was reisolated starting in 1979, and questions were raised as to the relationship of these new classical strains to the El Tor strains and to the pre-1973 classical strains. The VcA1 DNA (approximately 20 Mdaltons) hybridized to all classical strains examined from 1916 to 1982, giving similar, but not identical, Southern blot patterns. The El Tor strains from Asia did not hybridize with VcA1, but the U.S. Gulf Coast strains examined hybridized weakly, presumably due to the presence of VcA3 noted above. Thus, the probe data revealed the close relationship among classical strains isolated over the course of several decades and the greater divergence seen among El Tor strains over a shorter time period.

One other probe employed for *V. cholerae* is a *toxR* sequence which encodes a positive regulator of cholera toxin and possibly other products of *V. cholerae*. When both nontoxinogenic and toxinogenic strains of *V. cholerae* were probed, a 3.9-kb *Eco*RI fragment hybridized to *toxR* sequences in all strains, thus suggesting that an intact regulatory system for *ctx* genes is present in naturally occurring nontoxinogenic strains of *V. cholerae* as well as in toxinogenic strains of both biotypes.[38] The significance of these observations is currently under investigation.

III. DNA PROBES FOR *V. PARAHAEMOLYTICUS*

The mechanisms by which *V. parahaemolyticus* causes disease in man are not clear. The potential virulence factor of this species that has received the most attention is the thermostable direct hemolysin (TDH). This hemolysin is responsible for the so-called Kanagawa Phenomenon (KP), which is a β-type hemolysis on a special blood medium called Wagatsuma agar. Although there is no hard evidence confirming the role of TDH in disease, there is strong epidemiological evidence implicating this factor. Nearly all clinical isolates of *V. parahaemolyticus* produce TDH (that is, are KP^+), whereas nearly all environmental isolates of this species do not produce TDH.[9] Additional evidence is found in volunteer studies where KP^+ strains produced diarrhea when fed to volunteers, while KP^- strains did not.[39]

As part of ongoing studies in our laboratories to assess the role of TDH in disease due to *V. parahaemolyticus*, we have cloned and sequenced genes encoding TDH (*tdh* genes)[40,41] and have utilized these sequences in DNA probe studies. The first study[42] utilized an internal 406-bp *Taq*I-*Hin*fI fragment which represents 71% of the structural gene. To ease probe preparation, this fragment was ligated to *Pst*I linkers (yielding a 415-bp fragment) and cloned into pBR322 in multiple copies. pCVD518 contains six copies of this fragment in tandem, allowing high yields of the probe. In the first study, 141 *V. parahaemolyticus* strains were tested with the probe and compared to hemolysis on Wagatsuma agar (i.e., the KP test), and two immunological tests, a modified ELEK test and an enzyme-linked immunosorbent assay (ELISA). All KP^+ strains were gene positive. However, 86% of the strains that exhibited a weak KP reaction and 16% of KP^- strains also reacted with the gene probe. The immunological methods showed better correlation with the gene probe results. All probe-positive strains produced hemolysin detectable in the ELISA, although occasional strains showed only a weak reaction. The modified ELEK test was slightly less sensitive than the ELISA. All gene-negative strains were also negative in these immunological assays. The obvious question that arose was whether this probe was detecting strains which produced TDH at low levels (possibly indicating a mutation in the genes responsible for regulation) or strains which contained mutated *tdh* structural genes from which biologically less active or inactive TDH molecules are produced.

Southern blot analysis of representative strains with this gene probe helped to illuminate the situation. When chromosomal digests were examined for homology to the probe, all KP^+ *V. parahaemolyticus* strains were found to possess two chromosomal gene copies.[43] Each strain possssed a 1.3-kb *Hin*dIII fragment which was homologous to the probe and a second fragment which was usually, but not always, 2.5 kb in size. However, each KP^- or weakly KP^+ (KP^w) strain which reacted with the probe posessed only one chromosomal fragment homologous to the probe, which was never 1.3 kb in size. Interestingly, in one KP^- strain, a second gene copy was found on a 35-kb plasmid in addition to a single chromosomal copy.

Because of the inability of the 415-bp restriction fragment probe to distinguish definitively all of the KP^- strains, we next utilized synthetic oligodeoxyribonucleotide probes.[44] Four gene probes from 19 to 21 bp in length were synthesized, which represented four different regions of the *tdh* structural gene. These probes were hybridized to a variety of gene positive KP^+, KP^-, and

KP^{w} strains, as well as gene-negative strains, and the blots were washed at increasingly stringent temperatures. Under stringent washing conditions, one probe specific for a region near the carboxy-terminal end of the *tdh* gene clearly differentiated the KP^{+} *V. parahaemolyticus* from KP^{-} or KP^{w} strains which reacted with the fragment probe. Two other probes did not distinguish between the strains at any stringency, while a fourth probe gave intermediate results. Thus, the oligonucleotide probe with the sequence (5′ to 3′) GCATATGAGAGTGGTAGTGG is more suitable for definitive determinations than hybridization with the larger 415-bp gene probe.

The significance of the different *tdh* gene copies in *V. parahaemolyticus* is unknown. Unlike the situation in *V. cholerae* where the two *ctx* gene copies found in classical strains are identical, the two *tdh* gene copies are clearly not identical, as shown by the oligonucleotide results, and this finding has been confirmed by DNA sequence analysis of the different gene copies.[45] Mutants of *V. parahaemolyticus* deficient in one or both of these gene copies are being prepared to determine the contribution of each of these genes to pathogenicity.

The *tdh* gene probe has also been employed to study the presence of related sequences in other species of *Vibrio*. An initial survey of 20 *Vibrio* species with the 415-bp fragment probe revealed homologous sequences in all 11 strains of *V. hollisae* examined.[42] On Wagatsuma agar, *V. hollisae* is KP^{-}, and hybridization with the synthetic probes revealed homology with all four probes at low stringency, but no hybridization with any probe at high stringency.[44] Several strains of *V. mimicus* also contained sequences homologous to the 415-bp probe. In a preliminary study, 5 of 13 *V. mimicus* strains contained homologous sequences in the chromosome.[46] *V. vulnificus*, which has been reported to be KP^{+} on Wagatsuma agar,[47] was negative with the 415-bp probe and showed weak hybridization at low stringency with the synthetic probes. The initial survey of *V. cholerae* strains proved negative with the *tdh* probe,[42] but recently, two *V. cholerae* non-O1 strains were found to be positive for the production of TDH as shown by ELISA and to possess sequences homologous to the *tdh* gene fragment probe.[48] The homologous sequences were found, not on the chromosome, but on a 33-kb plasmid in these strains. These genes were cloned into pBR322, and the resulting *Escherichia coli* clones produced intracellular hemolysin. The finding of *tdh* genes on plasmids in *V. cholerae* non-O1 and at least one strain of *V. parahaemolyticus* could be quite significant. It is tempting to speculate that the transmission of the *tdh* gene is mediated by an extrachromosomal element which could explain the presence of this factor in several *Vibrio* species. We have cloned a variety of *tdh* genes, both chromosomal and plasmid-borne, from *V. parahaemolyticus, V. hollisae, V. mimicus,* and *V. cholerae* non-O1 and are comparing these different genes by DNA sequence analysis.

IV. DNA PROBE FOR *V. VULNIFICUS*

As with *V. parahaemolyticus*, the pathogenic mechanisms of *V. vulnificus* are still unclear. One putative virulence factor is a cytotoxin/hemolysin which may contribute to virulence, but does not appear to be a marker for pathogenicity as it is produced by both virulent and avirulent strains.[49] Wright et al.[50] reported the cloning of the gene encoding this factor (designated *vch*) and the use of a 3.2-kb *Hin*dIII-*Eco*RI fragment as a gene probe. All 54 *V. vulnificus* isolates tested possessed sequences homologous to this probe, whereas they were absent from 96 isolates representing 29 other bacterial species. In a second study,[51] this probe was used to screen 44 lactose-positive *Vibrio* strains isolated from the marine environment. (Lactose fermentation is a key characteristic of this species, but this property is also found in other *Vibrio* species.) All 29 marine isolates identified as *V. vulnificus* on the basis of numerical taxonomy and genomic DNA:DNA hybridization studies hybridized with the cytotoxin gene probe. Homologous gene sequences were not found in any other lactose-positive marine vibrio isolates.

The *vch* gene probe was also used to examine chromosomal digests of *V. vulnificus*. In a preliminary study,[51] at least three different RFLP patterns were seen, but it was not possible to correlate these patterns with any obvious characteristics such as isolate source, environmental

Table 2
APPLICATION OF DNA PROBES TO THE GENUS *VIBRIO*

	Probes						
					tdh		
Species	*ctx*	*hly*	VcA	*tox R*	Fragment	Oligonucleotide	*vvh*
V. cholerae O1 CT⁺	+	+	+/–	+	–	–	–
V. cholerae O1 CT⁻	–	+	+/–	+	–	–	–
V. cholerae non-O1 CT⁺	+	+	nd	nd	–	–	–
V. cholerae non-O1 CT⁻	–	+	nd	nd	+/–	+/–	–
V. mimicus CT⁺	+	–	nd	nd	–	nd	–
V. mimicus CT⁻	–	–	nd	nd	+/–	nd	–
V. parahaemolyticus KP⁺	–	–	nd	nd	+	+	–
V. parahaemolyticus KP⁻	–	–	nd	nd	+/–	–	–
V. hollisae	–	–	nd	nd	+	–	–
V. vulnificus	–	–	nd	nd	–	–	+

Note: Abbreviations: CT⁺, strains producing cholera toxin; KP⁺, strains producing the Kanagawa phenomenon on Wagatsuma agar. (See text for probe abbreviations.) + indicates sequences homologous to the probe are present in this species under stringent conditions; – indicates sequences are not present, +/– indicates that the sequences are present in only some strains of the indicated species; nd = no data yet reported.

vs. clinical isolate, etc. In contrast to the gene duplication seen in *V. cholerae* and *V. parahaemolyticus,* there was no clear evidence of duplication of the *vch* gene in the strains examined. This gene probe has alo been used to screen oyster samples for *V. vulnificus,* as oysters appear to be the major source of human infections. Shucked oyster meats (50 g) were homogenized in 450 ml of phosphate-buffered saline and seeded with various dilutions of *V. vulnificus* with concentrations ranging from 10^7 to 10^0 CFU/ml. Samples were spotted onto nitrocellulose filters, grown overnight on L-agar, and the filters were processed and hybridized to the probe. As few as 6×10^1 *V. vulnificus* were easily detectable with the probe with 6×10^0 *V. vulnificus* producing a faint signal on the autogradiograph.

Thus, it appears that while the *vch* probe will not distinguish virulent from avirulent strains, it provides a very sensitive and specific means of identifying both environmental and clinical isolates of *V. vulnificus*, giving results comparable to those obtained using considerably more involved techniques, such as numerical taxonomy and genomic DNA:DNA hybridization. Additional applications may include direct screening of food samples as well as Southern blot patterns for molecular epidemiology, but further research is needed to validate these uses.

V. SUMMARY AND CONCLUSIONS

DNA probes have been devised for those *Vibrio* species that are most significant in human disease. All probes so far reported, with the exception of the *V. cholerae* phage probes, contain sequences which encode putative virulence factors. While some probes hybridize only to the species from which they were cloned, other probes react with more than one species. A summary of the various probes and the species with which they react is given in Table 2. Some probes detect genes which are always expressed, such as cholera toxin, while other probes may also detect cryptic, inactive genes such as the *V. cholerae* hemolysin. These probes have been used for a variety of applications including identification, molecular epidemiology, and vaccine development. Because molecular pathogenicity studies are in their relative infancy for this genus, it is highly likely that many new probes will be developed in the near future for the genus *Vibrio.*

ACKNOWLEDGMENTS

Studies conducted in the authors' laboratories were supported by research grants AI19716, AI19165 (to J. B. Kaper), and AI22189 (to J. G. Morris, Jr.) from the National Institute of Allergy and Infectious Diseases and a subcontract (to J. G. Morris, Jr.) from Brooz, Allen, and Hamilton under Task XXVII, U.S. Food and Drug Administration contract 223-80-2295.

REFERENCES

1. **Morris, J. G., Jr. and Black, R. E.,** Cholera and other vibrioses in the United States, *N. Engl. J. Med.*, 312, 343, 1985.
2. **Morris, J. G., Jr.,** *Vibrio* and *Aeromonas*, in *Infectious Diarrhea*, Gorbach, S. L., Ed., Blackwell Scientific Publications, Boston, 1986, 101.
3. **Blake, P. A., Merson, M. H., Weaver, R. E., Hollis, D. G., and Heublein, P. C.,** Disease caused by a marine vibrio: clinical characteristics and epidemiology, *N. Engl. J. Med.*, 300, 1, 1979.
4. **Bode, R. B., Brayton, P. R., Colwell, R. R., Russo, F. M., and Bullock, W. E.,** A new *Vibrio* species, *Vibrio cincinnatiensis*, causing meningitis: successful treatment in an adult, *Ann. Intern. Med.*, 104, 55, 1986.
5. **Gangarosa, E. J. and Mosly, W. H.,** Epidemiology and surveillance of cholera, in *Cholera*, Barua, D. and Burrow, W., Eds., W.B. Saunders, Philadelphia, 1974, 381.
6. **Rogers, R. C., Cuffe, R. G. C. J., Cossins, Y. M., Murphy, D. M., and Bourke, A. T. C.,** The Queensland cholera incident of 1977. II. The epidemiological investigation, *Bull. W.H.O.*, 58, 665, 1980.
7. **Blake, P. A., Allegra, D. T., Snyder, J. D., et al.,** Cholera — a possible endemic focus in the United States, *N. Engl. J. Med.*, 302, 305, 1980.
8. **Miller, C. J., Feachem, R. G., and Drasar, B. S.,** Cholera epidemiology in developed and developing countries: New thoughts on transmission, seasonality, and control, *Lancet*, 2, 281, 1985.
9. **Joseph, S. W., Colwell, R. R., and Kaper, J. B.,** *Vibrio parahaemolyticus* and related halophilic vibrios, *CRC Crit. Rev. Microbiol.*, 10, 77, 1983.
10. **Shandera, W. X., Johnston, J. M., Davis, B. R., and Blake, P. A.,** Disease from infection with *Vibrio mimicus*, a newly recognized *Vibrio* species, *Ann. Intern. Med.*, 99, 169, 1983.
11. **Morris, J. G., Jr., Miller, H. G., Wilson, R., Tacket, C. O., Hollis, D. G., Hickman, S. W., Weaver, R. E., and Blake, P. A.,** Illness caused by *Vibrio damsela* and *Vibrio hollisae*, *Lancet*, 2, 1294, 1982.
12. **Tacket, C. O., Brenner, F., and Blake, P. A.,** Clinical features and an epidemiological study of *Vibrio vulnificus* infections, *J. Infect. Dis.*, 149, 558, 1984.
13. **Moseley, S. L. and Falkow, S.,** Nucleotide sequence homology between the heat-labile enterotoxin gene of *Escherichia coli* and *Vibrio cholerae* deoxyribonucleic acid, *J. Bacteriol.*,144, 444, 1980.
14. **Mekalanos, J. J.,** Duplication and amplification of toxin genes in *Vibrio cholerae*, *Cell*, 35, 253, 1983.
15. **Kaper, J. B., Moseley, S., and Falkow, S.,** Molecular characterization of non-toxigenic *Vibrio cholerae* and related species, *Infect. Immun.*, 32, 661, 1981.
16. **Levine, M. M., Black, R. E., Clements, M. L., Cisneros, L., Saah, A., Nalin, D. R., Giol, D. M., Craig, J. P., Young, C. R., and Ristaino, P.,** The pathogenicity of nonenterotoxigenic *Vibrio cholerae* serogroup O1 biotype El Tor isolated from sewage water in Brazil, *J. Infect. Dis.*, 145, 296, 1982.
17. **Kaper, J. B., Bradford, H. B., Roberts, N. C., and Falkow, S.,** Molecular epidemiology of *Vibrio cholerae* resident in the U.S. Gulf Coast, *J. Clin. Microbiol.*, 16, 129, 1982.
18. **Lockman, H. and Kaper, J. B.,** Nucleotide sequence analysis of the A2 and B subunits of *Vibrio cholerae* enterotoxin, *J. Biol. Chem.*, 258, 13722, 1983.
19. **Lockman, H. A., Galen, J. E., and Kaper, J. B.,** Further analysis of the *Vibrio cholerae* enterotoxin genes: nucleotide sequence analysis of the gene encoding the ADP-ribosylating subunit, *J. Bacteriol.*, 159, 1086, 1984.
20. **Mekalanos, J. J., Swartz, D. J., and Pearson, G. D. N., et al.,** Cholera toxin genes: nucleotide sequence, deletion analysis and vaccine development, *Nature (London)*, 306, 551, 1983.
21. **Roberts, N. C., Siebling, R. J., Kaper, J. B., and Bradford, H. B.,** Vibrios in the Louisiana Gulf Coast environment, *Microbiol. Ecology*, 8, 299, 1982.
22. **Blake, P. A., Wachsmuth, K., Davis, B. R., Bopp, C. A., Chaiken, B. P., and Lee, J. V.,** Toxigenic *Vibrio cholerae* O1 strain from Mexico identical to United States isolates, *Lancet*, 2, 912, 1983.

23. **Lin, F.-Y. C., Morris, J. G., Jr., Kaper, J. B., Gross, T., Michalski, J., Morrison, C., Libonati, J. P., and Israel, E.**, Persistence of cholera in the United States: isolation of *Vibrio cholerae* O1 from a patent with diarrhea in Maryland, *J. Clin. Microbiol*, 3, 624, 1986.
24. **Kaper, J. B., Nataro, J. P., Roberts, N. C., Siebeling, R. J., and Bradford, H. B.**, Molecular epidemiology of non-O1 *Vibrio cholerae* and *V. mimicus* in the U.S. Gulf Coast region, *J. Clin. Microbiol.*, 23, 652, 1986.
25. **Kaper, J. B., Lockman, H., Baldini, M. M., and Levine, M. M.**, Recombinant nontoxinogenic *Vibrio cholerae* strains as attenuated cholera vaccine candidates, *Nature*, 308, 655, 1984.
26. **Kaper, J. B., Lockman, H., Baldini, M. M., and Levine, M. M.**, A recombinant live oral cholera vaccine, *Bio/Technology*, 2, 245, 1984.
27. **Kaper, J. B., Levine, M. M., Lockman, H., Baldini, M. M., Black, R. E., Clements, M. L., and Morris, J. G.**, Development and testing of a recombinant live oral cholera vaccine, in *Vaccines 85. Molecular and Chemical Basis of Resistance of Viral, Bacterial, and Parasitic Diseases*, Lerner, R., Chanock, R., and Brown, F., Eds., Cold Spring Harbor Press, New York, 1985, 107.
28. **Yamamoto, K.**, personal communication, 1986.
29. **Goldberg, S. L. and Murphy, J. R.**, Molecular cloning of the hemolysin determinant from *Vibrio cholerae* El Tor, *J. Bacteriol.*, 160, 239, 1984.
30. **Manning, P. A., Brown, M. H., Heuzenroeder, M. W.**, Cloning of the structural gene (*hly*) for the hemolysin of *Vibrio cholerae* El Tor strain O17, *Gene*, 31, 225, 1984.
31. **Mobley, H. L. T., Michalski, J. M., and Kaper, J. B.**, Construction of a hemolysin-deficient *Vibrio cholerae* El Tor strain, *Abstr. 25th Intersci. Conf. Antimicrobial Agents and Chemotherapy*, Am. Soc. Microbiol., Washington, D.C., 1985, 299.
32. **Yamamoto, K., Iwanaga, M., Honda, T., Miwatani, T., Shinagawa, H., and Nakata, A.**, Genetical and biochemical studies on the hemolysin of *Vibrio cholerae* biotype El Tor, presented at the 22nd Jt. Conf. U.S.-Japan Coop. Med. Sci. Program Cholera Panel, July 20 to 23, 1986.
33. **Brown, M. H. and Manning, P. A.**, Haemolysin genes of *Vibrio cholerae*: presence of homologous DNA in non-haemolytic O1 and haemolytic non-O1 strains, *FEMS Microbiol. Lett.*, 30, 197, 1985.
34. **Kaper, J. B. and Michalski, J. M.**, unpublished data, 1985.
35. **Levine, M. M., Kaper, J. B., Morris, J. G., Herrington, D., Losonsky, G., and Tall, B.**, Reactogenicity, colonizing capacity, and immunogenicity of further attenuated, genetically-engineered *Vibrio cholerae* O1 vaccine strains, *Abstr. 21st Jt. Conf., U.S.-Japan Coop. Med. Sci. Program Cholera Panel*, Bethesda, Md., 1985, 37.
36. **Goldberg, S. and Murphy, J. R.**, Molecular epidemiological studies of United States Gulf Coast *Vibrio cholerae* strains: integration site of mutator vibriophage VcA-3, *Infect. Immun.*, 42, 224, 1983.
37. **Cook, W. L., Wachsmuth, K., Johnson, S. R., Birkness, K. A., and Samadi, A. R.**, Persistence of plasmids, cholera toxin genes, and prophage DNA in classical *Vibrio cholerae* O1, *Infect. Immun.*, 45, 222, 1984.
38. **Miller, V. L. and Mekalanos, J. J.**, synthesis of cholera toxin is positively regulated at the transcriptional level by *toxR*, *Proc. Natl. Acad. Sci. U.S.A.*, 81, 3471, 1984.
39. **Sanyal, S. C. and Sen, P. C.**, Human volunteer study on the pathogenicity of *Vibrio parahaemolyticus*, in Int. *Symp. Vibrio parahaemolyticus*, Fujino, T., Sakaguchi, G., Sakazaki, R., and Takeda, Y., Eds., Saikon Publ. Co., Tokyo, 1974, 227.
40. **Kaper, J. B., Campen, R. K., Seidler, R. J., Baldini, M. M., and Falkow, S.**, Cloning of the thermostable direct or Kanagawa phenomenon associated hemolysin of *Vibrio parahaemolyticus*, *Infect. Immun.*, 45, 290, 1984.
41. **Nishibuchi, M. and Kaper, J. B.**, Nucleotide sequence of *Vibrio parahaemolyticus tdh* gene (thermostable direct hemolysin gene), *J. Bacteriol.*, 162, 558, 1985.
42. **Nishibuchi, M., Ishibahi, M., Takeda, Y., and Kaper, J. B.**, Detection of the thermostable direct hemolysin gene and related DNA sequences in *Vibrio parahaemolyticus* and other vibrios by the DNA colony hybridization test, *Infect. Immun.*, 49, 481, 1985.
43. **Nishibuchi, M., Galen, J. E., and Kaper, J. B.**, The *tdh* gene (thermostable direct hemolysin gene of *Vibrio parahaemolyticus*), in *Advances in Research on Cholera and Related Diarrhoeas*, Vol. 4, Kuwahara, S. and Pierce, N. F., Eds. Martinus Nijhoff, The Hague, in press, 1988.
44. **Nishibuchi, M., Hill, W. E., Zon, G., Payne, W. L., and Kaper, J. B.**, Synthetic oligodeoxyribonucleotide probes to detect Kanagawa phenomenon-positive *Vibrio parahaemolyticus*, *J. Clin. Microbiol.*, 23, 1091, 1986.
45. **Nishibuchi, M. and Kaper, J. B.**, manuscript in preparation, 1988.
46. **Nishibuchi, M., Khaeomanee-iam, V., Honda, T., Miwatani, T., and Kaper, J. B.**, manuscript in preparation, 1988.
47. **Johnson, D. E. and Calia F. M.**, Hemolytic reaction of clinical and environmental strains of *Vibrio vulnificus*, *J. Clin. Microbiol.*, 14, 457, 1981.

48. **Honda, T., Nishibuchi, M., Miwatani, T., and Kaper, J. B.,** Demonstration of a plasmid-borne gene encoding a thermostable direct hemolysin in *Vibrio cholerae* non-O1 strains, *Appl. Environ. Microbiol.*, 52, 1218, 1986.
49. **Wood, P. K., Wright, A. C., and Morris, J. G.,** Association of virulence and iron responsiveness in *Vibrio vulnificus, Abstr. 25th Intersci. Conf. Antimicrobial Agents and Chemotherapy,* Am. Soc. Microbiol., Washington, D.C., 1985, 150.
50. **Wright, A. C., Morris, J. G., Jr., Maneval, D. R., Jr., Richardson, K., and Kaper, J. B.,** Cloning of the cytotoxin;dihemolysin of *Vibrio vulnificus, Infect. Immun.*, 50, 922, 1985.
51. **Morris, J. G., Jr., Wright, A. C., Roberts, D. M., Wood, P. K., Simpson, L. M., and Oliver, J. D.,** Identification of environmental *Vibrio vulnificus* isolates using a DNA probe for the cytotoxin-hemolysin gene, *Appl. Environ. Microbiol.*, 53, 193, 1987.

Chapter 7

NUCLEIC ACID PROBES FOR ROTAVIRUS DETECTION AND CHARACTERIZATION

Mary Kolb Estes and Tomoyuki Tanaka

TABLE OF CONTENTS

I. INTRODUCTION

Rotaviruses are the major cause of diarrheal disease in children and in the young of probably all mammalian and avian species throughout the world.[1-3] The discovery of the rotaviruses in 1963 in animals (mice and monkeys) and in 1973 in duodenal biopsies from infants and children hospitalized with acute gastroenteritis and the subsequent establishment that these viruses are responsible for severe (life-threatening) gastroenteritis in the young have represented major advances in elucidating the cause(s) of acute infectious nonbacterial gastroenteritis. It is now clear that rotavirus infections have a staggering impact in less-developed countries, where they constitute a major cause of mortality among the young. In the developing countries, diarrheal disease ranks first in both disease incidence and severity, and it has been estimated that in these regions 3 to 5 billion cases of diarrhea and 5 to 18 million diarrhea-associated deaths may occur annually.[4]

In developed countries, infectious gastroenteritis is second only to respiratory infections as the most common illness in families, and rotavirus infections result in significant morbidity and economic losses due to hospitalization costs and to time lost from work by parents caring for their ill children.[5,6] Nosocomial rotavirus infections are also a problem that may extend hospitalization for children admitted for other reasons; this is a particular problem for children with immune deficiencies, where rotavirus infections may be unusually prolonged.[7,8]

Rotaviruses are classified as members of the virus family Reoviridae. Their genome consists of 11 segments of double-stranded RNA. These discrete RNA segments are found within particles that possess a distinct morphologic appearance when observed by negative-stain electron microscopy. The name rotavirus (from the Latin *rota*, for wheel) was suggested because these particles have the appearance of a wheel with spokes radiating from a wide hub to the rim.[9] Three types of particles (double shelled, single shelled, and cores) can be observed in clinical or tissue-culture preparations of rotaviruses (Figure 1). The infectious particles are 70 nm in diameter, and the outermost margin of their double-shelled capsid has a well-defined, smooth, circular appearance. Single-shelled (60 nm) particles that have a rough outer layer and hexagonal (40 nm) core particles are often seen in viral preparations. Single-shelled particles can be produced in vitro from double-shelled particles by treating infectious particles with chelating agents; core particles can be produced by treating single-shelled particles with chaotropic agents.[10-12]

The proteins of rotavirus particles have been fairly well characterized (reviewed in References 1 to 3). The structural proteins found in virus particles are designated VP1 to VP9, and knowledge of their antigenic, biochemical, and functional properties is increasing.[1,13] The outer capsid of particles contains VP3, VP7, and possibly VP9. VP3 is the hemagglutinin,[14] and the cleavage of VP3 by trypsin into VP5* and VP8* is associated with activation of infectivity.[15] VP7 is a glycoprotein,[16] and it has been reported to be the cell attachment protein.[17] Recently, two glycoproteins (VP7[1] and VP7[2]) that are synthesized by the start of translation in a bicistronic mRNA at two distinct in-phase initiation codons in the neutralization gene have been reported in particles.[18] A minor amount of another protein — VP9, the protein product of genome segment 11 — has been reported in the outer virus capsid by some, but not all, investigators. The outer layer of the inner capsid is VP6, and this protein comprises 80% of the protein mass of particles.[11,19] Antigenic epitopes on VP6 are the major determinants detected by enzyme-linked immunosorbent assays (ELISAs) for virus particles.[13] VP1, VP2, and possibly VP4 are the proteins of viral core particles.

Rotavirus particles contain an endogenous RNA-dependent RNA polymerase; the polymerase is inactive in infectious double-shelled particles, but disruption and conversion of double-shelled particles to single-shelled particles by heat treatment or treatment with chelating agents activates the polymerase.[10,20,21] Activated polymerase will synthesize large amounts of single-stranded RNA transcripts representative of each template RNA genome segment in virus

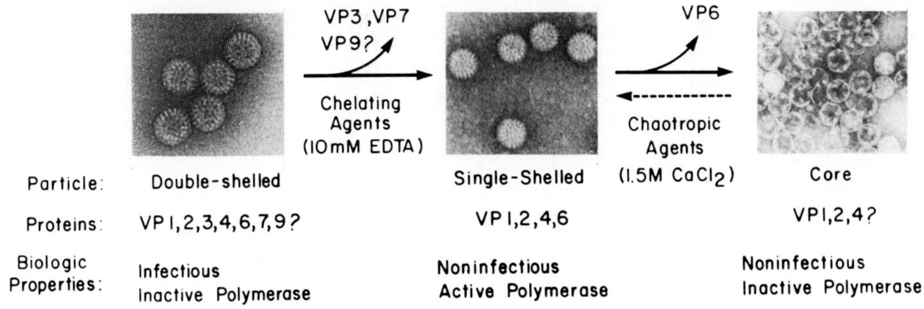

FIGURE 1. Properties of rotavirus particles. The viral proteins (VP1 to VP9), biologic properties, and appearance of rotavirus particles seen by negative-stain electron microscopy in clinial or tissue culture samples are shown. Methods to convert one virus form to another that are useful in immunologic procedures to detect common antigens or in biochemical procedures to produce transcripts for hybridization probes are also indicated.

particles.[20,22] Large amounts of transcripts can be made in vitro, and these have been useful (1) to determine the coding assignment of each genome segment,[23] (2) as templates to make cloned DNAs,[24] and (3) as probes for hybridization studies.[25-30]

Recognition of the significant mortality and morbidity caused by the rotaviruses has stimulated research on these viruses with the goals of developing effective methods of virus detection and ultimately of disease prevention and control. The need for rapid diagnosis of rotavirus infections in any clinical setting is well recognized in that it allows for prompt, appropriate treatment and the institution of infection-control measures in hospital environments. Rapid diagnosis and strain characterization are also becoming increasingly important to help evaluate efficacy in vaccine trials and to complete our understanding of the epidemiology and transmission of these viruses.

II. ROTAVIRUS CLASSIFICATION

An interesting new chapter in the history of rotavirus detection and epidemiology began in 1980, when viruses that possessed the characteristic rotavirus morphology but lacked the common (cross-reactive) group antigens located on several of the capsid proteins of prototype (group A) rotaviruses were described.[13,31,32] Rotaviruses with these characteristics were initially given various names, including atypical, antigenically distinct rotavirus, rotavirus-like agents, and pararotavirus, and their description emphasized our still incomplete knowledge of the diversity and biology of the rotaviruses. To date, these newly described rotaviruses have also possessed a distinct but characteristic pattern of migration of their genome segments (RNA electropherotype) after electrophoresis of extracted RNA in polyacrylamide gels (Figure 2). Observation of such unusual RNA patterns suggests a tentative identification of these rotaviruses, but confirmation requires serologic testing of isolates.

The description of an increasing number of rotaviruses from humans and animals with these distinct characteristics has resulted in the proposal that rotaviruses be subdivided into groups (to be designated groups A, B, C, D, etc.) with viruses in each group sharing cross-reactive antigens.[34-58] Five rotavirus groups (A to E) have now been proposed, based on comparative antigenic and genetic studies.[56] In this classification system, group A rotaviruses are those that are well characterized as causing infantile diarrhea. Further classification of group A rotaviruses involves serologic characterization of virus strains and identification of specific virus serotypes based on reactions in plaque-reduction (neutralization) assays with well characterized antiserum.[13,35]

The clinical significance of most of the nongroup A rotaviruses is not known, but annual epidemics of diarrhea among both adults and children in China have been attributed to group B

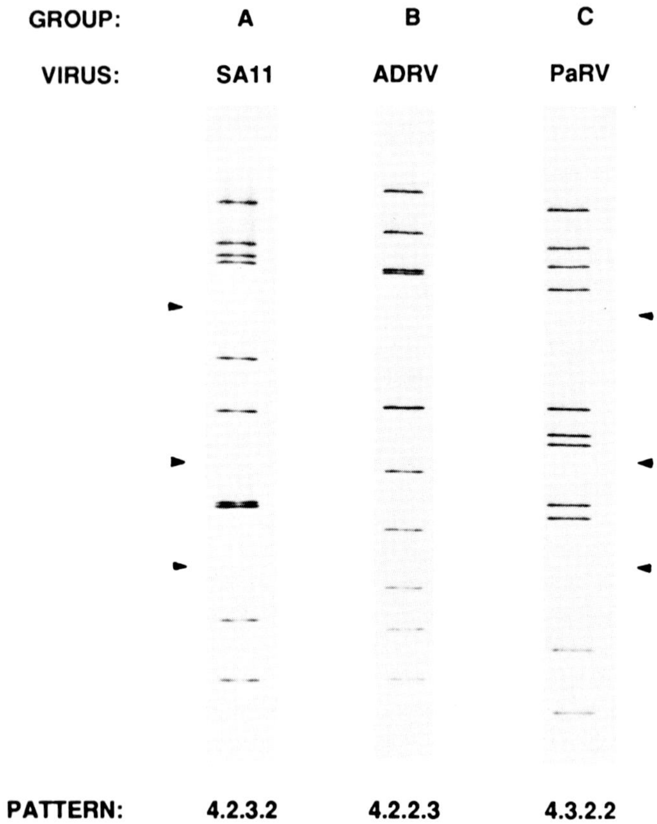

FIGURE 2. RNA electrophoretic patterns of rotavirus strains classified as group A, B, or C rotaviruses. The double-stranded RNA segments in three rotavirus strains (group A, simian rotavirus SA11; group B, adult diarrhea rotavirus, ADRV; and group C, porcine pararotavirus, PaRV) were visualized by silver staining after electrophoresis in 10% polyacrylamide gels. These rotaviruses do not share any detectable cross-reacting antigens. The RNA pattern of each virus is classified by comparing the distribution of the 11 RNA segments of the viruses relative to the 4:2:3:2 pattern for SA11, where segments 1 to 4, segments 5 and 6, segments 7 to 9, and segments 10 and 11 migrate, respectively, in separate groups as described by Lourenco et al.[33] The arrows indicate the empirical positions in the gel that separate the segments into different groups.

rotaviruses, also referred to as adult diarrhea rotaviruses (ADRV). Infections with ADRV strains have been extremely severe (cholera-like) in all age groups, and mortality (due to severe dehydration) was reported in some elderly patients.[55] Current seroepidemiologic studies have reported a low prevalence of antibody to ADRV, suggesting that widespread circulation of these viruses has not yet occurred outside China.[59,60] However, continued surveillance for the nongroup A rotavirus remains important because many people who lack detectable antibody would appear to be susceptible and group B infections in China have been extremely pathogenic. The data showing the importance of nongroup A rotaviruses as veterinary pathogens, based on transmission experiments and antibody prevalence, are more convincing.[39,40,47,48,50,53,56]

A combination of procedures is necessary to detect and characterize the nongroup A viruses; these rotaviruses are not detected with current ELISAs for group A rotaviruses. For example, samples that are positive by electron microscopy (EM^+) but are group A ELISA negative, or samples that contain viruses with an unusual RNA electropherotype, may contain nongroup A rotaviruses. Confirmation of the rotavirus group of any isolate currently requires serologic identification by a research group or reference laboratory with group-specific antisera that are not widely available. None of the nongroup A viruses have been successfully cultivated to allow easy preparation of large supplies of antisera for ELISAs. ELISAs to detect nongroup A rotaviruses have been described.[54,57] However, it has been suggested that these viruses may be excreted in lower amounts than group A rotaviruses so that the success of routine detection of nongroup A rotaviruses by EM and ELISA has been questioned.[57,58]

The synthesis and use of nucleic acid probes to these new groups of rotaviruses may be necessary to facilitate their detection if they continue to be refractory to cultivation. The use of nucleic acid hybridization to detect nongroup A virus isolates will not be considered further in this article. Instead, this chapter will focus on the use of nucleic acid probes to detect and characterize the group A rotaviruses. However, the principles summarized here from the existing work on the group A rotaviruses should be applicable to detection and characterization of rotaviruses in any of the other groups once the appropriate probes are available.

III. METHODS TO DIAGNOSE ROTAVIRUS INFECTIONS

Diagnosis of rotavirus infection by culture has not been feasible as a routine procedure because the majority of strains are difficult to isolate and some continue to be noncultivatable. Therefore, alternative methods of rapid detection have been developed. Presently, many techniques are available to detect rotaviruses, but rapid diagnosis (less than 1 day) requires visualization of virus by EM or immunologic detection of rotavirus antigens. ELISAs are the most popular detection method because they are simple enough to be performed in virtually any laboratory and they do not require expensive equipment. ELISAs that detect shared common antigens present in group A rotaviruses, regardless of their species of origin, are relatively sensitive tests, but false positive and false negative results do occur. In such cases, EM has remained the "gold standard" since it enables a rapid definitive diagnosis. In addition, EM gives unequivocal answers for questionable samples. However, both ELISA and EM appear to be approaching their limits of sensitivity, and these tests currently are not able to readily differentiate or characterize rotavirus strains.

In contrast to the limitations and difficulties with using classical virological methods to study rotaviruses, it has become increasingly easy to study rotavirus genomic RNAs, and the analysis of rotavirus nucleic acid is being exploited both for molecular epidemiology studies and for rapid diagnosis.[30,61-64] Diagnosis of rotavirus infections by analyzing genomic nucleic acid is accomplished by extracting nucleic acids from stool samples and then analyzing the RNA electropherotype observed after electrophoresis of the extracted nucleic acids in polyacrylamide gels and staining with ethidium bromide or silver nitrate (Figure 2). Rotavirus double-stranded RNAs migrate as discrete bands with a characteristic pattern, and this pattern is diagnostic because the rotaviruses are the only known viruses that possess 11 segments. Other mammalian viruses of the Reoviridae (reoviruses or orbiviruses) contain ten RNA segments, and the mobility of their segments is also sufficiently distinct to allow identification of any of these viruses by this method.

The analysis of RNA patterns of rotaviruses in diarrheal samples stimulated great interest because this method has resulted in the discovery of previously unrecognized rotaviruses, including new groups of rotaviruses (see above) and new rotavirus serotypes within group A viruses.[61,65] Detection of rotavirus double-stranded RNA in stool samples is also surprisingly simple, because the nucleic acids (double-stranded DNAs or single-stranded RNAs or DNAs) of other enteric viruses (enteric adenoviruses, Norwalk virus, caliciviruses, astroviruses, bacteriophage) or bacteria that might be present in diarrheal stools do not interfere with recognition of rotavirus double-stranded RNA; these other nucleic acids will not enter the polyacrylamide gels used to detect rotavirus RNAs. In addition, the double-stranded RNAs in stools have been found to be unexpectedly resistant to digestion with ribonucleases. It has been

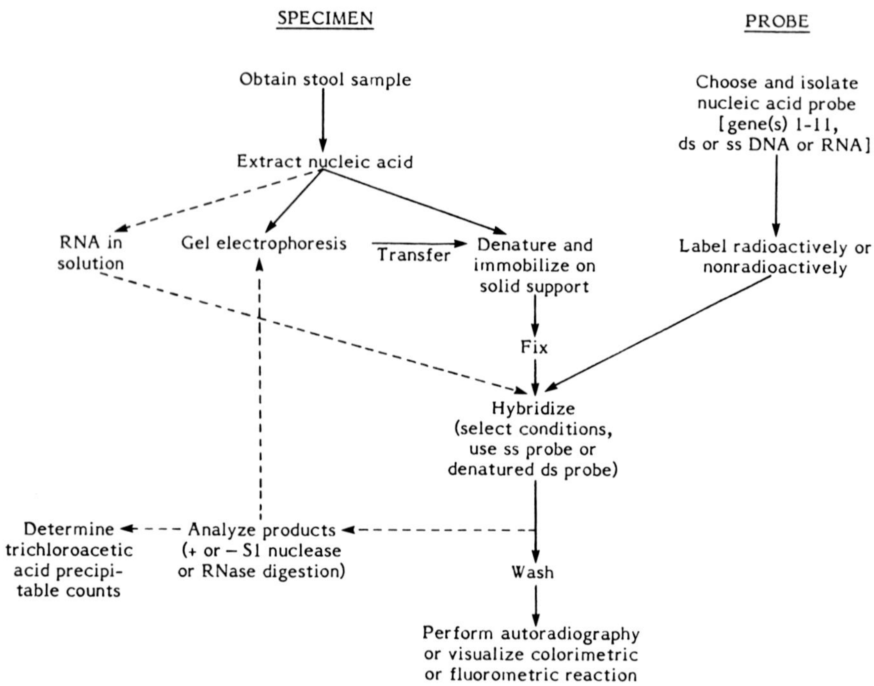

FIGURE 3. General steps in hybridization procedures to detect rotaviruses. The influence of choice of probe, labeling of probe, and test format on hybridization assays to detect rotaviruses are discussed in the text. The heavy arrows highlight the favored test format of hybridization on a solid (preferably nylon) support for greatest test sensitivity. ss = single-stranded; ds = double-stranded.

reported that the sample preparation procedure for RNA analysis can be further simplified by omitting purification of the nucleic acids;[66] however, caution in interpreting unusual patterns seen with this simplified method is recommended because residual salts in some unpurified samples may alter electrophoretic patterns.[62,66,67]

Recognition of the usefulness of detecting and analyzing rotavirus nucleic acids in clinical samples also has stimulated evaluation of other methods of molecular analysis (hybridization assays) as alternative approaches for virus detection.[30,68-70] This article briefly reviews the current status of using nucleic acid hybridization to detect and characterize rotaviruses, and the role of this technique in helping understand the biology and epidemiology of this important group of human and veterinary pathogens. Since rotaviruses are excreted in large amounts in stools (up to 10^{11} particles per gram), these viruses are also a useful system to evaluate these new methods of virus detection. The results from studies using rotavirus nucleic acid probes will continue to be helpful for extension of this new technology to other noncultivatable or fastidious enteric pathogens.

IV. OVERVIEW OF HYBRIDIZATION METHODS TO DETECT ROTAVIRUSES

The detection of rotaviruses by nucleic acid hybridization is based on the identification of rotavirus-specific nucleic acid sequences in a clinical or test sample. The general steps used in hybridization procedures are outlined in Figure 3. Viral sequences are extracted from specimens and detected after a specific reaction (hybridization) with a nucleic acid probe. The probe, which will be labeled radioactively or nonradioactively for subsequent detection, will contain sequences complementary (homologous) to the viral nucleic acid in the specimen. Hybridization tests can be performed using different formats (liquid or solid support), and positive reactions are seen by different procedures that detect the labeled, hybridized probe. These reactions succeed because the duplex rotavirus RNA strands in the test sample and the probe (RNA or DNA) nucleic acids dissociate into single-stranded molecules, and then complementary sequences in the separated test and probe strands will reassociate into detectable double-stranded molecules, depending on a number of physicochemical properties of both RNA and DNA that are known and used in establishing appropriate test conditions. Parameters that must be optimized for each test format include the base pair composition of the test and probe nucleic acids, the hybridization temperature, the ionic strength of the solution, the type of hybridization (liquid or solid support) being performed, the presence and concentration of denaturants or organics (formamide or dextran sulfate) in the reaction mixtures, and the type (single-stranded or double-stranded DNA or RNA molecules or oligonucleotides), size, and specific activity (and stability) of the probe. These parameters are of practical importance because they affect hybridization rates and hybrid stability, which in turn influence test speed, sensitivity, and specificity. Theoretical bases of each of these parameters have been reviewed elsewhere[71-74] (see also Chapter 1), and they will only be considered here as they relate to the design, optimization, or interpretation of hybridization tests for rotaviruses.

The requirement for sequence complementarity is not absolute for hybridization to occur, and some amount of mismatch can be tolerated and still allow duplexes to form. Therefore, nonidentical but related nucleic acids can participate in hybridization reactions under certain conditions. Nonidentical hybrids are less stable than identical hybrids, and they can be removed when subjected to heat in buffers of low ionic strength. This property enables one to discriminate between precisely and imprecisely matched duplexes. In practice, this is important because different amounts of hybridization of related but nonidentical sequences can be detected by performing hybridization reactions under different conditions of stringency. Hybridization stringency can be altered by adjusting the salt and/or formamide concentration and/or by changing the temperature. High stringency conditions indicate a high degree of specificity between the strands of the nucleic acid hybrids with large numbers of base pair matches; low stringency conditions indicate a relatively small number of base pair matches, yet enough to yield a double-stranded moiety. The conditions that affect the stringency of hybridization are determined somewhat empirically with each test format, and they will vary depending on the type of hybridization probe being used.

V. ROTAVIRUS GENE PROPERTIES AND HYBRIDIZATION TESTS

Table 1 summarizes the properties of rotavirus genes that should be considered in hybridization studies with these viruses. First, it should be noted that each rotavirus genome segment encodes a different protein product; therefore, each segment represents a distinct gene. This fact raises the following questions concerning hybridization with these viruses. Will test sensitivity or specificity, or both, be influenced by using a probe that represents only 1 or all 11 of the genome segments? Will variable G+C contents of individual genes influence optimal test conditions? Will cognate genes (i.e., genes that code for the same functional protein in different virus strains) possess sufficient sequence homology to allow detection of all heterologous virus strains? Will variations in the sequences of specific genes be useful for strain characterization? Will cognate genes possess identical migration rates on polyacrylamide gel electrophoresis (PAGE) so that a specific segment probe can be easily prepared from a new or previously uncharacterized virus strain? Will the 11 and 8 conserved nucleotide sequences found at the 5′ and 3′ ends of each of the 11 segments interfere with hybridization tests?

Sequence information is the most complete for the simian rotavirus SA11 genes; therefore, data on the SA11 genes are shown and have been used as a prototype. The numbering of the genome segments is based on the order of migration of the SA11 segments after electrophoresis

Table 1
PROPERTIES OF ROTAVIRUS GENES

Genome segment/ homologous gene comparisons	Protein product	Gene length	G+C content (%)	Nucleotide homology (%)	Has conserved sequences at ends 5′	3′
SA11 1	VP1	3300*	ND	ND	ND	ND
SA11 2	VP2	2600*	ND	ND	ND	ND
SA11 3	VP4?	2550*	ND	ND	ND	ND
SA11 4 (partial)	VP3	2370*	36.7[a]	ND	ND	ND
RS 5	NS53	1581	34	ND	+	+
SA11 6/WA 6	VP6	1356	38.6/37.0	78	+	+
SA11 7/UK 9	NS34	1104/1076	33.5/33.0	71	?+	+
SA11 8/UK 7	NS35	1059	35.5/35.9	88	+	+
SA11 9/UK 8	VP7[1], VP[2]	1062	35.9/37.6	78	+	+
SA11 10/UK 10	NS20	751	40.2/41.7	92	+	+
WA 11/UK 11	VP9?	663/667	37.9/39.7	86	+	+

Note: These data are based on published nucleotide sequences.[75-87] Data for the homologous genes of two group A rotavirus strains that are in different serotypes were compared. The protein coding assignments are as reported for SA11 by Mason et al. (References 18, 23, and unpublished data). The identity of the gene 3 and gene 11 products remains unclear.[1,2] The gene 3 product has also been reported as a structural protein and a nonstructural protein. The protein product of gene 11 has been reported to be an outer capsid protein, VP9, by some, but not all, investigators.[1-3] It is possible that the VP9 protein detected in particles was a cleavage product from an unidentified precursor; in this case, the protein product of genome segment 11 would be a nonstructural protein. Gene lengths (*) for genome segments 1, 2, 3, 4, and 5 are experimentally derived values from analysis of glyoxal-denatured RNA or cDNA copies with DND standards on denaturing gels.[2] ND = not determined. Each genome segment contains conserved terminal sequences. These sequences on the positive strand are 5′GGCUUU$^{A}_{U}$AAAG and $^{A}_{U}$UGUGACC3′.

[a] Calculated from partial gene 4 sequence data of 2243 base pairs.[75]

of the RNAs in 10% polyacrylamide gels. Complete nucleotide sequences are available for SA11 genes 6 to 10, and a partial sequence has been published for gene 4. The sequence of gene 11 of two other rotavirus strains is also available. Table 1 shows some comparative analyses of selected nucleotide sequences of SA11 and heterologous rotavirus strains that have been published[75-87] that help answer the above questions.

The first important fact about rotavirus gene structure is that the size of individual genome segments is distinct [ranging from 663 base pairs (bp) for gene 11 to ≈3300 bp for gene 1]. Since hybridization test sensitivity is roughly proportional to the length of the sequence from which the probe is prepared, theoretically maximum test sensitivity will be seen if the probe represents sequences of all 11 rotavirus segments. This principle has been demonstrated in two studies.[69,70] If individual probes representing each segment are not available, the use of combinations of probes (compared with the use of an individual probe) also will result in increased test sensitivity (see Figure 4).

The sequence data also reveal that there is a bias against CGN and NCC codons in rotavirus genes. This bias is reflected in the relatively low (33 to 41.7%) G+C content of individual genes as compared to other eukaryotic genes. Since the G+C content of nucleic acid influences the melting temperature of hybrids, one asks whether the variable G+C contents of the individual genes will affect their reactivities either as probes or as substrates. Will the expected different stabilities of the double-stranded duplexes formed during hybridization influence interpretation of results from studies performed with probes representing more than one genome segment?

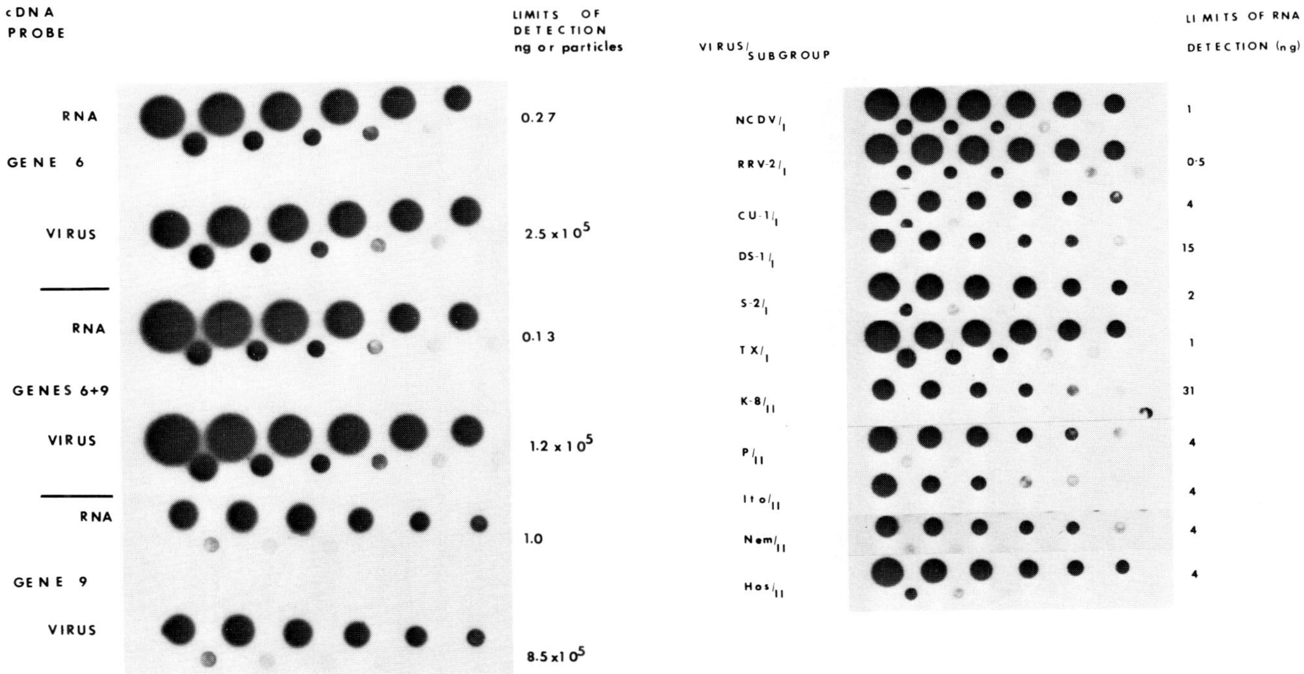

FIGURE 4. Sensitivity of the rotavirus dot-blot assay with individual or combined SA11 cDNA probes and with hterologous rotavirus strains. SA11 viral particles or double-stranded RNA extracted from [^{3}H]uridine-labeled SA11 virus (left-hand panel) or from 1 mg to 0.5 ng of double-stranded RNA extracted from heterologous NCDV, RRV-2, CI-1, DS-1, S-2, Tx-1258, K-8, P, Ito, nemoto, and Hosokawas strains (right-hand panel) were denatured by boiling, and serial twofold dilutions were applied to a nylon (Zetabind) filter. Hybridization was for 18 h at 65°C without formamide and with 0.5 ¥ 10^6 cmp/ml of SA11 gene 6 probe; autoradiography was for 12 h. The limits of virus or RNA detection are indicated in both panels. The results obtained with SA11 virus with different DNA probes are shown in the left-hand panel: (top) cDNA from gene 6, (bottom) cDNA from gene 9, and (center) cDNA from a combination of both gene 6 and gene 9. For heterologous strains (right-hand panel), the subgroup specificity and amount of RNA detected with the SA11 gene 6 probe are indicated.[69]

Will these different properties influence the specificity or sensitivity of results by changing the optimal conditions of hybridization to be used with probes representing either an individual or a combination of gene segments? Theoretically, hybridization with a genome segment 10 probe (which contains the highest G+C content) could be performed at higher temperatures, and this condition of hybridization might result in excellent specific reactions and lower nonspecific background reactions; however, hybridization using identical conditions and a gene 7 probe might fail to detect homologous nucleic acids in a test sample due to the conditions being too stringent. Answers to some of these questions were provided in the initial published studies that used hybridization with cloned DNA probes to detect rotaviruses.

Figure 4 illustrates typical results observed in a dot-blot assay using nylon membranes and ^{32}P-labeled SA11 cDNA probes to detect either homologous viral RNA or virus particles (left panel) or heterologous viral RNAs (right panel). Test sensitivity is clearly affected by the choice of the probe used for hybridization and by whether homologous or heterologous nucleic acids are being detected. Different conditions of hybridization were required for optimal test sensitivity to detect homologous RNAs using gene 6 (65°C) and gene 9 (56°C) probes.[69] Increased sensitivity was observed when two gene probes were combined. Finally, the limits of RNA detection were higher when rotavirus strains that were heterologous to the probe were detected[69] using optimal test conditions to detect homologous RNA.

The varied test sensitivity using individual genome segment probes is explained by the fact that the protein products encoded by each genome segment function either as structural or nonstructural proteins, and the extent of nucleotide sequence homology observed between the cognate genes differs, depending on the function of the individual gene product. For example, the sequences in genome segments 8 and 10 that clearly encode nonstructural proteins are the most conserved. In contrast, the sequences of genes 6 and 9 that encode the major viral structural protein VP6 and the outer capsid glycoprotein VP7 are the least conserved. Since it is presently unknown if the low nucleotide homology reported for the SA11 7/UK 9 genes reflects sequencing errors or true variation, these data have been excluded from this discussion. These theoretical points are supported by studies that show that genome segments encoding polypeptides comprising the ribonucleoprotein viral inner capsid (segments 1, 2, 3, and 6) and several nonstructural polypeptides (segments 7, 8, and 10) hybridized efficiently with a number of heterologous isolates, while probes of genome segments encoding outer capsid polypeptides (segments 4 and 9) varied in their efficiency of hybridization to individual isolates.[70] Detection of most strains can be achieved using the outer capsid protein gene 9 probe, but this requires use of different conditions of hybridization with lower stringency.[69,88]

Sequence data and gene-coding assignments determined from two or more virus strains have shown that cognate genes (i.e., SA11 6/WA 6, SA11 8/UK 7, or SA11 9/UK 8) are the same size (length). An exception to this general rule, again, is the large size difference reported for the SA11 gene 7 and the bovine rotavirus UK strain gene 9; as mentioned above, it is currently unknown whether this discrepancy reflects extreme variability in this gene. The fact that cognate genes in different virus strains are the same size, yet these genes often display different RNA electropherotypes, indicates that the heterogeneity observed in RNA patterns can reflect secondary structure that remains in the double-stranded RNAs during electrophoretic analysis in polyacrylamide gels. The gene sequence data also confirm previous biochemical and genetic data that had revealed that cognate genes may show a different order of migration in polyacrylamide gels. For example, the order of migration of the cognate genes of the SA11 and UK virus genome segments 7 to 9 is different. This means that cognate genes or probes of new virus strains cannot simply be identified by comparing the electrophoretic migration of rotavirus genes on gels. However, cognate genes with altered RNA patterns can be easily identified by hybridization.

The theoretical concern that the conserved nucleotide sequences found, respectively, at the 5′ and 3′ ends of each segment might result in false positive reactions with gene-specific probes has not been shown to cause problems in actual practice. This potential problem was initially tested using random primed cDNAs made from isolated segments as probes for hybridization with double-stranded RNA segments separated by electrophoresis on polyacrylamide gels and then electrophoretically transferred and covalently linked to diazobenzyloxymethyl paper.[89] Individual probes from one isolated segment hybridized only to their respective homologous genome segment under conditions of both low and high stringency.

In practice, many of the theoretical questions may only be critical for experiments where quantitative hybridization data are desirable to determine the relatedness of virus strains. Since rotaviruses can evolve by mechanisms of genome segment exchange (reassortment) and of accumulating point mutations and possibly by genome rearrangements,[8,89-91] the relatedness of virus strains can only be directly compared using individual probes that represent one of the 11 genes of one virus; otherwise, any homology data obtained between the genes of two virus strains represent an average of the homologies for all genome segments between the two strains. In other words, specific variation in any one genome segment cannot be reliably detected or quantitated without using individual genome segment probes. In addition, to maximize test sensitivity between heterologous virus strains, conditions of low or relatively low stringency of hybridization should be used.

In summary, the selection of a gene probe for hybridization with rotaviruses is important, and this selection will depend on the purpose of the experiment. Virus detection can be accomplished with any probe, but should be optimal with probes that represent the sequences in all 11 genes or with probes of the genes that encode the nonstructural proteins, since these show the least sequence variation. In contrast, screening virus strains for homologous genes that might have biologic or serologic significance will require the use of a more specific probe. For example, gene 4 probes might be used to compare homologies associated with virus virulence, hemagglutination, growth in tissue culture, or neutralization, and gene 9 probes could be useful to search for other homologies related to serotype specificity or to properties of cell attachment.

VI. HYBRIDIZATION STUDIES WITH ROTAVIRUSES

Relatively few studies using hybridization to detect rotaviruses in clinical specimens have been reported.[30,68-70] The published studies differ either in the test format used (liquid or solid support) or in the types of probes used. The following discussion will briefly consider the type of information gained from these studies and their potential future applications for rotavirus detection and characterization by comparing them according to the types of probe used.

A. Hybridization with Rotavirus Strains Using Transcript Probes

Transcript probes made with the endogenous viral RNA polymerase were used first to detect rotaviruses in dot-blot tests where the RNA in clinical samples was immobilized on nitrocellulose filters.[30] Subsequently, transcript probes have been used to identify the origin of specific gene segments in laboratory-derived reassortants,[27] using a combination of hybridization in solution followed by S1 nuclease digestion to remove partial hybrids and electrophoretic analysis of hybrid genome segments. The use of transcript probes for either procedure is relatively labor intensive because the virus must first be purified to make the ^{32}P- or ^{125}I-labeled probes. Therefore, hybridization with transcripts has not been used often for simple virus detection or for identifying gene reassortants unless the segments of the parental viruses were not able to be separated by the simpler procedure of merely analyzing the RNAs on long polyacrylamide gels.[92] However, detection of viral RNA with ^{125}I-labeled transcript probes is quite sensitive, and detection of as little as 8 pg of homologous (or 40 pg of heterologous) RNA has been reported in dot-blot tests.[30]

Transcript probes also have been used with hybridization experiments performed in solution to examine the relatedness of rotavirus strains.[26-29] Two methods to measure and quantitate the relatedness of the genes of double-stranded RNA viruses using hybridization in solution have been described for cell-culture-adapted reoviruses.[93,94] Both methods involve comparing heterologous hybrid genes that contain radiolabeled + strands and unlabeled – strands of genes from different serotypes with homologous hybrid genes (molecules formed by reannealing the separated + and – strands of the same gene) and have been well characterized. In the first technique, the amounts of radiolabeled probe in such hybrids sensitive to digestion with ribonuclease (or S1 nuclease) are compared and quantitated. In the second complementary method, the relative changes in the electrophoretic mobilities of hybrids (that are not digested with nuclease) are compared. Similar approaches were used for rotaviruses. However, when RNAs in clinical samples or from rotaviruses that grow poorly in cell culture were analyzed, it was often not possible to achieve optimal conditions. For example, it was not always possible to purify individual segments from each isolate or strain, and known amounts of excess cold RNA were not able to be used in the hybridization reactions. These limitations yielded semiquantitative results for rotaviruses, but the hybridization studies in solution have clearly shown that closer homology exists between some strains of human rotaviruses than between others, and little homology is generally observed between human and animal viruses. Due to the limitations and difficulties of these experiments, it may not be surprising that the levels of sequence homology reported between any of the human or animal viruses with these methods have not been completely confirmed by comparisons of direct sequence data. Estimates of sequence homology determined by solution hybridization studies have consistently underestimated the actual extent of homology observed by direct sequencing.

Although the precise reason for these discrepancies is not known, possible explanations include: (1) homologous hybridization of the positive and negative strands of the unlabeled test RNAs for each other may have outcompeted hybridization of the positive transcription probe for the heterologous negative strand and lowered the heterologous hybridization; (2) the hybridizations were not performed under conditions where the double-stranded RNA excess was sufficient to effect saturation conditions needed for quantitative hybridization; or (3) some of the heterologous hybrids were lost during the trichloroacetic acid precipitation such that some homology was not detected.

The rapidly recognized difficulties (discussed above) in performing and interpreting results if individual segment probes are not used for such studies, the labor requirements for separating individual genome segments or transcripts for use as probes, and the fact that such solution hybridizations allow assessment of the sequence relatedness of only one pair of virus isolates in each single reaction have not favored extensive use of this procedure. Fortunately, an alternative approach has also been developed to examine sequence homology among many isolates simultaneously by performing hybridization with double-stranded RNA genome segments electrophoretically separated and then transferred to a filter (Northern blots).

B. Hybridization with Rotavirus Strains Using Random Primed cDNA Probes

Northern blots were initially used with random primed cDNA probes to examine sequence diversity in individual RNA segments after hybridization with segments covalently linked to diazobenzyloxymethyl paper.[89] Hybridization was performed under conditions of both low and high stringency, and the results suggested that variation among rotavirus strains probably occurs by a process involving both drift (cumulative changes in nucleic acid sequences) and shift (the exchange of genomic segments). A probe from one isolated segment only hybridized to its respective homologous gene. Therefore, no cross-hybridization was seen between the 11 and 8 conserved nucleotide sequences that are found, respectively, at the 5′ and 3′ ends of each genome segment under the hybridization conditions used.

The sequence relationships of a number of cultivatable and noncultivatable human and animal rotavirus strains have been studied using this approach.[95] This work has shown that most rotavirus genome segments (except genes 4 and 5) exhibit partial homology under hybridization conditions of low stringency. Low stringency conditions in these initial studies were 52°C in the absence of formamide; these conditions of hybridization were reported to detect up to 34% of base mismatches. In contrast, few segments exhibited homology under more stringent conditions of hybridization (52°C, 50% formamide, reported to detect sequences having 8% base mismatches or less). These studies provided further support for the idea that both drift and shift are involved in the evolution of distinct rotavirus strains.

One problem in the interpretation of these hybridization results with Northern blots is that the cDNA reverse transcription of genome segments is not equimolar; therefore, the absolute differences in the intensities of hybridization signals observed on the blots cannot be precisely correlated with differing degrees of sequence homology. In addition, quantitative assessment of these experiments requires that the amounts of double-stranded RNA separated by PAGE are the same for each sample and that the efficiency of electrophoretic transfer and of binding of each genome segment to the filter matrix is the same.

Interpretation of results from hybridization on filters depends heavily on reconstruction experiments run in parallel using known quantities of reagents. In particular, with filter hybridizations it is important to assess experimental variation due to (1) the varying capacity of different filters or lots of filters to bind nucleic acids, (2) the variable transfer of nucleic acids of different molecular weights that can result in a variable proportion of the total immobilized nucleic acid being available for hybrid formation, (3) possible loss of formed hybrids from filters, and (4) some loss of fixed nucleic acid from the filters during long hybridization and during stringent washing procedures. Because of these possible variables, hybridization on filters must be controlled internally and cannot rely on mere comparisons with the published results of others. The greater simplicity of performance and versatility of blotting techniques, in terms of the numbers of samples that can be analyzed simultaneously or sequentially, are clear advantages of this method, and the limitations of blotting (if recognized) can be relatively easily controlled.

One should also note that the optimal conditions of hybridization on filters are somewhat empirical. Although published studies discuss calculations of hybridization stringency,[96] a complete biophysical understanding for these calculations, especially for RNA:RNA hybrids, is lacking. In comparing results between studies, one must consider whether the probe used was a DNA or RNA probe, and the relative different conditions of hybridization should be noted, as these factors may significantly affect the results.

It also should be noted that estimates of sequence homology made by hybridization studies performed on filters or in solution have differed.[29,94] In particular, sequence homology detected by solution hybridization studies with the rotavirus or reovirus systems was not always detected using filter hybridization studies.[29,97] This may reflect differences in the stringency of hybridization used in these studies. However, since the nucleotide sequence data for reovirus agree more closely with the results of the solution hybridization studies and the increased sensitivity and quantitative properties of hybridization in solution (compared to blotting techniques) are well documented, one must be cautious in assuming that negative results seen by Northern blots definitely reflect a lack of homology between two test samples.

C. Hybridization with Rotavirus Strains Using cDNA Probes

Following the use of random primed cDNA probes, the successful cloning of rotavirus genes renewed interest in the application of nucleic acid hybridization to detect and characterize rotaviruses.[68-70,92] The use of cDNA clones has the appealing feature that large amounts of probes for individual genes are able to be prepared quickly and relatively easily. The feasibility of detecting rotaviruses in tissue culture or clinical samples has now been clearly demonstrated using several test formats and different rotavirus cDNA probes (Table 2).

Cloned DNAs can clearly be used as probes to detect rotavirus in both tissue culture and clinical samples (see Figure 4). Although the number of samples tested in comparative studies

Table 2
DETECTION OF ROTAVIRUSES USING cDNA PROBES

Type of probe	Gene probes used	Probe excised from vector	Test format	Types of samples	Hybridization conditions	Test sensitivity	Comments	Ref.
Random-primed ^{32}P-labeled cDNA	All	Not applicable	Northern blots on DBM paper	Tissue culture and clinical samples	Varied	Not determined	Showed usefulness to compare genetic relatedness of strains	89, 95
Nick-translated, ^{32}P-labeled cDNA	6 or 9 or mixtures	Yes	Dot blot on nylon filters	Tissue culture and clinical samples	Varied	2.5×10^5 particles 0.27 ng homologous RNA 0.5–0.31 ng heterologous RNA	Good specificity with clinical samples; potential shown for strain characterization with individual probes	69
Nick-translated, ^{32}P-labeled cDNA	Each alone and mixtures	No	Dot blot on nitrocellulose filters	Tissue culture and clinical samples	Varied	1 ng homologous RNA	Good sensitivity and specificity compared with three other diagnostic tests; potential shown for differentiating strains with genes 6 and 9	70
Nick-translated, ^{32}P-labeled cDNA	2, 5-7	No	Dot blot on nitrocellulose filters	Tissue culture and field samples	Varied	3×10^8 particles 50 pg homologous RNA	Noted variable signals due to different concentrations of virus in feces; described problems when trying to analyze reassortants in field samples	68

with other diagnostic tests (EM, ELISA, and PAGE) is currently small, the available data show good sensitivity and specificity for hybridization tests using cDNA probes.

Lin et al.[70] reported good specificity in one comparative clinical study. This was somewhat surprising since these investigators did not purify their cDNA probes from the pBR322 vector DNA. The difficulties of false positive reactions between impure probes and bacterial DNA sequences in clinical specimens have been documented, and this potential problem cannot be overemphasized.[98] Lin et al.[70] recognized this possible problem and noted that it might have been seen with a larger sampling of specimens. Due to this, they recommended the use of purified cDNA inserts rather than the total recombinant plasmids as hybridization probes for clinical testing.

Dimitrov et al.[69] prepared gene 6 and gene 9 probes by purifying cDNA inserts from vector DNA and also found good test specificity (91%) when these probes were used to examine EM-positive samples. No false positive results were seen when 26 additional EM rotavirus-negative samples were tested. However, further work with these and other probes has shown that occasionally residual contaminating vector DNA may be found in probes purified by even two cycles of electrophoresis.[88,99] Therefore, probe specificity must be carefully evaluated by testing cross-hybridization with unlabeled pBR322 DNA. One solution to this problem is to use DNase digestion of extracted nucleic acid to decrease contamination of DNA sequences in clinical samples where RNA viral genomes are being sought.[100]

The exact limits of detection of either double-stranded RNA or virus particles have varied in the published studies, and the differences apparently reflect the use of different types of filters (nylon or nitrocellulose) to immobilize the double-stranded RNA, the use of different conditions of hybridization, and the use of different probes. Test sensitivity is reportedly better using nylon filters (compared with nitrocellulose filters) due to increased immobilization of the nucleic acids;[69] use of these filters has the added advantage that they can be reused after removal of the probe by boiling.

These initial studies have also demonstrated the feasibility of using nucleic acid hybridization for rotavirus strain characterization. The potential to differentiate rotaviruses into specific serotypes or subgroups using a combination of individual gene probes and different conditions of hybridization has been shown.[69,70] These studies also identified complicating factors that require further work to allow practical application of this approach to detect and characterize virus in clinical samples. First, the amount of virus and viral RNA in clinical samples varies considerably, so different dilutions of extracted nucleic acid must be tested; it would be optimal if a rapid and automated method to extract and quantitate the amount of nucleic acid from stools could be devised. Second, test sensitivity and specificity vary depending on whether the virus test strain is homologous or heterologous to the probe. Therefore, the clearest results for strain characterization (e.g., serotyping) will require hybridization with gene-specific probes that represent each serotype. Since at least five human rotavirus serotypes and many additional serotypes that include only animal viruses are recognized,[13] future applications for serotyping with cDNAs will necessitate the synthesis and characterization of additional gene-specific cDNA probes. Finally, shorter (<1 day) test times and a change to the use of nonradioactive probes will be required before these methods will become generally useful. Increasing success is being achieved with direct serotyping of rotaviruses in stool samples using ELISAs and monoclonal antibodies,[101,102] so it remains to be determined whether improved nucleic acid detection techniques will make this approach more useful. However, nucleic acid methods will clearly continue to be applied primarily to detect viruses that remain difficult to cultivate.

An interesting application of hybridization with rotavirus cDNAs has used gene-specific probes and Northern blots to identify cognate genes between different rotavirus strains.[88,103] Since these genes in different virus strains often migrate in different relative positions following PAGE, this procedure is critical to identify specific RNA segments that may be of interest for further cloning. Figure 5 illustrates results of Northern blots hybridized with SA11 nick-

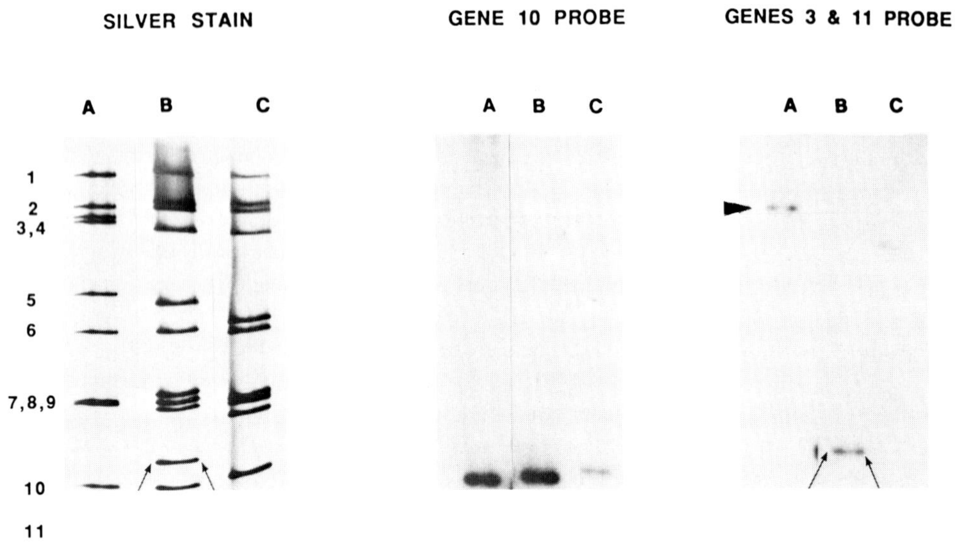

FIGURE 5. Hybridization to compare rotavirus genome segment homology using Northern blots. Double-stranded RNA segments of the simian rotavirus SA11 (A/SI/S.Africa/SA11/58) or two human rotavirus strains (A/Hu/USA, KA/1371/83; A/Hu/USA, W.Va./1389/83) were separated by electrophoresis in 10% polyacrylamide gels in lanes A, B, and C, respectively. The RNA patterns were visualized directly by staining with silver (left panel) or were electrophoretically transferred to a nylon membrane and hybridized with a ^{32}P-labeled cDNA probe specific for SA11 gene 10 (middle panel) or for genes 3 and 11 (right panel). The data shown here are from 24-h exposures of filters to

translated gene 10 alone or genes 3 and 11 cDNA probes with homologous virus (panel A) or with two human rotavirus strains (panels B, C). The results with the mixed gene 3 and 11 probes show that with the hybridization conditions used, only the homologous SA11 gene 3 was detected. In addition, the silver-stained gels showed that the RNA in one (panel B) of the human virus strains possessed an electropherotype called a short RNA pattern;[61] the hybridization results with the gene 10 and 11 probes show that the genome segment that corresponds to SA11 gene 11 has an altered and slower mobility in the short electropherotype. In contrast, the cognate gene 10's in all three virus strains have similar mobilities.

Further use of this technique to analyze the RNA profiles of viruses from children with immunodeficiencies and of virus passed serially in tissue culture at high multiplicities has demonstrated that the RNAs of some rotaviruses may contain genome rearrangements. In some cases where the RNA from plaque-purified viruses has been studied, unusual RNA patterns have been found where new bands that migrate between segments 1 and 7 are seen, and these new bands are derived from covalently linked rearrangements of genome segments of lower molecular weight.[8,103] These rearranged RNAs may be functional, and they may encode normal functional proteins. Sequence information from such rearranged RNA segments will be of interest to understand how these rearrangements are synthesized. Current evidence shows rearrangements only occur with the RNA within one segment and not between segments. The biological significance of these rearrangements remains unknown, but this is apparently a third mechanism (in addition to shift and drift) of evolution for the rotaviruses. The frequency of such rearrangements and their importance in the epidemiology of these viruses remain to be determined.

D. Hybridization with Rotavirus Strains Using Oligonucleotides or cDNA-Directed Transcript Probes

Two other types of probes, synthetic oligonucleotides and synthetic single-stranded RNA probes generated from cloned cDNAs inserted in transcription vectors, can also be used for rotavirus diagnosis.[99,105] Hybridization with cDNA probes is more sensitive than with oligonucleotide probes, but the ease of production and use of oligonucleotide probes makes them viable alternatives, particularly for viruses such as rotaviruses where the quantity of virus particles in most clinical specimens is high. In addition, hybridization with synthetic oligonucleotide probes allows quicker assays and the ability to easily choose and closely control specificity. For example, sequences that are highly type-specific or completely cross-reactive can theoretically be designed. The ability to synthesize such probes with linker arms also allows them to be labeled with various nonisotopic reporter groups,[106,107] and hybridization with these probes can be performed in less than 1 day.[99,105] The applications of such probes, that can be prepared from sequence data without the need for a molecular clone, remain to be fully explored.

The most sensitive hybridization tests reported to date utilized ^{125}I-labeled transcript probes synthesized using the endogenous viral polymerase in purified rotavirus particles.[30] The ability to insert cDNAs into transcription vectors adjacent to an RNA polymerase promoter that is recognized specifically by the polymerase encoded by either *Salmonella* phage SP6 or *Escherichia coli* phage T7 allows similar transcription probes to be synthesized very easily. Such single-stranded RNA probes have advantages for rotavirus RNA detection in that they can be used at high concentrations to shorten hybridization times, and the higher stability of RNA:RNA hybrids theoretically will permit higher test specificity while maintaining high sensitivity. Such probes can be synthesized easily and labeled nonradioactively (provided cDNA clones are available), and they can detect rotaviruses with sensitivities comparable to using ^{32}P-labeled, nick-translated probes.[99] However, their usefulness for clinical diagnosis remains untested; clinical evaluation with these probes is necessary since problems with test specificity have been reported in other systems.[74]

VII. SUMMARY

The detection of rotavirus nucleic acids in tissue culture or clinical samples has been shown to be a useful and specific method of diagnosis, and preliminary studies indicate sensitivities comparable to ELISAs. The importance of nucleic acid analysis by PAGE has become clear because the increased use of this method revealed new groups of rotaviruses, new serotypes within a group, and genome rearrangements. The ability to detect rotaviruses by hybridization with a number of different types of nucleic acid probes has also been clearly demonstrated, and the further use of specific gene probes will greatly facilitate studies on the evolution and transmission of virus strains. Although the available data do not show great increases in the sensitivity of detection of rotavirus with hybridization assays, compared with current immunologic assays, continued development and evaluation of hybridization probes and procedures can be expected. The development of rapid, automated methods of extraction of nucleic acid from stool samples and the expanded use of nonradioactive probes (biotin-labeled, enzyme-labeled, or chemically modified probes) are two improvements that will be needed to facilitate the use of these tests in clinical laboratories. The continued and expanded application of these improved test formats will then be indispensable in the acquisition of further epidemiologic and clinical information on rotavirus disease, such as the following: (1) What is the natural history of rotavirus disease following infections with specific viral serotypes? (2) Does recurrent disease reflect infections with viruses of different serotypes or reinfections with viruses of the same serotype? (3) Do specific rotavirus strains remain refractory to cultivation? (4) What is the prevalence of subclinical rotavirus infections in children and adults? (5) What is the importance of subclinical infections in disease transmission to children in homes, in the community, in day care centers, and in hospitals? (6) Do new rotavirus strains evolve by a mechanism of gene reassortment? Finally, this assay could be important in trials with live virus vaccines by providing a method of rapidly characterizing viruses excreted by vaccinees (often at low levels), of distinguishing these isolates from viruses circulating in the community, and of monitoring whether new virus strains emerge from vaccines by gene reassortment with circulating viruses.

ACKNOWLEDGMENTS

Research in this laboratory was supported in part by grants DK30144 and AI20649 from the National Institutes of Health and by grants from the World Health Organization and the National Science Foundation.

REFERENCES

1. **Estes, M. K., Palmer, E. L., and Obijeski, J. F.,** Rotaviruses: a review, *Curr. Top. Microbiol. Immunol.*, 105, 123, 1983.
2. **Holmes, I. H.,** Rotaviruses, in *The Reoviridae*, Joklik, W. K., Ed., Plenum Press, New York, 1983, 359.
3. **Kapikian, A. Z. and Chanock, R. M.,** Rotaviruses, in *Virology*, Fields, B. N., Knipe, D. M., Chanock, R. M., Melnick, J. L., Roizman, B., and Shope, R. E., Eds., Raven Press, New York, 1985, 863.
4. **Walsh, J. A. and Warren, K. S.,** Selective primary health care: an interim strategy for disease control in developing countries, *N. Engl. J. Med.*, 301, 967, 1979.
5. **Dingle, J. H., Badger, G. F., and Jordan, W. S.,** *Illness in the Home: a Study of 25,000 Illnesses in a Group of Cleveland Families*, Western Reserve University Press, Cleveland, Ohio, 1964, 19.
6. **Monto, A. S., Koopman, J. S., Longini, I. M., and Isaacson, R. E.,** The Tecumseh study. XII. Enteric agents in the community, 1976—1981, *J. Infect. Dis.*, 148, 284, 1983.
7. **Saulsbury, F. T., Winkelstein, J. A., and Yolken, R. H.,** Chronic rotavirus infection in immunodeficiency, *J. Pediatr.*, 97, 61, 1980.
8. **Pedley, S., Hundley, F., Chrystie, I., McCrae, M. A., and Desselberger, U.,** The genomes of rotaviruses isolated from chronically infected immunodeficient children, *J. Gen. Virol.*, 65, 1141, 1984.
9. **Flewett, T. H., Bryden, A. S., Davies, H., Woode, G. N., Bridger, J. C., and Derrick, J. M.,** Relation between viruses from acute gastroenteritis of children and newborn calves, *Lancet*, 2, 61, 1974.
10. **Cohen, J., Laporte, J., Charpilienne, A., and Scherrer, R.,** Activation of rotavirus RNA polymerase by calcium chelation, *Arch. Virol.*, 60, 177, 1979.
11. **Bican, P., Cohen, J., Charpilienne, A., and Scherrer, R.,** Purification and characterization of bovine rotavirus cores, *J. Virol.*, 43, 1113, 1982.
12. **Almeida, J., Bradburne, A. F., and Wreighitt, T. G.,** The effect of sodium thiocyanate on virus structure, *J. Med. Virol.*, 4, 269, 1979.
13. **Estes, M. K. and Graham, D. Y.,** Rotavirus antigens, in *Immunobiology of Proteins and Peptides — III*, Atassi, M. Z. and Bachrach, H. L., Eds., Plenum Press, New York, 1985, 201.
14. **Kalica, A. R., Flores, J., and Greenberg, H. B.,** Identification of the rotaviral gene that codes for hemagglutination and protease-enhanced plaque formation, *Virology*, 125, 194, 1983.
15. **Estes, M. K., Graham, D. Y., and Mason, B. B.,** Proteolytic enhancement of rotavirus infectivity: molecular mechanisms, *J. Virol.*, 39, 879, 1981.
16. **Ericson, B. L., Graham, D. Y., Mason, B. B., and Estes, M. K.,** Identification, synthesis and modifications of simian rotavirus SA11 polypeptides in infected cells, *J. Virol.*, 42, 825, 1982.
17. **Sabara, M., Gilchrist, J. E., Hudson, G. R., and Babiuk, L. A.,** Preliminary characterization of an epitope involved in neutralization and cell attachment that is located on the major bovine rotavirus glycoprotein, *J. Virol.*, 53, 58, 1985.
18. **Chan, W. K., Penaranda, M. E., Crawford, S. E., and Estes, M. K.,** Two glycoproteins are produced from the rotavirus neutralization gene, *Virology*, 151, 243, 1986.
19. **Gorziglia, M., Larrea, C., Liprandi, F., and Esparza, J.,** Biochemical evidence for the oligomeric (possibly trimeric) structure of the major inner capsid polypeptide (45K) of rotaviruses, *J. Gen. Virol.*, 66, 1889, 1985.
20. **Mason, B. B., Graham, D. Y., and Estes, M. K.,** In vitro transcription and translation of simian rotavirus SA11 gene products, *J. Virol.*, 33, 1111, 1980.
21. **Spencer, E. and Arias, M. L.,** In vitro transcription catalyzed by heat treated human rotavirus, *J. Virol.*, 40, 1, 1981.
22. **McCrae, M. A. and McCorquodale, J. G.,** Molecular biology of rotaviruses. V. Terminal structure of viral RNA species, *Virology*, 126, 204, 1983.

23. **Mason, B. B., Graham, D. Y., and Estes, M. K.,** Biochemical mapping of the simian rotavirus SA11 genome, *J. Virol.,* 46, 413, 1983.
24. **Cohen, J., Mason, B. B., and Estes, M. K.,** Molecular cloning of the simian rotavirus SA11, in *Double-Stranded RNA Viruses,* Compans, R. W. and Bishop, D. H. L., Eds., Elsevier/North Holland, Amsterdam, 1983, 65.
25. **Flores, J., Perez, I., White, L., Perez, M., Kalica, A. R., Marquina, R., Wyatt, R. G., Kapikian, A. Z., and Chanock, R. M.,** Genetic relatedness among human rotaviruses as determined by RNA hybridization, *Infect. Immun.,* 37, 648, 1982.
26. **Matsuno, S. and Nakajima, K.,** RNA of rotavirus: comparison of RNAs of human and animal rotaviruses, *J. Virol.,* 41, 710, 1982.
27. **Flores, J., Greenberg, H. B., Myslinski, J., Kalica, A. R., Wyatt R. G., Kapikian, A., Z., and Chanock, R. M.,** Use of transcription probes for genotyping rotavirus reassortants, *Virology,* 121, 288, 1982.
28. **Flores, J., Perez-Schael, I., Boeggeman, E., White, L., Perez, M., Purcell, R., Hoshino, Y., Midthun, K., Chanock, R. M., and Kapikian, A. Z.,** Genetic relatedness among human rotaviruses, *J. Med. Virol.,* 17, 135, 1985.
29. **Flores, J., Midthun, K., Hoshino, Y., Green, K., Gorziglia, M., Kapikian, A. Z., and Chanock, R. M.,** Conservation of the fourth gene among rotaviruses recovered from asymptomatic newborn infants and its possible role in attenuation, *J. Virol.,* 60, 972, 1986.
30. **Flores, J., Boeggeman, E., Purcell, R. H., Sereno, M., Perez, I., White, L., Wyatt, R. G., Chanock, R. M., and Kapikian, A. Z.,** A dot hybridisation assay for detection of rotavirus, *Lancet,* 1, 555, 1983.
31. **Saif, L. J., Bohl, E. H., Theil, K. W., Cross, R. F., and House, J. A.,** Rotavirus-like, calicivirus-like, and 23-nm virus-like particles associated with diarrhea in young pigs, *J. Clin. Microbiol.,* 12, 105, 1980.
32. **McNulty, M. S., Allan, G. M., Todd, D., McFerran, J. B., and McCracken, R. M.,** Isolation from chickens of a rotavirus lacking the rotavirus group antigen, *J. Gen. Virol.,* 55, 405, 1981.
33. **Lourenco, M. H., Nicolas, J. C., Cohen, J., Scherrer, R., and Bricout, F.,** Study of human rotavirus genome by electrophoresis: attempt at classification among strains isolated in France, *Ann. Virol.,* 132, 161, 1981.
34. **Pedley, S., Bridger, J. C., Brown, J. F., and McCrae, M. A.,** Molecular characterization of rotaviruses with distinct group antigens, *J. Gen. Virol.,* 64, 2093, 1983.
35. **Graham, D. Y. and Estes, M. K.,** Proposed working serologic classification system for rotaviruses, *Ann. Inst. Pasteur,* 136, 5, 1985.
36. **Hung, T., Wang, C., Fang, Z., Chou, Z., Chang, X., Liong, X., Chen, G., Yao, H., Chao, T., Ye, W., Den, S., and Chang, W.,** Waterborne outbreak of rotavirus diarrhoea in adults in China caused by a novel rotavirus, *Lancet,* 1, 1139, 1984.
37. **Wang, S., Cai, R., Chen, J., Li, R., and Jiang, R.,** Etiologic studies of the 1983 and 1984 outbreaks of epidemic diarrhea in Guangxi, *Intervirology,* 24, 140, 1985.
38. **Bridger, J. C., Clarke, I. N., and McCrae, M. A.,** Characterization of an antigenically distinct porcine rotavirus, *Infect. Immun.,* 35, 1058, 1982.
39. **Bohl, E. H., Saif, L. J., Theil, K. W., Agnes, A. G., and Cross, R. F.,** Porcine pararotavirus: detection, differentiation from rotavirus, and pathogenesis in gnotobiotic pigs, *J. Clin. Microbiol.,* 15, 312, 1982.
40. **Theil, K. W., Saif, L. J., Moorhead, P. D., and Whitmoyer, R. E.,** Porcine rotavirus-like virus (group B rotavirus): characterization and pathogenicity for gnotobiotic pigs, *J. Clin. Microbiol.,* 21, 340, 1985.
41. **Saif, L. J. and Theil, K. W.,** Antigenically distinct rotaviruses of human and animal origin, in *Infectious Diarrhoea in the Young: Strategies for Control in Humans and Animals,* Tzipori, S., Ed., Elsevier, Amsterdam, 1985, 208.
42. **Rodger, S. M., Bishop, R. F., and Holmes, I. H.,** Detection of a rotavirus-like agent associated with diarrhea in an infant, *J. Clin. Microbiol.,* 16, 724, 1982.
43. **Nicolas, J. C., Cohen, J., Fortier, B., Lourenco, M. H., and Bricout, F.,** Isolation of a human pararotavirus, *Virology,* 124, 181, 1983.
44. **Dimitrov, D. H., Estes, M. K., Rangelova, S. M., Shindarov, L. M., Melnick, J. L., and Graham, D. Y.,** Detection of antigenically distinct rotaviruses from infants, *Infect. Immun.,* 41, 523, 1983.
45. **Pereira, H. G., Leite, J. P., Azeredo, R. S., De Farias, V., and Sutmoller, F.,** An atypical rotavirus detected in a child with gastroenteritis in Rio de Janeiro, Brazil, *Mem. Inst. Oswaldo Cruz,* 78, 245, 1983.
46. **Espejo, R. T., Puerto, F., Soler, C., and Gonzalez, N.,** Characterization of a human pararotavirus, *Infect. Immun.,* 44, 112, 1984.
47. **Chasey, D. and Davies, P.,** Atypical rotaviruses in pigs and cattle, *Vet. Rec.,* 114, 16, 1984.
48. **Snodgrass, D. R., Herring, A. J., Campbell, I., Inglis, J. M., and Hargreaves, F. D.,** Comparison of atypical rotaviruses from calves, piglets, lambs and man, *J. Gen. Virol.,* 65, 909, 1984.
49. **Saif, L. J., Theil, K. W., and Redman, D. R.,** Detection and pathogenicity of an enteric bovine rotavirus-like agent (RVLA), in *63rd Annu. Meet. Conf. Res. Work in Anim. Dis.,* Chicago, Ill., November, 1982, 19.

50. **Chasey, D. and Banks, J.,** The commonest rotaviruses from neonatal lamb diarrhoea in England and Wales have atypical electropherotypes, *Vet. Rec.,* 115, 326, 1984.
51. **Vonderfecht, S. L., Huber, A. C., Eiden, J., Mader, L. C., and Yolken, R. H.,** Infectious diarrhea of infant rats produced by a rotavirus-like agent, *J. Virol.,* 52, 94, 1984.
52. **Chen, G.-M., Hung, T., Bridger, J. C., and McCrae, M. A.,** Chinese adult rotavirus is a group B rotavirus, *Lancet,* 2, 1123, 1985.
53. **Chasey, D., Bridger, J. C., and McCrae, M. A.,** A new type of atypical rotavirus in pigs, *Arch. Virol.,* 89, 235, 1985.
54. **Nakata, S., Estes, M. K., Graham, D. Y., Loosle, R., Hung, T., Wang, S., Saif, L. J., and Melnick, J. L.,** Antigenic characterization and ELISA detection of adult diarrhea rotaviruses, *J. Infect. Dis.,* 154, 448, 1986.
55. **Su, C., Wu, Y., Shen, H., Wang, D., Chen, Y., Wu, D., He, L., and Yang, Z.,** An outbreak of epidemic diarrhoea in adults caused by a new rotavirus in Anhui province of China in the summer of 1983, *J. Med. Virol.,* 19, 167, 1986.
56. **Pedley, S., Bridger, J. C., Chasey, D., and McCrae, M. A.,** Definition of two new groups of atypical rotaviruses, *J. Gen. Virol.,* 67, 131, 1986.
57. **Vonderfecht, S. L., Miskuff, R. L., Eiden, J. J., and Yolken, R. H.,** Enzyme immunoassay inhibition assay for the detection of rat rotavirus-like agent in intestinal and fecal specimens obtained from diarrheic rats and humans, *J. Clin. Microbiol.,* 22, 726, 1985.
58. **Eiden, J., Vonderfecht, S., and Yolken, R. H.,** Evidence that a novel rotavirus-like agent of rats can cause gastroenteritis in man, *Lancet,* 2, 8, 1985.
59. **Hung, T., Fan, R., Wang, C., Chen, G., Chou, D., Chang, J., McCrae, M. A., Wang, W., Se, W., Ran, R., and Ng, M. H.,** Seroepidemiology of adult rotavirus, *Lancet,* 2, 325, 1985.
60. **Nakata, S., Estes, M. K., Graham, D. Y., Wang, S., Gary, G. W., and Melnick, J. L.,** Detection of antibody to group B adult diarrhea rotaviruses in humans, *J. Clin. Microbiol.,* 25, 812, 1987.
61. **Estes, M. K., Graham, D. Y., and Dimitrov, D. H.,** The molecular epidemiology of rotavirus gastroenteritis, *Prog. Med. Virol.,* 29, 1, 1984.
62. **Rodger, S. M., Bishop, R. F., Birch, C., McLean, B., and Holmes, I. H.,** Molecular epidemiology of human rotaviruses in Melbourne, Australia, from 1973 to 1979, as determined by electrophoresis of genome ribonucleic acid, *J. Clin. Microbiol.,* 13, 272, 1981.
63. **Whitton, J. L., Hundley, F., O'Donnell, B., and Desselberger, U.,** Silver staining of nucleic acids: applications in virus research and in diagnostic virology, *J. Virol. Methods,* 7, 185, 1983.
64. **Clarke, I. N. and McCrae, M. A.,** A rapid and sensitive method for analyzing the genome of field isolates of rotavirus, *J. Virol. Methods,* 2, 203, 1981.
65. **Matsuno, S., Hasegawa, A., Mukoyama, A., and Inouye, S.,** A candidate for a new serotype of human rotavirus, *J. Virol.,* 54, 623, 1985.
66. **Dolan, K. T., Twist, E. M., Horton-Slight, P., Forrer, C., Bell, L. M., Jr., Plotkin, S. A., and Clark, H. F.,** Epidemiology of rotavirus electropherotypes determined by a simplified diagnostic technique with RNA analysis, *J. Clin. Microbiol.,* 21, 753, 1985.
67. **Estes, M. K.,** Simplified method of rotavirus RNA analysis: a word of caution, *J. Clin. Microbiol.,* 23, 806, 1986.
68. **Pedley, S. and McCrae, M. A.,** A rapid screening assay for detecting individual RNA species in field isolates of rotaviruses, *J. Virol. Methods,* 9, 173, 1984.
69. **Dimitrov, D. H., Graham, D. Y., and Estes, M. K.,** Detection of rotaviruses by nucleic acid hybridization with cloned DNA of simian rotavirus SA11 genes, *J. Infect. Dis.,* 152, 293, 1985.
70. **Lin, M., Imai, M., Bellamy, A. R., Ikegami, N., Furuichi, Y., Summers, D., Nuss, D. L., and Deibel, R.,** Diagnosis of rotavirus infection with cloned cDNA copies of viral genome segments, *J. Virol.,* 55, 509, 1985.
71. **Meinkoth, J. and Wahl, G.,** Hybridization of nucleic acids immobilized on solid supports, *Anal. Biochem.,* 138, 267, 1984.
72. **Minson, A. C. and Darby, G.,** Hybridization techniques. New developments in practical virology, in *Laboratory and Research Methods in Biology and Medicine,* Vol. 5, Howard, C., Ed., A. R. Liss, New York, 1982, 185.
73. **Edberg, S. C.,** Principles of nucleic acid hybridization and comparison with monoclonal antibody technology for the diagnosis of infectious diseases, *Yale J. Biol. Med.,* 58, 425, 1985.
74. **Richman, D. D.,** Nucleic acid probes to detect viral diseases, in *Concepts in Viral Pathogenesis II,* Notkins, A. L. and Oldstone, M. B. A., Eds., Springer-Verlag, Berlin, 1986, 305.
75. **Lopez, S., Arias, C. F., Bell, J. R., Strauss, J. H., and Espejo, R. T.,** Primary structure of the cleavage site associated with trypsin enhancement of rotavirus SA11 infectivity, *Virology,* 144, 11, 1985.

75a. **Bremont, M., Charpilienne, A., Chabanne, D., and Cohen, J.,** Nucleotide sequence and expression in *Escherichia coli* of the gene encoding the nonstructural protein NCDP2 of bovine rotavirus, *Virology,* 161, 138, 1987.

76. **Estes, M. K., Mason, B. B., Crawford, S., and Cohen, J.,** Cloning and nucleotide sequence of the simian rotavirus gene 6 that codes for the major inner capsid protein, *Nucleic Acids Res.,* 12, 1875, 1984.
77. **Both, G. W., Bellamy, A. R., and Siegman, L. J.,** Nucleotide sequence of the dsRNA genomic segment 7 of simian 11 rotavirus, *Nucleic Acids Res.,* 12, 621, 1984.
78. **Ward, C. W., Elleman, T. C., Azad, A. A., and Dyall-Smith, M. L.,** Nucleotide sequence of gene segment 9 encoding a nonstructural protein of UK bovine rotavirus, *Virology,* 134, 249, 1984.
79. **Ward, C. W., Azad, A. A., and Dyall-Smith, M. L.,** Structural homologies between RNA gene segments 10 and 11 from UK bovine, simian SA11, and human Wa rotaviruses, *Virology,* 144, 328, 1985.
80. **Dyall-Smith, M. L., Elleman, T. C., Hoyne, P. A., Holmes, I. H., and Azad, A. A.,** Cloning and sequence of the UK bovine rotavirus gene segment 7: marked sequence homology with simian rotavirus gene segment 8, *Nucleic Acids Res.,* 11, 3351, 1983.
81. **Imai, M., Richardson, M. A., Ikegami, N., Shatkin, A. J., and Furuichi, Y.,** Molecular cloning of double-stranded RNA virus genomes, *Proc. Natl. Acad. Sci. U.S.A.,* 80, 373, 1983.
82. **Both, G. W., Siegman, L. J., Bellamy, A. R., Ikegami, N., Shatkin, A. J., and Furuichi, Y.,** Comparative sequence analysis of rotavirus genomic segment 6 — the gene specifying viral subgroups 1 and 2, *J. Virol.,* 51, 97, 1984.
83. **Arias, C. F., Lopez, S., Bell, J. R., and Strauss, J. H.,** Primary structure of the neutralization antigen of simian rotavirus SA11 as deduced from cDNA sequence, *J. Virol.,* 50, 657, 1984.
84. **Elleman, T. C., Hoyne, P. A., Dyall-Smith, M. L., Holmes, I. H., and Azad, A. A.,** Nucleotide sequence of the gene encoding the serotype-specific glycoprotein of UK bovine rotavirus, *Nucleic Acids Res.,* 11, 4689, 1983.
85. **Both, G. W., Bellamy, A. R., Street, J. E., and Siegman, L. J.,** A general strategy for cloning double-stranded RNA: nucleotide sequence of the simian-11 rotavirus gene 8, *Nucleic Acids Res.,* 10, 7075, 1982.
86. **Both, G. W., Siegman, L. J., Bellamy, A. R., and Atkinson, P. H.,** Coding assignment and nucleotide sequence of simian rotavirus SA11 gene segment 10: location of glycosylation sites suggests that the signal peptide is not cleaved, *J. Virol.,* 48, 335, 1983.
87. **Both, G. W., Mattick, J. S., and Bellamy, A. R.,** Serotype-specific glycoprotein of simian 11 rotavirus: coding assignment and gene sequence, *Proc. Natl. Acad. Sci. U.S.A.,* 80, 3091, 1983.
88. **Tanaka, T., Nakata, S., and Estes, M. K.,** unpublished observations.
89. **Street, J. E., Croxson, M. C., Chadderton, W. F., and Bellamy, A. R.,** Sequence diversity of human rotavirus strains investigated by Northern blot hybridization analysis, *J. Virol.,* 43, 369, 1982.
90. **Urasawa, S., Urasawa, T., and Taniguchi, K.,** Genetic reassortment between two human rotaviruses having different serotype and subgroup specificities, *J. Gen. Virol.,* 67, 1551, 1986.
91. **Allen, A. M. and Desselberger, U.,** Reassortment of human rotaviruses carrying rearranged genomes with bovine rotavirus, *J. Gen. Virol.,* 66, 2703, 1985.
92. **Gombold, J. L., Estes, M. K., and Ramig, R. F.,** Assignment of simian rotavirus SA11 temperature-sensitive mutant groups B and E to genome segments, *Virology,* 143, 309, 1985.
93. **Ito, Y. and Joklik, W. K.,** Temperature-sensitive mutants of reovirus. I. Patterns of gene expression by mutants of groups C, D, and E, *Virology,* 50, 189, 1972.
94. **Gaillard, R. K., Jr. and Joklik, W. K.,** Quantitation of the relatedness of reovirus serotypes 1, 2, and 3 at the gene level, *Virology,* 123, 152, 1982.
95. **Schroeder, B. A., Street, J. E., Kalmakoff, J., and Bellamy, A. R.,** Sequence relationships between the genome segments of human and animal rotavirus strains, *J. Virol.,* 43, 379, 1982.
96. **Bodkin, D. K. and Knudson, D. L.,** Sequence relatedness of Palyam virus genes to cognates of the Palyam serogroup viruses by RNA-RNA blot hybridization, *Virology,* 143, 55, 1985.
97. **Bodkin, D. K. and Knudson, D. L.,** Assessment of sequence relatedness of double-stranded RNA genes by RNA-RNA blot hybridization, *J. Virol. Methods,* 10, 45, 1985.
98. **Ambinder, R. F., Charache, P., Staal, S., Wright, P., Forman, M., Hayward, S. D., and Hayward, G. S.,** The vector homology problem in diagnostic nucleic acid hybridization of clinical specimens, *J. Clin. Microbiol.,* 24, 16, 1986.
99. **Jiang, X., Metcalf, T. G., and Estes, M. K.,** unpublished data, 1986.
100. **Jiang, X., Estes, M. K., Metcalf, T. G., and Melnick, J. L.,** Detection of hepatitis A virus in seeded estuarine samples by hybridization with cDNA probes, *Appl. Environ. Microbiol.,* 52, 711, 1986.
101. **Shaw, R. D., Stoner-Ma, D. L., Estes, M. K., and Greenberg, H. B.,** Specific enzyme-linked immunoassay for rotavirus serotypes 1 and 3, *J. Clin. Microbiol.,* 22, 286, 1985.
102. **Urasawa, S., Taniguchi, K., Morita, Y., and Urasawa, T.,** Analysis of neutralizing antigen of group A human rotavirus by the use of neutralizing monoclonal antibodies, in *Working Conference on Viral Diseases,* The Japan-United States Cooperative Medical Science Program, Nagasaki, 1986, 25.
103. **Dyall-Smith, M. L., Azad, A. A., and Holmes, I. H.,** Gene mapping of rotavirus double-stranded RNA segments by Northern blot hybridization: application to segments 7, 8, and 9, *J. Virol.,* 46, 317, 1983.

104. **Hundley, F., Biryahwaho, B., Gow, M., and Desselberger, U.,** Genome rearrangements of bovine rotavirus after serial passage at high multiplicity of infection, *Virology,* 143, 88, 1985.
105. **Kerschner, J. H.,** Molecular BioSystems, San Diego, Calif., personal communication.
106. **Ruth, J. L.,** Chemical synthesis of non-radioactively-labeled DNA hybridization probes, *DNA,* 3, 123, 1984.
107. **Jablonski, E., Moomaw, E. W., Tullis, R. H., and Ruth, J. L.,** Preparation of oligodeoxynucleotide-alkaline phosphatase conjugates and their use as hybridization probes, *Nucleic Acids Res.,* 14, 6115, 1986.

Part IV: Nucleic Acid Probes for Respiratory Pathogens

Chapter 8

DNA PROBES FOR THE DETECTION OF *LEGIONELLA* SPECIES, *MYCOPLASMA PNEUMONIAE*, AND MEMBERS OF THE *MYCOBACTERIUM TUBERCULOSIS* COMPLEX

Sydney B. Shaw

TABLE OF CONTENTS

I. INTRODUCTION

Sensitivity and specificity are criteria by which diagnostic tests are evaluated. For tests based on nucleic acid hybridization, choice of the hybridization conditions and of the target nucleic acid are important determinants of the sensitivity and specificity that can be achieved. We have developed a nucleic acid hybridization procedure, useful for the detection of microorganisms, that targets ribosomal RNA (rRNA) and that utilizes in-solution hybridization and hybridization-rate accelerators. These principles are used in the sensitive and specific assays for the detection of *Legionella* species, *Mycoplasma pneumoniae,* and members of the *Mycobacterium tuberculosis* complex discussed in this chapter.

Eukaryotic and prokaryotic cells contain organelles called ribosomes that are complexes of RNA and proteins.[1] Ribosomes decipher the nucleic acid sequence of messenger RNAs and synthesize the proteins specified. Prokaryotic ribosomes contain three RNAs, denoted 5S, 16S, and 23S, based on their sizes determined using sedimentation gradients. Eukaryotic ribosomes also contain three RNAs, denoted 7S, 18S, and 28S. Selection of rRNA as the target nucleic acid presents distinct advantages in terms of sensitivity and specificity.

A typical bacterial cell contains one to several copies of any particular gene. In contrast, this bacterial cell contains about 10,000 ribosomes and, therefore, about 10,000 copies of each rRNA. In addition, with rRNA, there are no complementary strands present that would compete with the probe as would be the case with DNA. Thus, selection of rRNA as the target nucleic acid provides for increased sensitivity since one has a natural amplification of several thousandfold.[2]

The nucleic acid sequences of many 16S-like rRNAs have been determined. Analysis of these data indicates that the sequences of these rRNAs are highly conserved.[3-6] That is, many regions of the 16S and 18S rRNAs of evolutionarily distant organisms contain the same base sequences. Other regions, however, are less conserved and have diverged at varying rates throughout evolution. Figure 1 illustrates this point using the 16S rRNA molecule of *Escherichia coli.* The bold circles indicate nucleotides that are conserved throughout the bacterial kingdom. Other regions are conserved in members of the family Enterobacteriaceae, while other regions may be found only in members of the genus *Escherichia*. Still other regions may be specific for members of the species *E. coli*. Such family, genus, and species-specific sequences are found in the rRNA of all organisms.[3] Although fewer 23S-like rRNA sequences have been determined, conserved and divergent sequences are present.[3,5]

Thus, the choice of rRNA as target provides for a very flexible system that can be used for the detection of microorganisms. In the case of Legionnaires' disease, for example, a diagnostic probe test must be capable of detecting all members of the genus *Legionella*, while a diagnostic test for *Mycoplasma pneumoniae* must be specific for members of that species and not cross-react with any of the other mycoplasmas present in the respiratory tract.

The conditions of the hybridization reaction affect the rate and specificity of the hybridization. The temperature of hybridization and the concentration of the components in the hybridization reaction determine the rate at which hybridization occurs. Hybridizaion-rate accelerators can be used to drive the reaction to completion and, thus, allow for sensitive detection, even with low levels of target. The conditions of hybridization also affect the specificity of the reaction since the hybridization conditions determine how much mismatch between the probe and the target will be allowed.

Solution hybridization provides several advantages over the more widely used method of immobilizing the target nucleic acid on an inert support. The first advantage is one of availability of target. All of the target is available for hybridization in solution, while only a fraction of the target is available when it is immobilized on a filter. The second advantage is the 5- to 10-fold increase in the rate of the reaction when both the probe and the target are in solution.

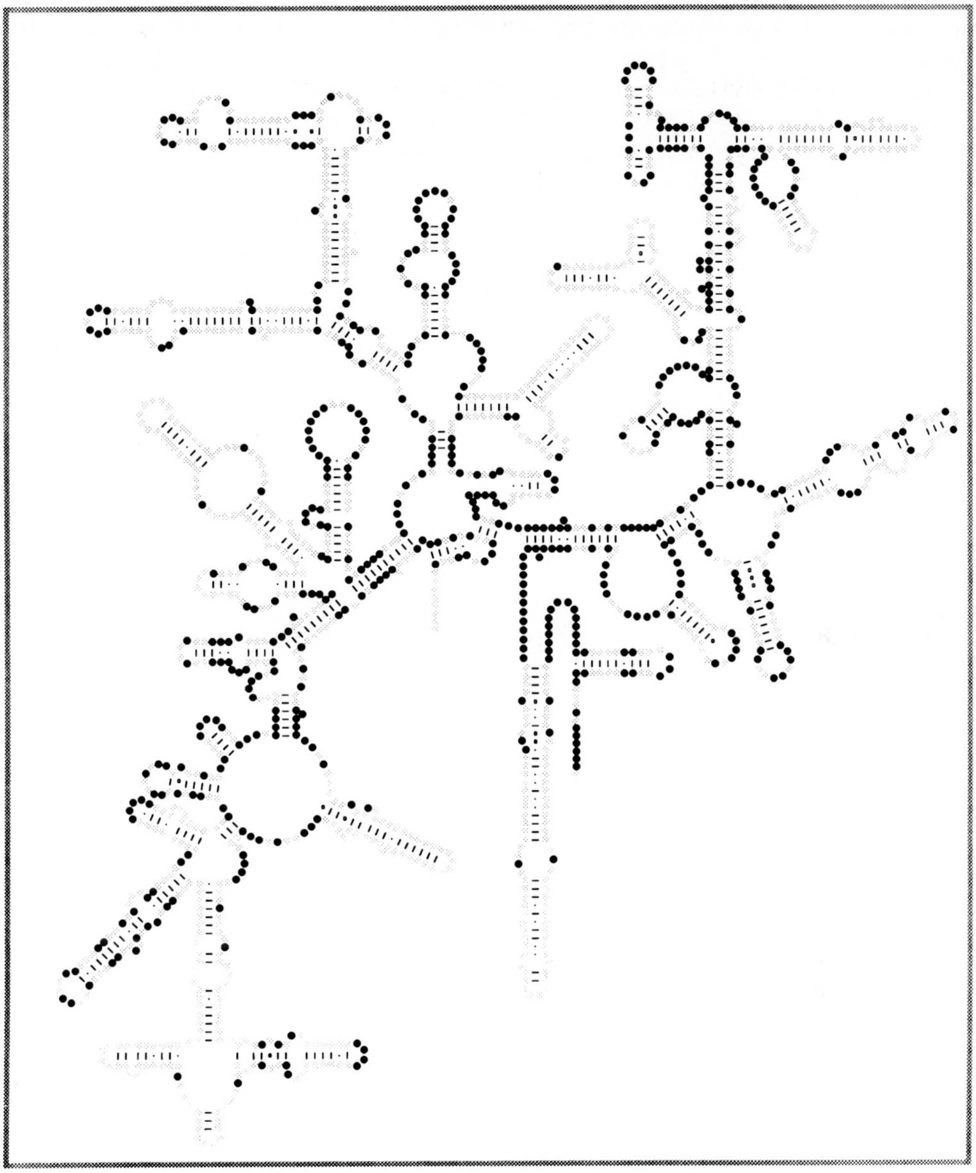

FIGURE 1. A schematic representation of the *Escherichia coli* 16S rRNA. The bold circles represent nucleotides that are conserved throughout the bacterial kingdom. (Courtesy of Robin Gutell, Gen-Probe, Inc., San Diego, Calif.)[4]

II. PRINCIPLES OF THE HYBRIDIZATION ASSAY

The hybridization procedure developed for the clinical laboratory tests discussed in this chapter involves three basic steps:

1. Hybridization
2. Separation
3. Measurement

For hybridization, a labeled probe is incubated with the sample in the appropriate salts and solubilizing agents at the appropriate temperature to give the required reactivity and specificity. If any of the target rRNA is present in the sample, the single-stranded probe will hybridize to it. The presence of hybridized probe thus indicates the presence of the target organism. Hybridization in this case occurs between target and probe molecules that are free in solution.

Hybridized probe is then separated from unhybridized probe using a separation agent called hydroxyapatite. Hydroxyapatite is a calcium phosphate crystal that will bind double-stranded nucleic acid but not single-stranded nucleic acid under certain conditions.[7] The hybridized probe, bound to the hydroxyapatite, is then separated from the unhybridized probe by centrifugation of the reaction mixture to collect the hydroxyapatite.

The probes used in the tests discussed in this chapter are labeled with ^{125}I. The amount of hybridized probe is quantitated using a gamma counter.

III. A DNA PROBE ASSAY FOR DIRECT IDENTIFICATION OF *LEGIONELLA* SPECIES

A. Introduction

At the 1976 American Legion Convention in Philadelphia, 221 attendees developed a pneumonia-like illness that resulted in 34 deaths. The etiologic agent responsible for this epidemic was a bacterium that was classified in a new genus and species, *Legionella pneumophila,* within the new family Legionellaceae. To date, 22 species of *Legionella* have been identified. New species continue to be identified using improved culture methods. *Legionella* species have been isolated from surface water, mud, thermally polluted lakes and streams, air conditioning cooling towers, evaporative condensers, showers, and nebulizers.[8-11]

All *Legionella* species are potentially pathogenic for man. Currently, *L. pneumophila,* serogroup 1, is responsible for 63% of *Legionella*-associated disease. *Legionella* species have been shown to be a significant cause of both Legionnaires'disease, an acute pneumonia, and Pontiac fever, an acute, nonpneumonic, febrile illness that is self-limiting. Without treatment, the fatality rate of Legionnaires' disease is 15% or higher. *Legionella* species have also been shown to be responsible for nosocomial pneumonia occurring predominantly among immunocompromised patients.

Legionellosis has been diagnosed on the basis of clinical symptoms, culture or biochemical procedures, and direct immunofluorescence antibody tests. Culture methods may require 2 to 3 weeks for isolation and identification to be completed. The presence of *Legionella* species is indicated when specimens produce growth on buffered charcoal yeast extract agar (BCYE) but not on blood agar plates. Suspected colonies are plated on selective BCYE media with and without L-cysteine, a necessary growth factor for *Legionella* species. Subsequent biochemical and serological tests are used for further identification. Rapid, direct immunofluorescent antibody assays are available for detection of *Legionella* species; however, the sensitivity of these tests vary widely from laboratory to laboratory (25 to 80%), suggesting that considerable skill is required for performance of the tests. The usefulness of these tests is further limited by the availability of reagents for only some of the disease-causing species and strains of *Legionella.*[11,12]

Rapid, specific diagnosis of Legionnaires' disease is important because while the disease may be rapidly fatal, prompt treatment is effective in the majority of cases. Recently, Kohne et al. demonstrated that a DNA probe complementary to the ribosomal RNAs of the *Legionella* species could be used to specifically detect and identify all 22 members of the genus *Legionella.*[13] A clinical laboratory test was then developed that uses this DNA probe and in-solution hybridization to directly detect the organism in patient specimens. The procedure used in this test and the results of the first clinical evaluation of this test are discussed in this section.[14]

B. Test Procedure

Fresh or frozen specimens to be tested using the Gen-Probe Rapid Diagnostic System for Legionella are first liquified with a solubilizing agent. Homogenized lung tissue, sputa, endotracheal suctions, and transtracheal aspirates suspected of containing *Legionella* species have been tested. A portion of the liquified sample (100 μl) is transferred to a reaction tube containing lysing solution and glass beads and sonicated to release rRNA from the bacterial cells. After the addition of 2 ml of a solution containing the ^{125}I-labeled probe specific for *Legionella* rRNA, the sample is hybridized during a 1-h incubation at 72°C. Hybridized probe is separated from unhybridized probe using hydroxyapatite and is quantitated in a gamma counter. A positive sample is indicated when the ratio of sample counts to negative control counts is greater than or equal to 4.0.

C. Specificity of the *Legionella* Probe Test

The specificity of the DNA probe test was evaluated using strains of bacteria obtained from the American Type Culture Collection, Rockville, Md. Included among the strains tested were representatives of all 22 known *Legionella* species and 109 bacteria from other genera. The percent of the input probe that hybridized to each strain tested is presented in Tables 1 and 2. The results demonstrated that the probe detects all 22 of the named *Legionella* species and does not hybridize significantly with any of the other strains of bacteria tested. The specificity of the *Legionella* DNA probe has been confirmed in two independent studies.[15,16]

D. Comparison of the DNA Probe Test with Culture for Direct Detection of *Legionella* Species in Clinical Specimens

The performance of the DNA probe test was evaluated in a retrospective study of 342 specimens collected over an 8-year period.[14] The relatively low incidence of Legionnaires' disease precluded the performance of a prospective evaluation. The 342 specimens tested included sputa, transtracheal aspirates, endotracheal suctions, nasotracheal suctions, bronchial washes, pleural fluids, and lung and autopsy tissues. These specimens had been cultured upon collection using reference methods. The DNA probe assay detected 63 out of 112 of the originally culture-positive specimens and 228 out of the 230 culture-negative specimens. When compared to the culture originally performed on these specimens at the time of collection, the DNA probe assay showed overall specificity of 99.1% and sensitivity of 56.7%.

Repeat culture was performed whenever possible on the discrepant specimens using six plates of selective and nonselective media. Of the original 51 discrepant specimens, 34 were available for reculture. Of the initially culture-positive specimens, 70% (23/33) had low titers (1+) at the time they were initially tested, and 61% (20/33) had been stored frozen for more than 4 years. These results indicated that these specimens may have suffered losses of titer during the prolonged storage period. Of the 33 originally culture-positive, discrepant specimens that were available for repeat culture, 11 showed no growth of *Legionella* species. One of the two specimens, that were initially culture negative but DNA probe test positive, was available for repeat culture. All six repeat culture plates for this specimen, both selective and nonselective, were heavily overgrown with contaminating flora. The original culture plates had also been heavily overgrown. These results made it impossible to determine whether or not *Legionella* species were present in the specimen.

If the results of the repeat culture are taken into account, the results obtained are as shown in Table 3. Results from the 17 specimens unavailable for repeat culture and the one originally culture-negative specimen, that yielded heavily overgrown cultures, were dropped from the evaluation. Analysis of the remaining samples to determine sensitivity and specificity of the DNA probe assay with respect to the reference culture method indicated 74% sensitivity and 100% specificity. Overall agreement was 93%.

Table 1
REACTIVITY OF THE *LEGIONELLA* DNA PROBE WITH *LEGIONELLA* SPECIES

Sample number	Sample identification	% Hybrid
1	*L. pneumophila* 1	51.3
2	*L. pneumophila* 2	44.7
3	*L. pneumophila* 3	50.4
4	*L. pneumophila* 4	41.6
5	*L. pneumophila* 5	39.8
6	*L. pneumophila* 6	44.2
7	*L. pneumophila* 7	42.7
8	*L. pneumophila* 8	39.7
9	*L. pneumophila* 9	47.9
10	*L. bozemanii*	30.7
11	*L. longbeachae* 1	17.3
12	*L. longbeachae* 2	21.8
13	*L. gormanii*	34.7
14	*L. micdadei* 1	31.1
15	*L. micdadei* 2-Tatlock	38.4
16	*L. dumoffii*	32.1
17	*L. wadsworthii*	29.6
18	*L. jordanis*	21.4
19	*L. oakridgensis*	18.6
20	*L. anisa*	19.6
21	*L. maceachernii*	28.7
22	*L. rubrilucens*	31.0
23	*L. erythra*	20.6
24	*L. hackeliae*	28.9
25	*L. parisiensis*	30.0
26	*L. cherrii*	35.4
27	*L. jamestowniensis*	33.6
28	*L. sainthelensi*	27.3
29	*L. steigerwaltii*	48.2
30	*L. spiritensis*	26.1
31	*L. feeleii* 2	40.1
32	*L. santicrucis*	37.9

Table 2
REACTIVITY OF THE *LEGIONELLA* DNA PROBE WITH NON-*LEGIONELLA* SPECIES

Sample identification	% Hybrid	Sample identification	% Hybrid
Acetobacter aceti	0.1	*Methanobacterium thermoautotrophicum*	0.2
Acinetobacter calcoaceticus	1.0	*Methylomonas thalassica*	0
Actinobacillus actinomycetemcomitans	0	*Micromonospora rubra*	0
Actinoplanes italicus	0.8	*Moraxella osloensis*	0.7
Aeromonas hydrophila	0.8	*Mycobacterium gordonae*	0.6
Alcaligenes denitrificans subsp. *xylosoxydans*	0.4	*Mycoplasma hominis*	0.5
		M. pneumoniae	0
A. faecalis	0	*Myxococcus fulvus*	0.4
Alteromonas rubra	0.6	*Neisseria meningitidis*	0
Archangium gephyra	0.3	*Nitrobacter species*	0.6
Arthrobacter oxydans	0	*Nocardia asteroides*	0
Azomonas macrocytogenes	0.3	*Obesumbacterium proteus*	0.9
Bacillus subtilis	0.1	*Oceanospirillum japonicum*	0.3

Table 2 (continued)
REACTIVITY OF THE *LEGIONELLA* DNA PROBE WITH NON-*LEGIONELLA* SPECIES

Sample identification	% Hybrid	Sample identification	% Hybrid
Bacteroides fragilis	0	*Paracoccus denitrificans*	0.5
Beggiatoa alba	0	*Pasteurella multocida*	0.7
Beijerinckia indica	0.8	*Peptococcus magnus*	0
Branhamella catarrhalis	1.3	*Photobacterium phosphoreum*	0.1
Campylobacter jejuni	0	*Planctomyces staleyi*	0
Cardiobacterium hominis	0	*Polyangium cellulosum*	0
Caryophanon latum	0.4	*Propionibacterium acnes*	0.7
Cedecea lapagei	1.4	*Proteus mirabilis*	1.1
Cellulomonas cartae	0	*Pseudomonas aeruginosa*	0.8
Chlorobium thiosulfatophilum	0	*P. cepacia*	0
Chromatium vinosum	1.2	*P. diminuta*	0
Chromobacterium violaceum	0.8	*P. fluorescens*	0
Citrobacter freundii	1.1	*P. maltophilia*	0.9
Clostridium perfringens	0	*Rahnella aquatilis*	0
Cystobacter fuscus	0.3	*Rhizobium lupini*	0
Cytophaga johnsonae	0	*Rhodopseudomonas blastica*	0
Deinococcus radiodurans	0.5	*Rhodospirillum rubrum*	0.7
Dermatophilus congolensis	0.3	*Salmonella enteritidis*	0.3
Derxia gummosa	0.9	*Serpens flexibilis*	0.1
Desulfovibrio desulfuricans	0	*Serratia marcescens*	1.3
Edwardsiella tarda	0.5	*Shigella sonnei*	0.1
Eikenella corrodens	0.2	*Simonsiella muelleri*	0.4
Enterobacter cloacae	0	*Sphaerotilus species*	0.2
Escherichia coli	0.3	*Spirosoma linguale*	1.0
Flavobacterium meningosepticum	1.1	*Staphylococcus aureus*	0.4
Flectobacillus major	0	*Streptobacillus moniliformis*	0.7
Flexibacter sancti	0.7	*Streptococcus pneumoniae*	0
Francisella novicida	1.0	*S. pyogenes*	0.2
Frankia species	0	*Streptomyces griseus*	0.2
Fusobacterium nucleatum	0	*Tatumella ptyseos*	0
Gardnerella vaginalis	0.5	*Thermomicrobium roseum*	0.2
Haemophilus influenzae	1.7	*Thermus aquaticus*	1.1
Halobacterium saccharovorum	0.7	*Thiobacillus intermedius*	0.4
Halomonas elongata	0.2	*Thiomicrospira denitrificans*	1.2
Hyphomicrobium indicum	1.1	*Treponema denticola*	0
Janthinobacterium lividum	0.4	*Veillonella atypica*	0.6
Kingella kingae	0.1	*Vibrio parahaemolyticus*	0.1
Klebsiella pneumoniae	0.7	*Vitreoscilla stercoraria*	0
Kluyvera ascorbata	0.1	*Xanthomonas campestris*	0
Lactobacillus acidophilus	0	*Xenorhabdus nematophilus*	0
Leucothrix mucor	0	*Yersinia enterocolitica*	1.5
Meniscus glaucopis	0.4	*Zymomonas mobilis*	0

IV. A DNA PROBE ASSAY FOR DIRECT DIAGNOSIS OF *MYCOPLASMA PNEUMONIAE* INFECTION

A. Introduction

Mycoplasma pneumoniae[17,18] is one of the leading causes of pneumonias and is responsible for about 10% of the pneumonias in the general population. The incidence of infection with this organism is higher among populations in prolonged, close contact such as armed forces personnel and college students where, in some cases, more than 50% of pneumonias have been caused by *M. pneumoniae*. Each year, 8 to 15 million cases of *M. pneumoniae* are estimated to

Table 3
RESULTS OF THE COMPARISON OF THE *LEGIONELLA* DNA PROBE TEST WITH CULTURE FOR DIRECT DETECTION OF *LEGIONELLA* SPECIES IN CLINICAL SPECIMENS

		Culture	
		+	−
Gen-Probe	+	63	0
assay	−	22	239

occur. Infection with this organism can be asymptomatic, or produce illness ranging from mild upper respiratory infection to severe bronchitis and pneumonia.

M. pneumoniae is a member of the order Mollicutes. Organisms in this group lack cell walls and are very fastidious in their growth requirements. *M. pneumoniae* is particularly fastiduous, and strains adapted to culture have generation times of about 16 h.

Diagnosis of *M. pneumoniae* infection has, until now, been based upon isolation of the organism or demonstration of a rise in specific antibody titer. These procedures take 1 to 3 weeks to complete, and the results are useful only retrospectively. Detection of cold agglutinins has also been used for diagnosis of *M. pneumoniae,* but this method lacks specificity and sensitivity.

A DNA probe test has been developed that allows direct diagnosis of *M. pneumoniae* infection using throat swab specimens. The principles involved in this test have been described earlier in this chapter. Specifics of the test procedure and results of an evaluation of the test will be discussed here.[19]

B. Test Procedure

Throat swab specimens are collected in mycoplasma transport medium using dacron, rayon, or cotton swabs and stored at 4°C. Hanks' Balanced Salt Solution containing 0.5% gelatin or Tryptic Soy Broth containing 0.5% bovine serum albumin are recommended for transport of mycoplasmas. The throat swab specimens to be tested using the Gen-Probe Rapid Diagnostic System for Mycoplasma pneumoniae are concentrated by centrifugation in a microcentrifuge at $12{,}000 \times g$ for 10 min. The pellet is resuspended in 150 μl of lysing solution that contains components that lyse the organisms and inhibit nucleases. Probe solution (150 μl) is added and hybridization of the ^{125}I-labeled probe to any *M. pneumoniae* rRNA occurs in solution during a 1-h incubation at 72°C. Hybridized probe is separated from unhybridized probe using a hydroxyapatite suspension and quantitated using a gamma counter. A ratio of sample counts to negative control counts greater than or equal to 3.0 is considered positive for *M. pneumoniae.*

C. Specificity of the *M. pneumoniae* Probe Test

The specificity of the DNA probe test was demonstrated using 57 isolates obtained from the American Type Culture Collection, Rockville, Md. For these experiments, more than 10^7 organisms were tested in each assay. The isolates chosen for study represented 28 mycoplasmas, including all of the known human mycoplasmas; organisms found in the respiratory tract as pathogens or as normal flora; organisms thought to be related to *M. pneumoniae;* and genera across the phylogenetic tree. Of the 57 species tested, only *M. pneumoniae* and *M. genitalium* produced positive results (Table 4). Hybridization of the probe to *M. genitalium* is less than that to *M. pneumoniae.*

D. Sensitivity of the *M. pneumoniae* Probe Test

Various amounts of purified *M. pneumoniae* rRNA or *M. pneumoniae* organisms were added

Table 4
REACTIVITY OF THE *MYCOPLASMA PNEUMONIAE* DNA PROBE

ATCC identification	DNA probe assay: Ratio	DNA probe assay: +/–
Acholeplasma axanthum	1.3	–
A. laidlawii	1.4	–
Actinomyces israelii	1.0	–
Bacteroides fragilis	1.3	–
Bifidobacterium breve	1.3	–
Bordetella bronchiseptica	1.4	–
Clostridium innocuum	1.0	–
C. pasteurianum	2.7	–
C. perfringens	1.1	–
C. ramosum	1.5	–
Corynebacterium xerosis	1.1	–
Erysipelothrix rhusiopathiae	1.1	–
Escherichia coli	1.0	–
Flavobacterium meningosepticum	1.1	–
Haemophilus influenzae	1.2	–
Klebsiella pneumoniae	1.0	–
Lactobacillus acidophilus	1.2	–
Legionella pneumophila	1.0	–
Listeria monocytogenes	1.0	–
Moraxella osloensis	1.1	–
Mycobacterium tuberculosis	1.9	–
Mycoplasma arginini	1.6	–
M. arthritidis	1.1	–
M. bovigenitalium	1.0	–
M. bovis	2.5	–
M. buccale	1.4	–
M. californicum	1.3	–
M. capricolum	1.3	–
M. columbinasale	1.4	–
M. columborale	1.2	–
M. faucium	1.7	–
M. fermentans	1.5	–
M. gallisepticum	1.9	–
M. gallopavonis	1.8	–
M. genitalium	20.8[a]	+
M. hominis	1.3	–
M. hyorhinis	1.4	–
M. lipophilum	1.7	–
M. orale	1.7	–
M. pneumoniae, FH strain	30.0	+
M. primatum	1.3	–
M. pulmonis	1.3	–
M. salivarium	1.6	–
M. synoviae	1.2	–
Neisseria meningitidis	1.0	–
Pasteurella multocida	1.4	–
Peptococcus magnus	1.0	–
Propionibacterium acnes	1.0	–
Pseudomonas aeruginosa	1.1	–
Spiroplasma citri	1.3	–
S. mirum	1.2	–
Staphylococcus aureus	1.0	–
Streptococcus faecalis	1.2	–

Table 4 (continued)
REACTIVITY OF THE *MYCOPLASMA PNEUMONIAE* DNA PROBE

ATCC identification	DNA probe assay Ratio	+/–
S. mitis	1.3	–
S. pneumoniae	1.4	–
S. pyogenes	1.0	–
Ureaplasma urealyticum	1.1	–

[a] *M. genitalium* has been isolated on rare occasions from the human urogenital tract. Pathogenicity of this organism in humans is unknown.[25,26]

to throat swab material that had been concentrated by centrifugation in order to determine the sensitivity of the DNA probe test. The results, shown in Table 5, indicate that the DNA probe assay detected 30 pg of *M. pneumoniae* rRNA and 50,000 *M. pneumoniae* organisms.

E. Comparison of the DNA Probe Test with Culture for Direct Detection of *M. pneumoniae* in Clinical Specimens

Pharyngeal swabs, collected in two studies from pediatric patients (aged 4 to 17 years) or from university students, were assayed blindly for *M. pneumoniae* using the DNA probe test. These specimens were from frozen repositories and had been previously cultured for *M. pneumoniae* using a reference culture method. A total of 388 specimens were tested using the DNA probe assay. Half of the specimens tested were culture positive for *M. pneumoniae* and half were culture negative. The DNA probe assay demonstrated complete agreement with culture. The results are presented in Tables 6 and 7. Culture-positive specimens produced ratios in the DNA probe assay between 11.8 and 29.7. The DNA probe assay ratios of the culture-negative specimens ranged from 0.8 to 1.6. Thus, the DNA probe test clearly distinguishes between positive and negative specimens.

V. A DNA PROBE ASSAY FOR THE IDENTIFICATION OF MEMBERS OF THE *MYCOBACTERIUM TUBERCULOSIS* COMPLEX

A. Introduction

Mycobacteria[20-22] are ubiquitous in nature, and several species cause disease in man. Among the mycobacteria, the most important causes of human disease are members of the *Mycobacterium tuberculosis* complex (TB complex) which includes *M. tuberculosis, M. bovis, M. africanum,* and *M. microti. M. tuberculosis* infection may produce the disease tuberculosis which causes the deaths of approximately three million people worldwide each year. The incidence of tuberculosis reported in the U.S. in 1983 was 10.2 per 100,000 population. While the incidence of this disease had been declining in this country, an increase has now been observed that is attributable to the large number of cases seen among patients with Acquired Immune Deficiency Syndrome (AIDS). *M. bovis* is now rarely seen as a cause of human disease in the U.S., but this organism is occasionally isolated from immigrants and from individuals infected by ingestion of contaminated milk products or by exposure to infected cattle. *M. africanum* is occasionally isolated as a cause of pulmonary tuberculosis in Africa. The remaining member of the TB complex, *M. microti,* is an animal pathogen that mainly infects rodents. Members of the TB complex may actually be subspecies, biovars, or pathovars of a single species.

Tuberculosis is transmitted via airborne droplet nuclei from sputa of patients with infectious

Table 5
DETERMINATION OF THE MINIMUM AMOUNT OF *M. PNEUMONIAE* rRNA AND THE MINIMUM NUMBER OF *M. PNEUMONIAE* ORGANISMS DETECTED BY THE DNA PROBE ASSAY

M. pneumoniae rRNA

Amount tested	DNA probe assay ratio
0	1.0
10 pg	1.8
30 pg	3.5
100 pg	8.2
1 ng	15.1

M. pneumoniae organisms

Number of organisms	DNA probe assay ratio
0	1.0
10,000	1.2
20,000	1.9
50,000	3.5
100,000	4.8

Table 6
COMPARISON OF THE DNA PROBE TEST WITH A REFERENCE CULTURE METHOD FOR DETECTION OF *M. PNEUMONIAE* IN THROAT SWAB SPECIMENS FROM PEDIATRIC PATIENTS

		DNA probe assay ratio		
***M. pneumoniae* culture results**	**No. of specimens**	**Mean**	**SD**	**Range**
Positive	94	22.8	3.8	17.5—29.7
Negative	94	1.1	0.1	0.8—1.5

Table 7
COMPARISON OF THE DNA PROBE TEST WITH A REFERENCE CULTURE METHOD FOR DETECTION OF *M. PNEUMONIAE* IN THROAT SWAB SPECIMENS FROM UNIVERSITY STUDENTS

		DNA probe assay ratio		
***M. pneumoniae* culture results**	**No. of specimens**	**Mean**	**SD**	**Range**
Positive	100	19.6	3.1	11.8—27.7
Negative	100	1.2	0.2	0.9—1.6

tuberculosis. Effective antimicrobial therapy is available which reduces communicability promptly. Rapid diagnosis of members of the TB complex is, therefore, extremely important to prevent spread of the disease.

Preliminary identification of mycobacteria may be made by examination of smears of the patient specimen for bacilli that retain dye after exposure to acid-alcohol. Mycobacteria are currently identified using cultural and biochemical procedures which may take 2 to 3 months to complete. Specimens are either plated directly onto specialized culture media or are processed, to reduce the levels of other organisms and to concentrate the mycobacteria, prior to plating. Lowenstein-Jensen and Middlebrook 7H10 or 7H11 media are the most commonly used for primary isolation. Incubation periods of 4 to 6 weeks or longer may be required. Identification of any mycobacteria isolated may take an additional 4 weeks and is based on observations of their growth rate, colonial morphology and pigmentation, and biochemical properties.

A DNA probe test has been developed that shortens the time required for identification of members of the *M. tuberculosis* complex to a few hours. The procedure used and the results of evaluations of this test are presented here. DNA probe tests have also been developed for the identification of *M. avium* and *M. intracellulare* and for the direct detection of members of the genus *Mycobacterium* in clinical specimens.

B. Test Procedure

The Gen-Probe Rapid Diagnostic System for the Mycobacterium TB Complex confirms the identification of members of the *M. tuberculosis* complex. Organisms growing on standard Lowenstein-Jensen or Middlebrook media or in broth are removed from the media and suspended in 2 ml of sterile, distilled water. The turbidity of the suspension is adjusted to match a No. 1 McFarland nephalometer barium sulfate standard which corresponds to approximately 300 million bacteria per ml.

The bacterial suspension (100 μl) is transferred to a tube containing lysing reagents and glass beads and sonicated in a water bath sonicator for 15 min. Probe solution (1 ml) containing the ^{125}I-labeled probe is then added to each tube. The sample is hybridized for 1 h at 72°C. Hydroxyapatite is used to separate hybridized probe from unhybridized probe. The amount of hybridized probe in each sample is quantitated using a gamma counter. An aliquot of the probe solution is also counted to determine the total counts of probe used for each sample. The percent of the probe that hybridized is then calculated by dividing the sample count by the total number of counts of probe used. A percent hybridization greater than or equal to 10% is positive for members of the TB complex.

C. Specificity of the *M. tuberculosis* Complex Probe

The specificity of the TB complex probe was tested in two studies. The first study evaluated the specificity of the probe using 105 reference strains from the American Type Culture Collection, Rockville, Md. and 20 additional isolates obtained from other institutions.[23] The 125 strains tested included 21 members of the TB complex, 31 species of mycobacteria of which 30 were clinically significant species, and 92 species from 41 other genera that included respiratory pathogens and other clinically significant bacteria. The results of this study are presented in Table 8. The 21 TB complex isolates tested gave percent hybridization values that ranged from 40.4 to 68.1% and had a mean hybridization value of 57.2%. Organisms outside the TB complex gave results ranging from 0.5 to 3.4%, with a mean of 1.2%. Thus, all of the TB complex isolates tested produced positive results, and all of the isolates outside the TB complex produced negative results using the DNA probe test.

The second study determined the specificity of the TB complex probe when reacted with 240 clinical isolates referred to the Center for Disease Control (CDC), Atlanta, Ga. for identification.[24] The 240 isolates tested by the CDC included 126 different TB complex isolates and 114 species of other mycobacteria, *Nocardia* and *Rhodococcus*. The results obtained with the DNA

Table 8
SPECIFICITY OF THE TB COMPLEX PROBE

Sample identification	% Hybrid	Results +/–[a]
Acinetobacter calcoaceticus	1.1	–
Actinomadura madurae	1.0	–
Actinoplanes italicus	1.0	–
Arthrobacter oxydans	1.0	–
Bacillus subtilis	0.9	–
Bacteroides fragilis	1.1	–
Branhamella catarrhalis	1.4	–
Brevibacterium linens	1.4	–
Campylobacter jejuni	1.0	–
Chromobacterium violaceum	1.0	–
Clostridium perfringens	1.2	–
Corynebacterium aquaticum	1.0	–
C. diphtheriae	1.9	–
C. genitalium	1.9	–
C. haemolyticum	1.9	–
C. matruchotii	2.0	–
C. minutissimum	2.0	–
C. pseudodiphtheriticum	2.0	–
C. pseudogenitalium	2.0	–
C. pseudotuberculosis	1.9	–
C. pyogenes	2.0	–
C. renale	2.0	–
C. striatum	2.0	–
C. xerosis	1.6	–
C. xerosis	0.7	–
Deinococcus radiodurans	1.0	–
Dermatophilus congolensis	1.1	–
Derxia gummosa	1.0	–
Enterobacter aerogenes	1.0	–
Escherichia coli	1.0	–
Fusobacterium nucleatum	0.5	–
Haemophilus influenzae	0.9	–
Klebsiella pneumoniae	0.7	–
Legionella pneumophila	0.8	–
Microbacterium lacticum	1.0	–
Mycobacterium africanum	68.1	+
M. asiaticum	3.4	–
M. asiaticum	1.7	–
M. avium	0.9	–
M. avium 1	1.3	–
M. avium 2	1.3	–
M. avium 2	1.2	–
M. avium I	1.3	–
M. avium II	1.2	–
M. avium III	1.2	–
M. avium complex	1.9	–
M. avium complex	1.9	–
M. avium complex	1.2	–
M. avium complex	1.6	–
M. avium complex	1.5	–
M. bovis	63.1	+
M. bovis (BCG) Copenhagen	40.4	+
M. bovis (BCG) TMC 1011	57.7	+
M. chelonae	1.1	–

Table 8 (continued)
SPECIFICITY OF THE TB COMPLEX PROBE

Sample identification	% Hybrid	Results +/–[a]
M. flavescens	0.9	–
M. fortuitum	1.1	–
M. gastri	0.8	–
M. gordonae	1.1	–
M. gordonae	1.3	–
M. haemophilum	0.8	–
M. intracellulare	1.1	–
M. intracellulare 12	1.6	–
M. intracellulare 14	1.0	–
M. kansasii	1.3	–
M. kansasii	2.3	–
M. leprae	1.6	–
M. malmoense	0.9	–
M. marinum	1.1	–
M. microti	51.2	+
M. nonchromogenicum	1.1	–
M. phlei	1.3	–
M. scrofulaceum	1.1	–
M. shimoidei	1.0	–
M. simiae	1.2	–
M. smegmatis	0.9	–
M. szulgai	1.1	–
M. terrae	1.0	–
M. thermoresistibile	1.0	–
M. triviale	1.2	–
M. tuberculosis	66.2	+
M. tuberculosis	62.4	+
M. tuberculosis	54.5	+
M. tuberculosis	62.5	+
M. tuberculosis	54.4	+
M. tuberculosis	52.2	+
M. tuberculosis	53.2	+
M. tuberculosis	47.2	+
M. tuberculosis	58.9	+
M. tuberculosis	60.0	+
M. tuberculosis	60.8	+
M. tuberculosis	63.2	+
M. tuberculosis	61.9	+
M. tuberculosis	49.6	+
M. tuberculosis lung	53.4	+
M. tuberculosis sputum	59.9	+
M. ulcerans	0.9	–
M. vaccae	0.8	–
M. xenopi	2.6	–
Mycoplasma hominis	0.8	–
M. pneumoniae	0.6	–
Neisseria meningitidis	0.6	–
Nocardia asteroides	1.1	–
N. brasiliensis	1.0	–
N. otitidiscaviarum	1.2	–
Nocardiopsis dassonvillei	1.1	–
Oerskovia turbata	1.2	–
O. xanthineolytica	1.2	–
Propionibacterium acnes	0.6	–

Table 8 (continued)
SPECIFICITY OF THE TB COMPLEX PROBE

Sample identification	% Hybrid	Results +/–[a]
Proteus mirabilis	1.0	–
Pseudomonas aeruginosa	0.9	–
P. cepacia	2.2	–
Rahnella aquatilis	1.6	–
Rhodococcus aichiensis	1.1	–
R. bronchialis	1.1	–
R. chubuensis	1.0	–
R. equi	1.2	–
R. obuensis	1.0	–
R. sp.	1.2	–
R. sputi	1.1	–
Rhodospirillum rubrum	0.8	–
Staphylococcus aureus	0.5	–
Streptococcus mitis	0.9	–
S. pneumoniae	0.6	–
Vibrio parahaemolyticus	0.8	–
Yersinia enterocolitica	0.9	–

[a] +/– = a sample is positive by the Gen-Probe assay if the percent hybridization is ≥10%. A sample is negative if the percent hybridization is <10%.

probe test agreed with the identification determined using standard culture and biochemical techniques for all but one isolate. This isolate was an unidentified mycobacteria from outside the TB complex.

D. Performance of the DNA Probe Test in 20 Clinical Laboratories

A total of 898 clinical isolates were tested by 20 different institutions from across the U.S. that included state and county health department laboratories, university hospital laboratories, Veterans Administration medical centers, a military hospital, private hospitals, and a private reference laboratory. Each site identifed their isolates using their standard culture and identification procedures. The isolates tested included 593 members of the TB complex, 290 other mycobacteria, and 15 bacteria from other genera.

None of the bacteria from outside of the genus *Mycobacterium* produced a positive result with the DNA probe. Of the 290 mycobacteria outside the TB complex that were tested, all but one gave a negative result. One institution obtained a hybridization value of 14% for *M. asiaticum.* The two other institutions in this study that tested this organism obtained negative results. All but 3 of the 593 TB complex isolates tested gave positive results with the DNA probe test. The four isolates that gave discrepant results are still under investigation.

Thus, the DNA probe test demonstrated 99.6% agreement, 99.5% sensitivity, and 99.7% specificity when compared to the conventional identification procedures used in these clinical laboratories.

VI. SUMMARY

The DNA probe tests described here illustrate that hybridization and detection procedures convenient for the clinical laboratory can be developed that provide results within 2 or 3 h. The use of rRNA as the target molecule in these tests has allowed for clinically useful levels of sensitivity and for flexibility in achieving the desired specificity.

DNA probes have tremendous potential for the diagnosis of infectious diseases. Application of probe technology to the identification of organisms that are difficult to culture or that require long incubation periods for isolation and identification should allow for the prompt initiation of specific therapy for many diseases that had, heretofore, been treated solely on clinical presentation. In addition, more rapid identification will allow for better patient management in the control of spread of infection.

REFERENCES

1. **Chambliss, G., Craven, G. R., Davies, J., Davis, K., Kahan, L., and Nomura, M.,** *Ribosomes: Structure, Function, and Genetics,* University Park Press, Baltimore, Md., 1980.
2. **Kohne, D. E.,** The application of DNA probe technology in the clinical laboratory: is it the diagnostic wave of the future?, *Infect. Dis.,* 15, 4, 1985.
3. **Noller, H. F.,** Structure of ribosomal RNA, *Annu. Rev. Biochem.,* 53, 119, 1984.
4. **Gutell, R. R., Weiser, B., Woese, C. R., and Noller, H. F.,** Comparative anatomy of 16S-like ribosomal RNA, in *Progr. Nucleic Acid Research and Molecular Biology,* Vol. 32, Academic Press, New York, 1985, 155.
5. **Brimacombe, R. and Steige, W.,** Structure and function of ribosomal RNA, *Biochem. J.,* 229, 1, 1985.
6. **Woese, C. R.,** Bacterial evolution, *Microbiol. Rev.,* 51, 221, 1987.
7. **Kohne, D. E. and Britten, R. J.,** Hydroxyapatite techniques for nucleic acid reassociation, in *Proc. Nucleic Acid Res.,* Vol. 2, Harper & Row, New York, 1971, 500.
8. **Fraser, D. W., Tsai, T. F., Orenstein, W., Parkin, W. E., Beecham, H. J., Sharrar, R. G., Harris, J., Mallison, G. F., Martin, S. M., McDade, J. E., Shepard, C. C., Brachman, P. S., and the Field Investigation Team,** Legionnaires' disease: description of an epidemic of pneumonia, *N. Engl. J. Med.,* 297, 1189, 1977.
9. **McDade, J. E., Shepard, C. C., Fraser, D. W., Tsai, T. R., Redus, M. A., Dowdle, W. R., and the Laboratory Investigation Team,** Legionnaires' disease: isolation of a bacterium and demonstration of its role in other respiratory disease, *N. Engl. J. Med.,* 297, 1197, 1977.
10. **Brenner, D. J., Steigerwalt, A. G., and McDade, J. E.,** Classification of the Legionnaires' disease bacterium: *Legionella pneumophila,* genus novum, species nova, of the family *Legionellaceae,* familia nova, *Ann. Intern. Med.,* 90, 656, 1979.
11. **Myerowitz, R. L.,** *Legionella pneumophila* and other legionella species: an update, *Lab. Med.,* 13, 611, 1982.
12. **Edelstein, P. H., Meyer, R. D., and Finegold, S. M.,** Laboratory diagnosis of Legionnaires' disease, *Am. Rev. Respir. Dis.,* 121, 317, 1980.
13. **Kohne, D. E., Steigerwalt, A. G., and Brenner, D. J.,** Nucleic acid probe specific for members of the genus *Legionella,*in *Legionella: Proc. 2nd Int. Symp.,* Thornsberry, C., Balows, A., Feeley, J. C., and Jakubowski, W., Eds., American Society for Microbiology, Washington, D.C., 1984, 107.
14. **Edelstein, P. H., Bryan, R. N., Enns, R. K., Kohne, D. E., and Kacian, D. L.,** Retrospective study of Gen-Probe Rapid Diagnostic System for detection of Legionellae in frozen clinical respiratory tract samples, *J. Clin. Microbiol.,* 25, 1022, 1987.
15. **Wilkinson, H. W., Sampson, J. S., and Plikaytis, B. B.,** Evaluation of a commercial gene probe for identification of *Legionella* cultures, *J. Clin. Microbiol.,* 23, 217, 1986.
16. **Edelstein, P. H.,** Evaluation of the Gen-Probe DNA probe for detection of *Legionellae* in culture, *J. Clin. Microbiol.,* 23, 481, 1986.
17. **Kenny, G. E.,** Mycoplasmas, in *Manual of Clinical Microbiology,* 4th ed., Lennette, E. H., Balows, A., Hausler, W. J., Jr., and Shadomy, H. J., Eds., American Society for Microbiology, Washington, D.C., 1985, 407.
18. **Velleca, W. M., Bird, B. R., and Forrester, F. T.,** *Laboratory Diagnosis of Mycoplasma Infections,* Course 8226-C, Public Health Service, U.S. Department of Health, Education and Welfare, Atlanta.
19. **Shaw, S. B., Weiner, L. B., Fultz, T. J., Lawrence, T. G., Boas, L. A., Putnam, J. G., Cabanas, D. K., McMillan, J. A., Kohne, D. E., and Kacian, D. L.,** A DNA probe assay for direct diagnosis of *Mycoplasma pneumoniae* infections, submitted for publication, 1987.
20. **Sommers, H. M. and Good, R. C.,** Mycobacterium, in *Manual of Clinical Microbiology,* 4th ed., Lennette, E. H., Balows, A., Hausler, W. J., Jr., and Shadomy, H. J., Eds., American Society for Microbiology, Washington, D.C., 1985, 216.
21. **Wayne, L. G.,** Microbiology of tubercle bacilli, *Am. Rev. Respir. Dis.,* 125(Suppl.), 31, 1982.

22. **Benenson, A. S., Ed.,** *Control of Communicable Diseases in Man,* 14th ed., American Public Health Association, Washington, D.C., 1985, 409.
23. **Dean, E. D., Murphy, K. A., McDonough, S. H., Smith, R. D., Daly, W. J., Cabanas, D. K., Holden, A. L., Bhatt, R., and Kohne, D. E.,** Rapid assay detection and identification of species in the *Mycobacterium tuberculosis* complex by specific cDNA probes, presented at the 1987 Am. Soc. Microbiol. Annu. Meet., Atlanta, March, 1987.
24. **Silcox, V. A., Floyd, M. M., and Woodley, C. L.,** Rapid identification of *Mycobacterium tuberculosis* compex, using a DNA probe, presented at the 1987 Am. Soc. Microbiol. Annu. Meet., Atlanta, March, 1987.

Chapter 9

IDENTIFICATION OF ADENOVIRUS DNA IN CLINICAL SPECIMENS

Timo Hyypiä and Ulf Pettersson

TABLE OF CONTENTS

I. INTRODUCTION

A. Classification and Clinical Importance of Adenoviruses

Human adenoviruses consist of 41 known serotypes on the basis of the neutralization test. These serotypes are classified into subgenera according to the hemagglutination properties and oncogenicity in newborn hamsters. Biochemical analysis of viral DNA and proteins has further been used for more precise classification of adenovirus strains.

Subgenus A consists of serotypes 12, 18, and 31, which have tropism for the alimentary tract and have been detected in patients with diarrhea. Their etiological role in gastroenteritis, however, remains to be confirmed. Members of subgenus B (serotypes 3, 7, 11, 14, 16, 21, 34, 35) and subgenus C (1, 2, 5, 6) are causative agents of respiratory infections and are found also in persistent infections of kidney and lymphoid tissue. Adenovirus tonsillitis is characterized by prolonged high fever and often is confused with streptococcal tonsillitis.

Subgenus D consists of more than 20 serotypes which are known to cause keratoconjunctivitis, while subgenus E (serotype 4) is responsible for conjunctivitis and respiratory symptoms. Recently discovered subgenus F consists of serotypes 40 and 41, which are a common cause of acute infantile gastroenteritis.

B. Diagnosis of Adenovirus Infections

Members of subgenera A, B, C, D, and E grow in a variety of human epithelial cell lines, but some types require a 2- to 3-week incubation and often a blind passage. Specific cell lines are needed for propagation of enteric adenoviruses (types 40 and 41). Typing of the isolated viruses can be carried out by neutralization with specific antisera. Because the isolation technique is laborious and time consuming, several approaches have been used for demonstration of adenoviruses or their structural components directly in clinical specimens.

Stool samples of patients often contain sufficient amounts of virus particles to be visualized by electron microscopy. This method has been widely used for the diagnosis of acute gastroenteritis. Since electron microscopy requires special training and facilities, and because analysis of large numbers of specimens is tedious, immunoassays are becoming the most commonly used modern diagnostic method in adenovirus infections. The sensitivity of immunoassays in detecting adenoviruses in stool specimens is at least two times that of electron microscopy. Immunoassays are either broad reacting among the adenoviruses or more specific for a subgenus. They have been successfully used for the detection of viruses in both stool and nasopharyngeal specimens.

Serology is also used to diagnose acute adenovirus infections, but because this requires serum pairs taken at 2 to 3 week intervals, it is not convenient as a modern approach for viral diagnosis. Recently, direct identification of adenovirus DNA in clinical specimens and restriction enzyme analysis of isolated strains have become a useful complement to the other diagnostic methods, and they will be discussed in more detail below.

C. Molecular Biology of Adenoviruses

Adenoviruses are nonenveloped, icosahedral viruses which replicate in the nucleus of the infected cell. The capsid consists of 252 capsomers; 240 hexons form the facets of the icosahedron, and 12 pentons are located at the corners of the virus particle (Figure 1). The pentons consist of the penton base which is noncovalently linked to the fiber. After attachment of the virion to the cell, the viral DNA is transported to the nucleus, where early transcription takes place. The incoming viral DNA is associated with histone-like proteins that are encoded by the viral genome. Late after infection, the viral DNA, together with viral core proteins, aided by a specific sequence located at the left end of the genome, are encapsidated in a preformed head.

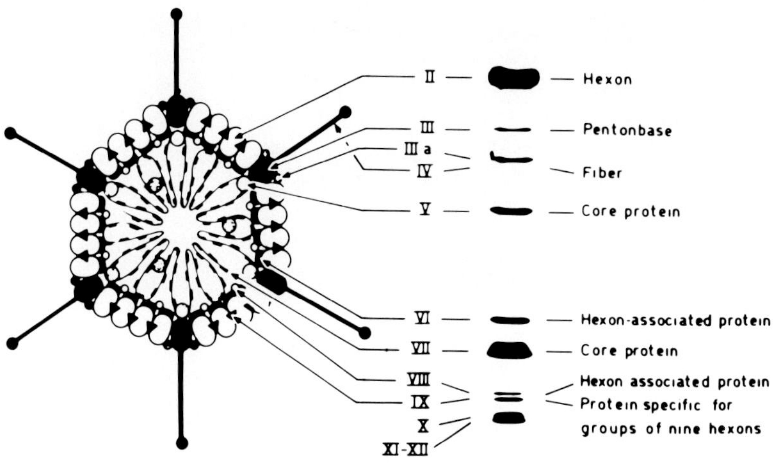

FIGURE 1. A model of an adenovirus particle. The double-stranded DNA genome is surrounded by proteins which form a regular icosahedral structure. (From Persson, H. and Philipson, L., *Curr. Top. Microbiol. Immunol.*, 97, 159, 1982. With permission.)

The genomes of adenoviruses are linear, double-stranded DNA molecules with molecular weights of 20 to 25 $\times$ 10^6. The genome of adenovirus type 2 has a length of 35,937 base pairs. Both ends of the viral genome function as origins for DNA replication and contain a protein which is covalently linked to the 5′ end of each DNA strand. The ends of the viral DNA contain an inverted terminal repetition with a length of approximately 100 base pairs.

Adenovirus gene expression is regulated at multiple levels. The lytic cycle is traditionally divided into two phases: an early phase which precedes viral DNA replication and a late phase during which the structural proteins are synthesized. Both strands of the genome are utilized for transcription. Widely separated regions of the genome are expressed during the early phase. The primary transcripts from the early regions are further processed by differential splicing. The region which is located at the left-hand end contains genes which regulate early viral transcription and also specifies the genes required for cell transformation.

The adenovirus genome encodes two low molecular weight RNAs designated virus-associated RNA (VA RNA) I and II. Both RNAs, which are approximately 160 nucleotides in length, are transcribed by RNA polymerase III. VA RNAI is known to function during the initiation of protein synthesis by maintaining the catalytic activity of initiation factor eIF-2.

Late after infection, transcription initiates predominantly at the major late promoter, and a long primary transcript is initiated from this promoter. It becomes polyadenylated at one of five locations along the genome, generating five RNA families with coterminal 3′ ends. The 3′ ends are generated by endonucleolytic cleavage of the nuclear RNA precursor. Following the selection by endonucleolytic cleavage, the primary transcript is processed by splicing in such a way that each mature mRNA gains a common set of three short 5′ leader segments. This tripartite leader, which is 201 nucleotides long, is thus joined to at least 15 different downstream acceptor splice sites. Among the gene products coded by this late region are structural polypeptides IIIa, III, V, pVII, pVI, hexon, pVIII, and fiber (Figure 1). The structure and function of the adenovirus 2 genome have recently been reviewed by Akusjärvi et al.[1]

II. ADENOVIRUS DNA IN CLINICAL SPECIMENS

Because the molecular biology of adenovirus is well known, it has been one of the model pathogens in modern approaches to viral diagnostics. Adenoviruses were also among the first viruses which were detected directly in clinical specimens by hybridization.

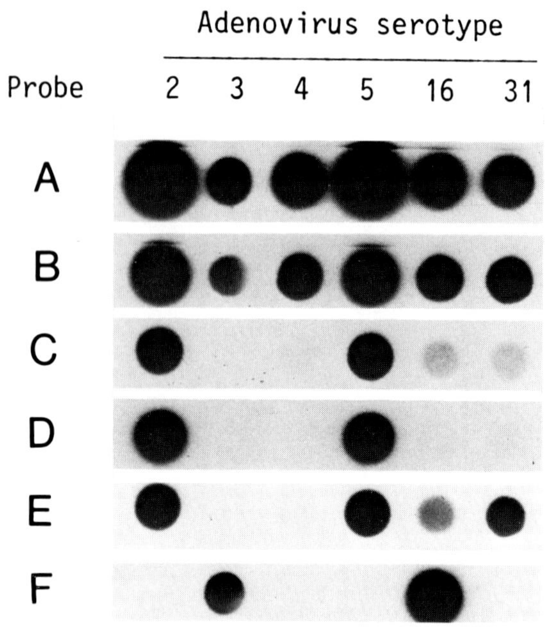

FIGURE 2. Specificity of various hybridization probes in detection of adenovirus DNA in cells infected with different virus strains. (A) Total Ad2 DNA labeled with ^{32}P using nick translation; (B) *Hind*II A fragment of Ad2 DNA cloned in pBR322; (C) XbaI/*Hind*III fragment of Ad2 DNA containing the VA gene region cloned in pSP65; (D) ^{32}P-labeled RNA probe generated by in vitro transcription of the VA gene region from the pSP65 construct; (E) fragment of Ad2 DNA cloned in a M13 vector used as a single-stranded probe and further detected by nick-translated replicative form of M13 vector DNA; and (F) synthetic oligonucleotide (20-mer) presenting an Ad3 DNA sequence.

A. Nucleic Acid Hybridization for the Detection of Adenovirus DNA

In order to evaluate the applicability and specificity of various hybridization probes, we have tested a selection of adenovirus strains (Figure 2). The cells were infected with the viruses, and when a cytopathic effect was observed in more than 50% of cells, they were treated with proteinase K; DNA was denatured, and the sample was applied onto membrane filters. The strains included serotypes 31 (from subgenus A); 3, 16 (B); 2, 5 (C); and 4 (E).

Each of the strains tested gave a signal which was most intense with serotypes 2 and 5 when whole adenovirus 2 was labeled with ^{32}P in a nick-translation reaction and used as a hybridization probe (Figure 2A). A similar result was obtained when a cloned *Hind*III-A fragment of adenovirus 2 was used (Figure 2B).

The use of vectors containing bacterial promoters makes it possible to prepare labeled RNA probes. We used a construction containing the SP6 promoter followed by a region coding for the VA genes from the adenovirus 2 genome. The in vitro transcription reaction was carried out in the presence of ^{32}P-dUTP. This probe showed preference for adenovirus types 2 and 5, but also gave a slight signal with some other serotypes (Figure 2D). Corresponding results were observed when the same probe was used after labeling by nick translation (Figure 2C).

We also used a fragment of adenovirus 2 cloned in an M13 vector which can be used to produce single-stranded DNA probes. The single-stranded probe bound to target DNA was detected with nick-translated replicative form vector DNA. This experiment, again, showed preference against types 2 and 5, but types 16 and 31 were positive as well (Figure 2E).

An interesting possibility to make specific probes for any microorganism with known nucleic acid sequence is the use of synthetic oligonucleotides. We used a 20-mer representing a region coding for the 20.1K polypeptide of adenovirus 3. When this oligonucleotide was labeled with ^{32}P by using polynucleotide kinase and used as a probe, a signal was observed with serotypes 3 and 16, both belonging to subgenus B (Figure 2F).

B. Detection of Adenovirus DNA in Clinical Specimens

Direct demonstration of adenoviruses by nucleic acid hybridization in stool and nasopharyngeal specimens has been reported by several groups. Both spot and sandwich hybridization assays have been employed, either directly or after phenol extraction of the samples.

Stanlhandske et al.[2] have used a spot hybridization assay with radioactivity labeled probes including whole adenovirus 2 (Ad2) DNA and a cloned fragment of Ad41 genome. The stool samples were treated with proteinase K, phenol extracted, and ethanol precipitated. DNA was then denatured and applied onto nitrocellulose filters and hybridized using probes labeled with ^{32}P in a nick-translation reaction. The filters were washed, and the results were revealed by autoradiography on X-ray films. Forty selected specimens from hospitalized children with gastroenteritis were analyzed, and the hybridization results were compared with those obtained by a radioimmunoassay (RIA) for adenovirus hexon antigen. Whole Ad2 DNA probe detected 15 of the 18 RIA-positive specimens and 1 RIA-negative specimen. The cloned fragment of Ad41 gave a signal with five samples, all of which were also positive with the Ad2 probe. Typing of the viruses was not done and, therefore, the specificity of the Ad41 probe in this study remains unknown.

A similar assay has been described by Kidd et al.[3] for enteric adenovirus types 40 and 41 (subgenus F). Cloned restriction fragments were used as ^{32}P-labeled probes in a spot hybrdization test, allowing detection of picogram quantities of adenovirus DNA in stool suspension. The specimens were tested either directly or after phenol extraction. Without extraction, 61 of the 76 specimens known to contain enteric adenoviruses on the basis of restriction enzyme analysis were positive. After phenol extraction, each of the specimens was positive. The strains were directly typed as Ad40 and Ad41 with 100% correlation to the results of restriction enzyme analysis.

Two other groups have also reported construction of specific reagents for the detection of enteric adenoviruses. Allard et al.[4] describe cloning of probes from Ad41 DNA, which were used either to detect representatives from all human adenovirus subgenera or specifically for detection of enteric adenoviruses, in particular Ad41. Similar results have also been reported by Takiff et al.[5] who used a cloned DNA fragment of an enteric adenovirus with minimal cross-hybridization to represent strains from other adenovirus groups.

The applicability of nucleic acid hybridization to detect adenovirus DNA in nasopharyngeal specimens has also been demonstrated. In a comparative study of radioactive and enzymatic probe systems, Hyypiä[6] showed that they both can be used for diagnosis of adenovirus infections, and the results correlate well with those obtained by immunoassays. The sensitivity of the radioactive and enzymatic probe systems was 10 and 100 pg of adenovirus DNA, respectively. Twenty-four selected nasopharyngeal mucus aspirates were treated with proteinase K, phenol extracted, and the DNA was denatured and applied onto nitrocellulose filters. Sixteen specimens were positive when ^{32}P-labeled whole adenovirus DNA and a cloned restriction fragment were used as probes. Equivalent results were obtained when a biotinylated whole adenovirus DNA probe was detected by alkaline-phosphatase conjugated streptavidin using an enzyme reaction. The same 16 specimens were also positive by EIA for adenovirus hexon antigen which was used as a reference test.

A sandwich hybridization assay for adenoviruses has been introduced by Virtanen et al.[7] In the test, the sample DNA mediates the binding of the labeled probe to a catching DNA reagent on the filter. The reagents included cloned fragments from the genomes of Ad2 and Ad3. The catching DNA reagent and the probe were cloned in plasmid pBR322 and phage M13, respectively. The hybridization results were compared with RIA in detection of adenoviruses from nasopharyngeal specimens of children with acute respiratory infections. Of the 25 RIA-positive specimens, 22 were also positive in the sandwich hybridization assay, and all the 22 RIA-negative samples were negative by hybridization. In addition, the test could discriminate between adenovirus subgenera B and C. An advantage of this method is the low background which allows testing of mucus specimens directly without phenol extraction. Recently, Lehtomăki and co-workers[8] have shown that the sandwich hybridization test can also be used in respiratory infections of young adult men, and the sensitivity also in these cases is comparable to that of EIA and immunofluorescence. Virus isolation, however, was the most sensitive method for demonstration of adenoviruses in this study.

C. Molecular Epidemiology of Adenoviruses

Restriction enzyme (RE) analysis of adenovirus DNA makes it possible to study relatedness between viral genomes. The approach is not restricted on the capsid proteins as is the neutralization test. Comparison of RE patterns has yielded interesting data on the epidemiology and variation of adenoviruses.

In a recent analysis of the global distribution of Ad7 by Wadell and co-workers,[9] seven distinct genome types were identified using restriction endonucleases *Bam*HI and *Sma*I. Types 7b and 7c, for instance, were detected in Europe and in Australia, with a shift in Europe from 7c to 7b in 1969 and in Australia in 1975. Type 7c was never detected among isolates from North or Latin America, and in Japan, only 7b was found. The subtype 7b can further be subdivided into three distinct genome subgroups by digestion with additional enzymes as shown by Bailey and Richmond.[10] These variants appeared in similar frequency and were found in similar clinical illnesses.

Recently, Adrian et al.[11] have analyzed RE patterns of all human adenovirus prototypes with five enzymes. These data can now be used as references for strain identification, thus providing us with tools for sophisticated epidemiology and comparative analysis of adenoviruses.

D. Conclusions

The examples described above clearly show that nucleic acid hybridization assays give comparable results with reference methods in the identification of adenoviruses in clinical samples. The whole sequence of Ad2 genome is already known and the knowledge of the other serotypes is rapidly increasing, which allows construction of defined reagents for the detection of these viruses. Therefore, the adenoviruses serve as a good model system for the evaluation of hybridization probes, including synthetic oligonucleotides, for diagnostic purposes. In addition, there are reliable rapid reference tests available for the evaluation of the specificity and sensitivity of various probe systems.

ACKNOWLEDGMENTS

We thank KabiGen AB for the synthetic oligonucleotide, Mrs. Marita Maaronen for technical assistance, Mrs. Taina Kivela for typing the manuscript, and Professor Pekka Halonen for critical comments.

REFERENCES

1. **Akusjärvi, G., Pettersson, U., and Roberts, R. J.,** Structure and function of the adenovirus-2 genome, in *Adenovirus DNA*, Doerfler, W., Ed., Martinus Nijhoff Publ., Boston, Mass., 1986.
2. **Stålhandske, P., Hyypiä, T., Allard, A., Halonen, P., and Pettersson, U.,** Detection of adenoviruses in stool specimens by nucleic acid spot hybridization, *J. Med. Virol.*, 16, 213, 1985.
3. **Kidd, A. H., Harley, E. H., and Erasmus, M. J.,** Specific detection and typing of adenovirus types 40 and 41 in stool specimens by dot-blot hybridization, *J. Clin. Microbiol.*, 22, 934, 1985.
4. **Allard, A. K., Wadell, G., Evander, K. M., and Lindman, G. K. K.,** Specific properties of two enteric adenovirus 41 clones mapped within early region 1A, *J. Virol.*, 54, 145, 1985.
5. **Takiff, H. E., Seidin, M., Krause, P., Rooney, J., Brandt, C., Rodriguez, W., Yolken, R., and Straus, S. E.,** Detection of enteric adenoviruses by dot-blot hybridization using a molecularly cloned viral DNA probe, *J. Med. Virol.*, 16, 107, 1985.
6. **Hyypiä, T.,** Detection of adenovirus in nasopharyngeal specimens by radioactive and nonradioactive DNA probes, *J. Clin. Microbiol.*, 21, 730, 1985.
7. **Virtanen, M., Palva, A., Laaksonen, M., Halonen, P., Söderlund, H., and Ranki, M.,** Novel test for rapid viral diagnosis: detection of adenovirus in nasopharyngeal mucus aspirates by means of nucleic-acid sandwich hybridization, *Lancet*, 1, 381, 1983.
8. **Lehtomäki, K., Julkunen, I., Sandelin, K., Salonen, J., Virtanen, M., Ranki, M., and Hovi, T.,** Rapid diagnosis of respiratory adenovirus infections in young adult men, *J. Clin. Microbiol.*, 24, 108, 1986.
9. **Wadell, G., Cooney, M. K., da Costa Linhares, A., de Silva, L., Kennett, M. L., Kono, R., Gui-Fant, R., Lindman, K., Nascimento, J. P., Schoub, B. D., and Smith, C. D.,** Molecular epidemiology of adenoviruses: global distribution of adenovirus 7 genome types, *J. Clin. Microbiol.*, 21, 403, 1985.
10. **Bailey, A. S. and Richmond, S. J.,** Genetic heterogeneity of recent isolates of adenovirus Types 3, 4, and 7, *J. Clin. Microbiol.*, 24, 30, 1986.
11. **Adrian, Th., Wadell, G., Hierholzer, J. C., and Wigand, R.,** DNA restriction analysis of adenovirus prototypes 1 to 41, *Arch. Virol.*, 91, 277, 1986.
12. **Persson, H. and Philipson, L.,** Regulation of adenovirus, *Curr. Top. Microbiol. Immunol.*, 97, 157, 1982.

Chapter 10

NUCLEIC ACID PROBES FOR THE DETECTION OF EPSTEIN-BARR VIRUS

John W. Sixbey and Pamela S. Shirley

TABLE OF CONTENTS

I. INTRODUCTION

The application of nucleic acid probes to the area of rapid diagnostics in infectious diseases provides striking new information at several levels. The biology peculiar to each infectious agent may require novel hybridization strategies for detection in clinical materials, the diversity of which is illustrated throughout this volume. Appreciation of these varied approaches promises not only to provide us with a diagnostic wherewithall previously unavailable in clinical pathology, but leads us, if unwittingly, to a broader understanding of disease mechanisms at epidemiologic, cellular, and molecular levels. The specific requirements imposed by the biology of the Epstein-Barr virus (EBV) on the clinical application of nucleic acid hybridization analysis are the focus of this discussion.

II. COMMENSAL OR PATHOGEN: CHANGING PERSPECTIVE ON EBV INFECTION

EBV exists as a widespread infection in virtually all human communities. Most people carry the virus for life in what has been termed a "latent" form.[1] The nature of the asymptomatic carrier state is of particular interest, first, because of the central role postulated for the B lymphocyte, a cell type for which EBV possesses potent transforming activity in vitro;[2,3] second, because of the malignant nature of what could be termed "reactivation" diseases: Burkitt's lymphoma, lymphoproliferative disorders of the immunocompromised host, and nasopharyngeal carcinoma.

It has generally been assumed that EBV persists as a latent infection in virus-transformed B lymphocytes which intermittently reactivates to allow viral replication at permissive sites such as the oropharynx.[1] Supporting this notion is the observation that lymphocytes derived from the blood or lymph node of EBV-seropositive donors will occasionally transform spontaneously into EBV genome bearing lymphoblastoid cell lines.[4] Although the intermittent shedding of virus from the oropharynx has been considered more of epidemiologic than clinical significance, elucidation of a link between viral excretion and EBV-induced disease has been hampered by the cumbersome lymphocyte transformation assay[3] used to identify virus in saliva. The diagnostic utility of any assay for detection of EBV is similarly confounded by the universal and persistent nature of EBV infection.

The view that EBV persistence is maintained through foci of virus-transformed B cells resident in lymphoid tissue is increasingly at odds with work from several laboratories.[5-13] First, there is clear evidence for a potent virus-specific cytotoxic T cell surveillance system by which infected B cells are rapidly eliminated.[5,6] Indeed, the EBV genome cannot routinely be detected by hybridization analysis in peripheral lymphoid tissue from healthy individuals.[7] Second, in vivo replication of the virus in lymphocytes has not been demonstrated, despite identification of the oropharynx as site of viral shedding. By contrast, epithelial cells of the oropharynx appear to support chronic EBV replication.[8-10] Moreover, widespread distribution of a putative EBV receptor on epithelium from various tissue sites supports a primary role for this cell type in EBV infection.[11] Finally, what appeared to be a sporadic phenomenon — viral excretion by asymptomatic seropositive carriers — is now reported to occur in the majority of healthy subjects.[12] Of even greater importance is what seems to be a direct relationship between the level of virus excretion in the throat and the proportion of virus-infected B cells in the peripheral blood.[12,13]

These findings suggest that chronic, low-grade replication of the virus in the pharynx is, in fact, crucial to the mechanism of virus persistence in the human host. Involvement of peripheral lymphoid tissue may be a secondary phenomenon representing ongoing recruitment of B cells by infection with endogenous virus. Unlike the rather static notion of latency, this view of EB virus persistence reflects the dynamic balance between host immune surveillance and the

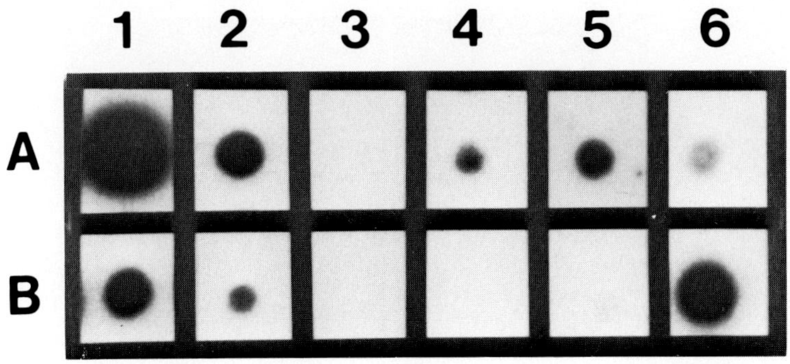

FIGURE 1. Epstein-Barr virus (EBV) DNA by dot-blot hybridization. Test cells or DNA was applied to a nylon filter and hybridized with a ^{32}P-oligolabeled *Bam*H1 W DNA probe (2×10^8 cpm/100 ng) and exposed to Kodak XAR 5 film for 72 h.[7,22,23,25] The figure illustrates results from the productively infected B95-8 cell line at 1×10^5 and 1×10^4 cells per spot (1A, 1B); latently infected Raji cells at 1×10^5 and 1×10^4 cells per spot (2A, 2B); EBV-negative celll ine BJAB at 1×10^5 and 1×10^4 cells per spot (3A, 3B); 0.2 ml of unconcentrated, EBV-positive throat washing from a leukemia patient (4A); an EBV-negative throat washing (4B); DNA (0.5 μg) from an EBV-positive liver biopsy (5A); DNA (5.0 μg) from normal liver (5B); lymphoma cell suspension (1×10^5 cells) from a patient with the X-linked lymphoproliferative (XLP) syndrome (6A); nasopharyngeal carcinoma biopsy DNA, 1.5 μg (6B).

continual replenishment of infected B cells via epithelial-derived virus, a view which immediately offers parameters by which to gauge virologic activity. The concordance between virus replication in the throat and virus-infected B cells in the blood, even more striking in the immunosuppressed patient,[13] may provide markers of disturbance in host:virus equilibrium which precede the development of EBV-related lymphoproliferative disorders. Both viral shedding and tissue burden represent indices of the EB virus carrier state suitable for analysis by rapid quantitative assays.

III. NUCLEIC ACID HYBRIDIZATION

A. Basic Concepts

The basis for nucleic acid hybridization centers on the ability of single strands of nucleic acid to recognize and form stable hybrids with homologous strands of either DNA or RNA having complementary sequences. Double-stranded DNA is first denatured before hybridization is attempted. Once a hybrid is formed, the duplex of probe and target sequence can then be analyzed, quantitated, or visualized, based on the label incorporated into the probe strand.

Maximum information can be obtained by tailoring hybridization methodology to the clinical material tested. Sample DNA containing the target nucleotide sequence may be extracted from cells and fixed to a membrane filter (Figure 1). Preparations affixed to filters are sufficiently durable to allow sequential hybridizations with different probes, a feature of some utility for clinical diagnostics.[14] Alternatively, DNA may be left unextracted in a tissue preparation and hybridization procedures carried out *in situ*. By preserving cellular morphology, cytohybridization allows localization of target sequences to specific subsets of cells within tissue or to a defined subcellular region (Figure 2A). Membrane hybridization provides no information with respect to localization of target nucleic acid sequences throughout tissue. In addition, *in situ* cytohybridization allows detection of targeted nucleotides in the rare cell, whereas biochemical extraction of DNA from total sample would dilute viral sequences beyond the level of detection.

B. Probe Selection

Restriction enzyme analysis and recombinant DNA technology have provided abundant,

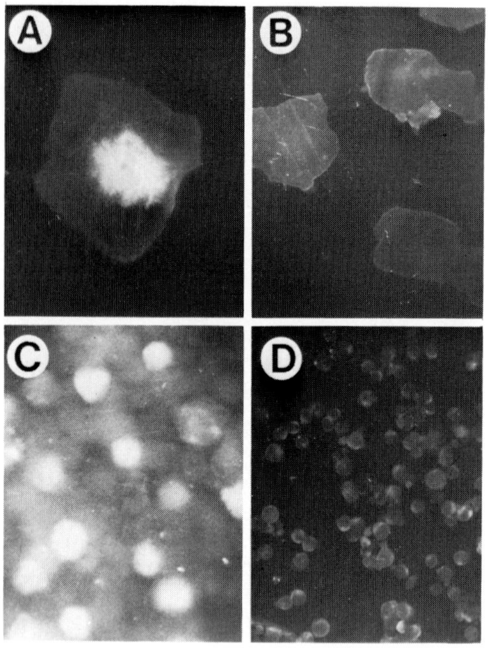

FIGURE 2. Cytohybridization for Epstein-Barr virus (EBV) RNA with biotynylated EBV DNA fragment *Bam*H1 H.[15] Fluorescein isothiocyanate (FITC)-avidin detects bound probe denoting EBV RNA in the nucleus of an oropharyngeal epithelial cell (A) as well as in 30% of superinfected Raji cell controls (C). Epithelial cells from an EBV-seronegative donor (B) as well as ribonuclease-treated super-infected Raji cells (D) did not hybridize for viral-specific RNA. Detection of an EBV-specific RNA in epithelial cells from the throat confirms the intracellular location of this DNA virus (as opposed to nonspecific adherence to cell surfaces).

well-defined subgenomic sequences of viral DNA for use as probes. Furthermore, these technologies have permitted detailed analysis of EBV structure to ascertain what viral sequences may be optimal as targets.[15] Because cells infected by EBV retain the complete viral genome either as extrachromosomal episomes or linear DNA integrated into cellular sequences, selection as probe of a particular fragment of EBV DNA over another does not seem critical.[16] A complete representation of the EBV genome is available within infected cells as target. Yet, strategic decisions remain which may optimize detection of viral sequences.

EBV consists of double-stranded DNA with a molecular weight of 115×10^6 daltons (170 kilobase pairs). The viral genome has a complex structure of several regions of repetitive DNA interspersed with unique sequences.[15] A 5 to 15 tandem reiteration of one 3-kbp sequence within the EBV genome provides a sensitive target for hybridization analysis. Selection as probe of a restriction fragment of EBV DNA that is within this internal repeat might favor detection of EBV in clinical materials by the repetitive binding of probe to the reiterated sequence. Use of fragments that encode an Epstein-Barr virus nuclear antigen that is invariably expressed when the EBV genome is present represents another approach.[17] Alternatively, a combination of fragments, which together represent a large portion of the genome, may be used as probe.

Transcriptional products of viral DNA may provide an amplified target for DNA probes in the form of abundant, complementary RNA from which presence of the viral genome can be inferred. Use of DNA-RNA hybridization analysis may be particularly useful in the EBV system where most cells harboring the virus are latently infected. Viral replication is stringently controlled, and the amount of viral DNA present may be insufficient for detection by some hybridization techniques. An abundantly transcribed region of the viral genome could provide RNA sequences far in excess of persisting viral DNA copy number. Since relatively few regions of the EBV genome are actively transcribed in latent infection, careful selection of fragments of DNA for use as probe is required.[18]

As detailed analysis of transcriptional patterns in both activated and latent EBV states are reported, opportunities increase for selection as probe of DNA fragments which identify target sequences of biologic significance in host-virus interactions. Specifically, virus-derived RNA might originate from a limited region of the genome, the abundant expression of which is necessary for the maintenance of the latent state. Selected fragments have been utilized in reports from several laboratories to detect transcripts which suggest an activated viral state in clinical materials (Figure 2A).[9,19] Wang and associates have recently reported that expression of a gene for an EBV-determined membrane antigen can confer tumorigenicity on a variety of cell types.[20] Its transcriptional product represents an approximately 60-fold amplification of the gene and is one of the more abundant EBV RNAs present in latently infected cells.

In preparation of probes, cloned DNA fragments can be excised from their cloning vector DNA sequences or used in combination with vector DNA. There is evidence to suggest that vector sequences may, in fact, amplify the hybridization signal by providing loose tails on which to build further hybrids, thereby increasing detection of a lower abundance target.[21] However, this approach is accompanied by increased nonspecific background. The probes described below were derived from cloned EBV DNA excised from the plasmid vector.

IV. HYBRIDIZATION TECHNIQUES FOR RAPID SCREENING OF CLINICAL SPECIMENS

A. Dot Blot

A simple nucleic acid hybridization method that allows rapid quantitatve screening of large numbers of samples for viral DNA without the necessity of first purifying viral or cellular DNA has been described by Brandsma and Miller.[22] Whole cells or body fluids are spotted on nylon filters, and their DNA is denatured and fixed to the filter. The DNA spots are then hybridized to radiolabeled probe DNA, and the degree of hybridization is assessed by autoradiography or scintillation counting. Serial dilutions of test sample can be compared to a known EBV-positive standard, such as latently infected Raji cells, to permit estimation of genome content.

The feasibility of dot blots for monitoring oropharyngeal viral excretion, as well as for screening tissue samples for presence of EBV, has been reported (see Figure 1).[7,23] The semiquantitative assay yields results in approximately 3 days, as opposed to the 3- to 8-week period required for cord blood transformation assays. The dot blot is somewhat less sensitive than the transformation assay for detection of cell-free infectious virus. Ambinder et al. reported a hybridization sensitivity of 75% and a specificity of 97% when the dot blot and lymphocyte transformation assay were compared using identical oropharyngeal washings.[23] However, concentration of samples by ultracentrifugation,[12,24] together with use of probes with a higher specific activity,[25] may resolve these difficulties. Of note is the fact that the dot-blot assay provides, for the first time, a means for detecting the EBV genome in specimens unsuitable for the lymphocyte transformation assay because of toxicity to lymphocytes (e.g., semen). Dot-blot hybridization also allows identification of virus in samples where infectivity is lost secondary to improper handling or in specimens containing defective or nontransforming strains of EBV.

B. *In Situ* Cytohybridization

We have utilized both biotin-labeled and radiolabeled EBV DNA probes to identify an apparent productive infection by EBV in epithelial cells of the pharynx, and more recently, the

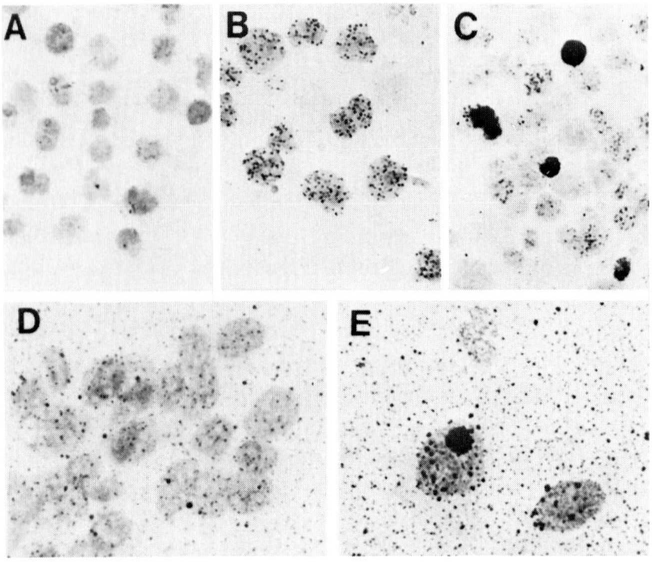

FIGURE 3. Epstein-Barr virus (EBV) DNA in latently infected or permissive cell types. The ABC detection system[28] labels bound biotinylated probe (*Bam*H1 W EBV DNA)[15] with granules of peroxidase precipitate. Latently infected Raji cells (B), which contain approximately 60 to 100 EBV genome equivalents, bear a mean of 48 granules per cell. Productively infected HR_1 cells (C) are heavily labeled as opposed to the latently infected majority which contain sparse granules. The EBV-negative cell line BJAB did not hybridize (A). EBV DNA in malignant epithelial cells from a tissue imprint of nasopharyngeal carcinoma (E); touch prep of healthy tissue adjacent to the tumor (D).

uterine cervix.[9,26] Figure 2 depicts EBV RNA in oropharyngeal epithelial cells after hybridization with a biotinylated *Bam*H1 H fragment of EBV DNA (a 3.9×10^6-dalton fragment generated by digestion of EBV DNA with the restriction endonuclease *Bam*H1[15]). Virus-specific RNA would be anticipated only in those cells with replicative mechanisms activated for synthesis of viral products. In that two abundant early RNAs are transcribed from the *Bam*H1 H region of the genome in lymphoid cultures actively replicating viral DNA, detection of transcripts in epithelial cells implies an activated, not latent, viral state.[27] Hybridization to RNA was not detected in control cells pretreated with RNAse (Figure 2D) or in oropharyngeal epithelial cells from EBV-seronegative donors (Figure 2B).

Figure 3 demonstrates the hybridization signal in a number of cell types, the intensity of which varies in direct relation to their EBV DNA content. Biotin-labeled EBV DNA fragment *Bam*H1 W, a 1.9×10^6-dalton fragment which recognizes the large internal repeat within the EBV genome, was used as probe.[15] With the avidin-biotin-peroxidase complex detection system of Hsu and colleagues,[28] biotinylated probe is shown bound to EBV sequences not only in the subset of HR_1 cells which actively replicate virus, but in the latently infected population as well. No signal was detected over the EBV-negative BJAB cell line. EBV DNA was demonstrated in touch preparations from latently infected nasopharyngeal carcinoma tissue (Figure 3E), whereas imprints of nonmalignant tissue adjacent to the tumor did not hybridize (Figure 3D).

V. CONCLUSIONS

Conventional laboratory diagnosis of EBV is made by one of three techniques: serology, Epstein-Barr virus nuclear antigen (EBNA) staining of tissue preparations, or lymphocyte transformation assays. Of these approaches, serology is the most readily available. This method assumes an intact immune system, which may be absent or impaired in the more puzzling EBV-related syndromes.[29] Although detection of EBNA in tissue imprints is rapid, the anticomplement immunofluorescence assay is subject to varied interpretations and, when applied to clinical materials, requires extensive controls. Detection of cell-free infectious virus by its transforming activity on umbilical cord blood lymphocytes is labor intensive, requires regular access to cord blood, and results are rarely finalized while still clinically relevant.

These difficulties make the broader application of hybridization analysis for detection of EBV in clinical samples particularly inviting. Moreover, against a background of almost universal seropositivity to EBV in adult populations, hybridization analysis provides supportive evidence of an etiologic role for EBV in its more perplexing disease associations — specifically those arising from reactivation of persistent virus, not primary infection. The EBV genome is regularly detected in several diseases classically linked to EBV, as well as in some emerging clinical entities: nasopharyngeal carcinoma,[30,31] African Burkitt's lymphoma,[30,32] polyclonal B cell lymphoproliferations in immunocompromised hosts,[33] Burkitt's lymphoma in adult AIDS,[34,35] pulmonary lymphoid hyperplasia of pediatric AIDS,[36] and a syndrome of severe EBV disease in young children (organomegaly, failure to thrive, pneumonitis, hypergammaglobulinemia, T cell abnormalities).[7,37] Viral DNA is clearly not found as a matter of routine in a wide variety of normal and diseased lymphoid tissue,[7] an argument against the view of EBV as a simple passenger and of all lymphoid tissue as foci of viral latency. Finally, due to the ease with which hybridization techniques can be performed on multiple samples in a rapid, semiquantitative manner, these assays provide a useful approach for evaluating the impact of various therapeutic interventions on EBV shedding and viral activity, particularly in immunocompromised patients where disruption of virus-host balance may have an adverse clinical consequence.

REFERENCES

1. **Epstein, M. A. and Achong, B. G.,** Introduction: discovery and general biology of the virus, in *The Epstein-Barr Virus,* Epstein, M. A. and Achong, B. G., Eds., Springer-Verlag, Berlin, 1979, 1.
2. **Henle, W., Diehl, V., Kohn, G., Zur Hausen, H., and Henle, G.,** Herpes-type virus and chromosomal marker in normal leukocytes after growth with irradiated Burkitt cells, *Science,* 157, 1064, 1967.
3. **Pope, J. H., Horne, M. K., and Scott, W.,** Transformation of foetal human leukocytes *in vitro* by filtrates of a human leukaemic cell line containing herpes-like virus, *Int. J. Cancer,* 3, 857, 1968.
4. **Nilsson, K., Klein, G., Henle, W., and Henle, G.,** The establishment of lymphoblastoid lines from adult and from foetal human lymphoid tissue and its dependence on EBV, *Int. J. Cancer,* 8, 443, 1971.
5. **Moss, D. J., Rickinson, A. B., and Pope, J. H.,** Long-term T-cell-mediated immunity to Epstein-Barr virus in man. I. Complete regression of virus-induced transformation in cultures of seropositive donor leukocytes, *Int. J. Cancer,* 22, 662, 1978.
6. **Rickinson, A. B., Moss, D. J., Wallace, L. E., Rowe, M., Misko, I. S., Epstein, M. A., and Pope, J. H.,** Long-term T-cell-mediated immunity to Epstein-Barr virus, *Cancer Res.,* 41, 4216, 1981.
7. **Andiman, W., Gradoville, L., Heston, L., Neydorff, R., Savage, M. E., Kitchingman, G., Shedd, D., and Miller, G.,** Use of cloned probes to detect Epstein-Barr viral DNA in tissues of patients with neoplastic and lymphoproliferative diseases, *J. Infect. Dis.,* 148, 967, 1983.
8. **Lemon, S. M., Hutt, L. M., Shaw, J. E., Li, J.-L. H., and Pagano, J. S.,** Replication of EBV in epithelial cells during infectious mononucleosis, *Nature (London),* 268, 268, 1977.
9. **Sixbey, J. W., Nedrud, J. G., Raab-Traub, N., Hanes, R. A., and Pagano, J. S.,** Epstein-Barr virus replication in oropharyngeal epithelial cells, *N. Engl. J. Med.,* 310, 1225, 1984.
10. **Greenspan, J. S., Greenspan, D., Lennette, E. T., Abrams, D. I., Conant, M. A., Peterson, V., and Freese, U. K.,** Replication of Epstein-Barr virus within the epithelial cells of oral "hairy" leukoplakia, an AIDS-associated lesion, *N. Engl. J. Med.,* 313, 1564, 1985.
11. **Sixbey, J. W., Davis, D. S., Young, L. S., Hutt-Fletcher, L., Tedder, T. F., and Rickinson, A. B.,** Human epithelial cell expression of an Epstein-Barr virus receptor, *J. Gen. Virol.,* 68, 805, 1987.

12. **Yao, Q. Y., Rickinson, A. B., and Epstein, M. A.,** A re-examination of the Epstein-Barr virus carrier state in healthy seropositive individuals, *Int. J. Cancer,* 35, 35, 1985.
13. **Yao, Q. Y., Rickinson, A. B., Gaston, J. S. H., and Epstein, M. A.,** *In vitro* analysis of the Epstein-Barr virus:host balance in long-term renal allograft recipients, Int. *J. Cancer,* 35, 43, 1985.
14. **Gatti, R. A., Concannon, P., and Salser, W.,** Multiple use of Southern blots, *BioTechniques,* 1, 148, 1984.
15. **Dambaugh, T., Beisel, C., Hummel, N., King, W., Fennewald, S., Cheung, A., Heller, M., Raab-Traub, N., and Kieff, E.,** Epstein-Barr Virus (B95-8) DNA VII: Molecular cloning and detailed mapping, *Proc. Natl. Acad. Sci. U.S.A.,* 77, 2999, 1980.
16. **Sugden, B.,** Comparison of Epstein-Barr viral DNAs in Burkitt lymphoma biopsy cells and in cells clonally transformed *in vitro, Proc. Natl. Acad. Sci. U.S.A.,* 74, 4651, 1977.
17. **Summers, W. P., Grogan, E. A., Shedd, D., Robert, M., Liu, C.-R., and Miller, G.,** Stable expression in mouse cells of nuclear neoantigen after transfer of a 3.4 megadalton cloned fragment of Epstein-Barr virus DNA, *Proc. Natl. Acad. Sci. U.S.A.,* 79, 5688, 1982.
18. **King, W., Van Santen, V., and Kieff, E.,** Epstein-Barr virus RNA VI. Viral RNA in restringently and abortively infected Raji cells, *J. Virol.,* 38, 649, 1981.
19. **Raab-Traub, N., Hood, R., Yang, C.-S., Henry, B., II, and Pagano, J. S.,** Epstein-Barr virus transcription in nasopharyngeal carcinoma, *J. Virol.,* 48, 580, 1983.
20. **Wang, D., Liebowitz, D., and Kieff, E.,** An EBV membrane protein expressed in immortalized lymphocytes transforms established rodent cells, *Cell,* 43, 831, 1985.
21. **Singer, R. H., Lawrence, J. B., and Villnave, C.,** Optimization of in situ hybridization using isotopic and non-isotopic detection methods, *BioTechniques,* 4, 230, 1986.
22. **Brandsma, J. and Miller, G.,** Nucleic acid spot hybridization: rapid quantitative screening of lymphoid cell lines for Epstein-Barr viral DNA, *Proc. Natl. Acad. Sci. U.S.A.,* 77, 6851, 1980.
23. **Ambinder, R. F., Wingard, J. R., Burns, W. H., Hayward, S. D., Saral, R., Perry, H. R., Santos, G. W., and Hayward, G. S.,** Detection of Epstein-Barr virus DNA in mouthwashes by hybridization, *J. Clin. Microbiol.,* 21, 353, 1985.
24. **Buffone, G. J., Schimbor, C. M., Demmler, G. J., Wilson, D. R., and Darlington, G. J.,** Detection of cytomegalovirus in urine by nonisotopic DNA hybridization, *J. Infect. Dis.,* 154, 163, 1986.
25. **Feinberg, A. P. and Vogelstein, B.,** A technique for radiolabeling DNA restriction endonuclease fragments to high specific activity, *Anal. Biochem.,* 132, 6, 1983.
26. **Sixbey, J. W., Lemon, S. M., and Pagano, J. S.,** A second site for Epstein-Barr virus shedding: the uterine cervix, *Lancet,* 2, 1122, 1986.
27. **Hummel, M. and Kieff, E.,** Epstein-Barr virus RNA. VIII. Viral RNA in permissively infected B95-8 cells, *J. Virol.,* 43, 262, 1982.
28. **Hsu, S.-M., Raine, L., and Fanger, H.,** Use of avidin-biotin-peroxidase complex (ABC) in immunoperoxidase techniques: a comparison between ABC and unlabeled antibody (PAP) procedures, *J. Histochem. Cytochem.,* 29, 577, 1981.
29. **Fackler, J. G., Nagel, J. E., Adler, W. H., Mildvan, P. T., and Ambinder, R. F.,** Epstein-Barr virus infection in a child with acquired immunodeficiency syndrome, *Am. J. Dis. Child.,* 139, 1000, 1985.
30. **Zur Hausen, H., Schulte-Holthausen, H., Klein, G., Henle, W., Henle, G., Clifford, P., and Santesson, K.,** EBV DNA in biopsies of Burkitt tumors and anaplastic carcinomas of the nasopharynx, *Nature (London),* 228, 1056, 1970.
31. **Wolf, H., Zur Hausen, H., and Becker, V.,** EB viral genomes in epithelial nasopharyngeal carcinoma cells, *Nature New Biol.,* 244, 245, 1973.
32. **Lindahl, T., Klein, G., Reedman, B. M., Johansson, B., and Singh, S.,** Relationship between Epstein-Barr virus (EBV) DNA and the EBV-determined nuclear antigen (EBNA) in Burkitt lymphoma biopsies and other lymphoproliferative malignancies, *Int. J. Cancer,* 13, 764, 1974.
33. **Hanto, D. W., Frizzera, G., Purtilo, D. T., Sakamoto, K., Sullivan, J. L., Saemundsen, A. K., Klein, G., Simmons, R. L., and Najarian, J. S.,** Clinical spectrum of lymphoproliferative disorders in renal transplant recipients and evidence for the role of Epstein-Barr virus, *Cancer Res.,* 41, 4253, 1981.
34. **Ziegler, J., Drew, W. L., Miner, R. C., Mintz, L., Grenshaw, J., Rosenbaum, E., Lennette, E. T., Greenspan, J., Shillitoe, E., Beckstead, J., Casavant, C., and Yamamoto, K.,** Outbreak of Burkitt's-like lymphoma in homosexual men, *Lancet,* 2, 631, 1982.
35. **Snider, W. D., Simpson, D. M., Aronyk, K. E., and Nielsen, S. L.,** Primary lymphoma of the nervous system associated with acquired immunodeficiency syndrome (letter), *N. Engl. J. Med.,* 308, 45, 1983.
36. **Rubinstein, A., Morecki, R., Silverman, B., Charytan, M., Krieger, B. Z., Andiman, W., Ziprkowski, M. N., and Goldman, H.,** Pulmonary disease in children with acquired immune deficiency syndrome and AIDS-related complex, *J. Pediatr.,* 108, 498, 1986.
37. **Roberts, W. M., Wotherspoon, R., Herrod, H. G., and Sixbey, J. W.,** Morbidity of Epstein-Barr virus infection in children, in *Epstein-Barr Virus and Human Disease,* Levine, P. H., Ablashi, D. V., Nonoyama, M., Pearson, G. R., and Glaser, R., Eds., Humana Press, Clifton, N.J., 1987, 101.

Part V: Nucleic Acid Probes for the Detection and Identification of Viral Pathogens and Chlamydia

Chapter 11

DNA PROBES FOR THE DETECTION OF HUMAN CYTOMEGALOVIRUS

**Stephen A. Spector, Wayne M. Dankner,
Frank Denaro, and Deborah H. Spector**

TABLE OF CONTENTS

I. INTRODUCTION

Human cytomegalovirus (HCMV) infects approximately 1% of infants born in the U.S. and results in over 4000 neurologically impaired children annually.[1-3] HCMV is also a common pathogen of immunocompromised individuals, particularly organ transplant recipients and individuals with Acquired Immunodeficiency Syndrome (AIDS).[4-7] In the immunocompromised host, the virus causes a broad range of clinical diseases, including pneumonia, retinitis, colitis, hepatitis, and encephalitis. In healthy individuals, HCMV infection is usually mild and self-limited but may result in Epstein-Barr virus negative mononucleosis. In AIDS patients, the virus has been further implicated as a possible cofactor in progressive immune deficiency and in the development of Kaposi's sarcoma.[8]

Diagnosis of HCMV infection is established currently through both serologic tests and viral growth in tissue culture which may require several weeks. However, the complex clinical settings in which the virus is frequently a pathogen, and recent advances in antiviral therapy, necessitate the development of a more rapid and efficient means for identification of HCMV in clinical specimens. Additionally, a better understanding of the routes of viral transmission and the mechanisms of viral infection, persistence, and latency are required so that effective means for HCMV prevention and treatment can be developed.

Although the principles of nucleic acid hybridization have been known for over 20 years, the application of this technology to clinical medicine is only now in its infancy. However, advances in molecular biology promise to supply the necessary tools to provide solutions to many of the problems and questions relating to viral infections. The studies summarized below demonstrate how cloned HCMV DNA fragments have been used as probes to study the molecular biology, diagnosis, treatment, and pathogenesis of HCMV infections.

II. HCMV GENOME STRUCTURE AND ORGANIZATION

HCMV consists of a double-stranded DNA genome, 240 kbp in length, divided into two covalently linked components, designated L (approximately 198 kbp) and S (approximately 42 kbp), each bounded by inverted repeats.[9,10] The genome structure allows for the L and S components to be in any one of four different configurations which are represented in approximately equal amounts.[11,12]

Molecular analysis of different HCMV strains, using restriction enzyme digests, indicates that there is considerable heterogeneity of DNA sequence, while viruses from individuals who transmitted HCMV from one to the other (e.g., a mother and her congenitally infected infant, or sexual partners) have identical viruses.[13,14] These molecular differences among different strains have been exploited to help identify the epidemiology of HCMV infections.[14-17] As will be discussed below, DNA probes can be used to simplify performing such comparisons.

We have used cloned DNA probes as hybridization reagents for the analysis of RNA transcription patterns in human fibroblasts permissively infected with the AD169 HCMV strain.[18] During permissive infection, HCMV viral gene transcription and translation are subject to temporal, quantitative, and post-transcriptional controls. The immediate early gene products are synthesized shortly after virus entry into the cell, and the major protein has been found to be autoregulatory.[19] Early RNA and protein synthesis begin before viral DNA replication and require some immediate early gene products. The late gene products are encoded by a large portion of the viral genome and are abundantly synthesized following initiation of viral DNA synthesis. DNA probes are currently in use in several laboratories to determine the structure and exact genomic locations of specific RNA transcripts and to define functions to their protein products.

III. USE OF DNA PROBES FOR DETECTION OF HCMV IN CLINICAL SPECIMENS

A. Selection of Probe

Prior to using cloned HCMV DNA fragments as probes to detect HCMV nucleic acid in clinical specimens, it was necessary to demonstrate homology between the cloned fragments and DNA obtained from clinical isolates and to determine if there was any cross-hybridization between the potential probes and human cell DNA or other herpesvirus DNA.[20] To determine which of the 32 *Eco*RI AD169 cloned fragments hybridized to DNA extracted from HCMV clinical isolates or human DNA, 100 ng of DNA were dotted onto nitrocellulose filters and hybridized to each of the 32 clones. Of the 32 fragments, 31 hybridized to the DNA extracted from six clinical isolates. Only fragment h (0.85 kbp, mapping at the junction of the long repeat and long unique regions) failed to hybridize to the clinical isolates. In similar experiments, 16 of the 32 clones hybridized to human DNA. Two *Eco*RI fragments, R and b, hybridized strongly to human DNA, while fragments A, C, E, I, M, O, P, Q, V, c, d, and e, and the L-S junction fragments F (WL) and H (WN) hybridized at low levels.

For our original studies, we selected fragment B (17 kbp) from the S segment and fragment D (17 kbp) from the L segment. Both clones fail to hybridize with DNA from other herpesviruses, including herpes simplex virus type 1, herpes simplex virus type 2, varicella-zoster virus, and Epstein-Barr virus. In more recent studies, we have used only fragment D as our probe. Reconstruction experiments indicate that at least 10 pg of HCMV genomic DNA or 1 pg of DNA homologous to either fragment B or D can be detected using dot/slot-blot hybridization.

B. Detection of HCMV in Urine Specimens

Initial studies of clinical specimens were designed to determine if HCMV DNA could be detected in urine specimens obtained from infants with cytomegalic inclusion disease.[21] Urine specimens obtained from these infants ranged in titer from 10^4 to 10^6 $TCID_{50}$/ml. When DNA was extracted from as little as 1 ml of urine from these infants, strong signals were obtained using our ^{32}P-labeled probe. In subsequent studies, we have found using dot/slot-blot hybridization that over 90% of culture-positive urine specimens from a wide range of patients are hybridization positive. Approximately 10^4 virus particles are required for detection. However, since both infectious and noninfectious viruses generate signals, we have consistently identified HCMV urine samples with titers of 10^1 to 10^2 $TCID_{50}$/ml. We have also identified a few urine specimens that have been culture negative but hybridization positive, and, in most cases, evidence existed suggesting that the hybridization findings were true positives. Importantly, we have found the nucleic acid hybridization test to be substantially more rapid than routine tissue culture. For example, during our studies, we have found that urine cultures performed in our viral diagnostic laboratory require a mean of 12 days before being identified as positive, compared with 48 to 72 h required for the DNA-DNA hybridization assay. Studies by other investigators have verified our initial results and indicate that nucleic acid hybridization can be useful for identification of HCMV in urine.[22,23]

C. Detection of HCMV in Blood Specimens

The use of nucleic acid hybridization for identification of HCMV in buffy-coat specimens has proven to be less straightforward than for urines. In our studies in bone marrow transplant recipients, we found that when 5 μg of DNA were spotted onto a nitrocellulose filter, approximately 75% of culture-positive buffy-coat specimens were hybridization positive.[24] On the other hand, in this high-risk population, we found as many as 82 (31%) of 264 culture-negative specimens gave signals over background. A review of these 82 specimens indicated that 76 were from patients who had a positive culture for HCMV prior to and/or following the buffy-coat specimen that was culture negative but hybridization positive. Additionally, in specimens where sufficient DNA has been available, we have been able, in some specimens, to document by Southern hybridization the presence of authentic HMCV bands. However, it is clear that some positive hybridization results are false positive, indicating that methodologic improvement is necessary before dot/slot-blot hybridization can be applied with confidence to buffy-coat specimens.

Recent studies performed by other investigators reflect on this problem. Schrier et al.[25] have identified the presence of HCMV RNA in eight of eight seropositive individuals. They have found the presence of RNA in from 0.03 to 2% mononuclear cells and have localized apparent viral RNA preferentially in T helper lymphocytes. However, 1 (8%) of 12 seronegative individuals tested positive by their *in situ* hybridization assay, indicating either the presence of HCMV in individuals with no detectable antibody or nonspecific hybridization.

Using a ^{35}S-labeled probe consisting of the 23 cloned fragments of HCMV strain AD169 known not to hybridize to normal human DNA, we have initiated studies to identify HCMV in peripheral blood cells of individuals with active infection. In these preliminary studies, we have consistently found that the polymorphonuclear leukocytes are the cell population most frequently containing HCMV nucleic acid (Figure 1). We have detected HCMV RNA, as well as DNA, indicating that virus is not only present in these cells, but is actively replicating. We are currently determining to what degree viral replication occurs in these actively infected cells.

D. Detection of HCMV in Tissue Specimens by Nucleic Acid Hybridization

Nucleic acid hybridization is playing an increasingly important role in identification of HCMV in tissue specimens for diagnosis and pathogenetic studies. We have extracted DNA and/or RNA from tissue specimens including brain, lung, liver, kidney, adrenals, spleen, etc. and, using dot/slot-blot hybridization, identified HCMV-specific nucleic acid. These findings can be documented by either direct culture or Southern hybridization for DNA.

In situ cytohybridization has been particularly useful for identifying the presence of HCMV in tissue. A major problem in HCMV detection in clinical specimens by dot/slot or Southern hybridization is that a heterogenous mixture of infected and uninfected cells are invariably present. Since there is usually a substantially larger amount of uninfected cellular nucleic acid present than viral DNA or RNA, a dilutional effect can decrease the sensitivity of the assay. *In situ* cytohybridization can circumvent this problem by determining in which specific cells viral nucleic acid is present. Using biotinylated probes, this procedure has been used for the rapid identification of HCMV in tissue specimens.[26] Most notably, the detection of HCMV in lung specimens has been useful for clinical purposes in bone marrow transplant recipients with suspected HCMV pneumonia.[27]

We have used *in situ* cytohybridization to examine specimens of Kaposi's sarcoma from patients with AIDS for the presence of HCMV.[20,28] Frozen sections cut from Kaposi's sarcoma lesions of AIDS patients were hybridized with a mix of ^{3}H-labeled or ^{35}S-labeled cloned HCMV fragments (representing those fragments not hybridizing to normal cell DNA) under conditions for detection of both DNA and RNA. All specimens were also hybridized separately with the plasmid vector pACYC184 as a control for nonspecific hybridization or for specific adherence of the probe to bacterial or plasmid-related sequences that might be present in the specimens. In these assays, we were able to detect HCMV nucleic acid in Kaposi's tumor cells from 11 of 14 specimens. Two additional specimens hybridized equally well to both HCMV probe and the control plasmid probe, making interpretation of hybridization data impossible. Although the role of HCMV in the development of Kaposi's sarcoma remains unclear, *in situ* hybridization and double-labeling studies will be critical to determine the importance of this association.

In situ hybridization has been used to detect HCMV-infected cells in the brain.[29] However, this does not give specific information as to the cell type(s) which are infected. To overcome this problem, we have used a double-labeling technique to identify the specific brain cells infected

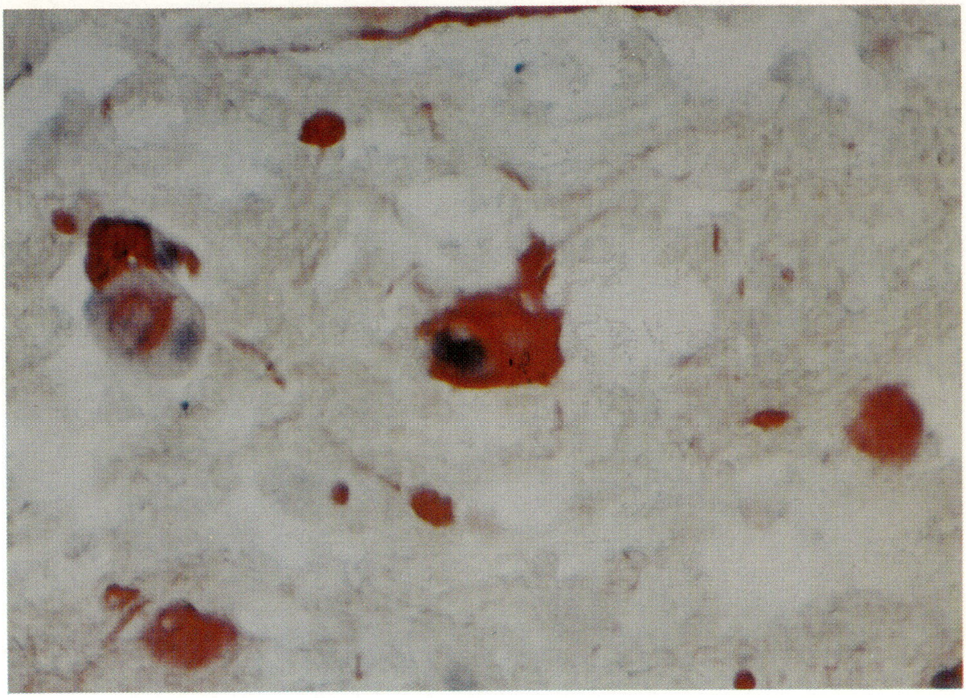

PLATE 1. **Double label for HCMV (*in situ* hybridization) GFAP (immunocytochemistry).** Brain obtained at postmortem from an AIDS patient was fixed in 10% buffered formaldehyde and embedded in paraffin. Ten-μm thick sections were collected on acid-cleaned slides and incubated overnight at 30°C. Sections were deparaffinized with xylene and hydrated. Endogenous peroxidase activity was blocked with hydrogen peroxide (final concentration 0.5%) in methanol, rinsed, and digested with proteinase K. The specimen was first hybridized with HCMV-specific DNA probe nick translated with biotin and visualized with avidin-alkaline phosphatase. The chromagen used was Fast Blue BB. Rabbit antiserum to GFAP was used with a peroxidase antiperoxidase procedure to identify glial cells (red chromagen). Cell in center of field shows strongly red-staining glial cell with HCMV-specific nucleic acid appearing blue.

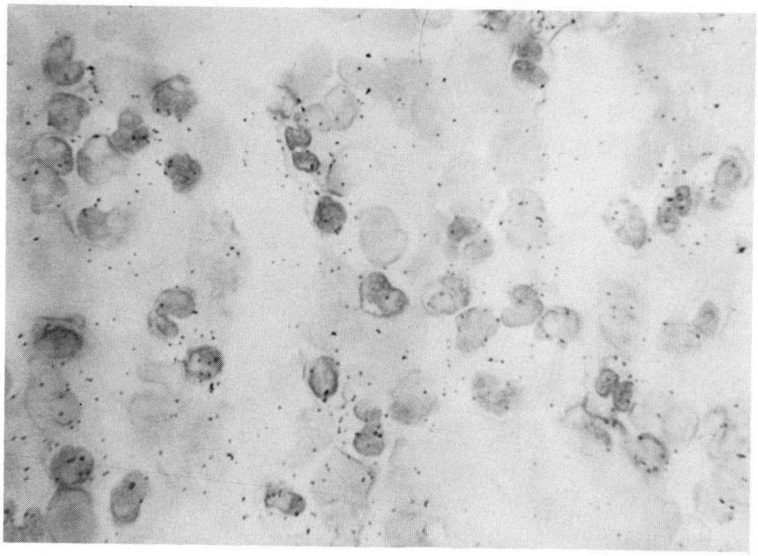

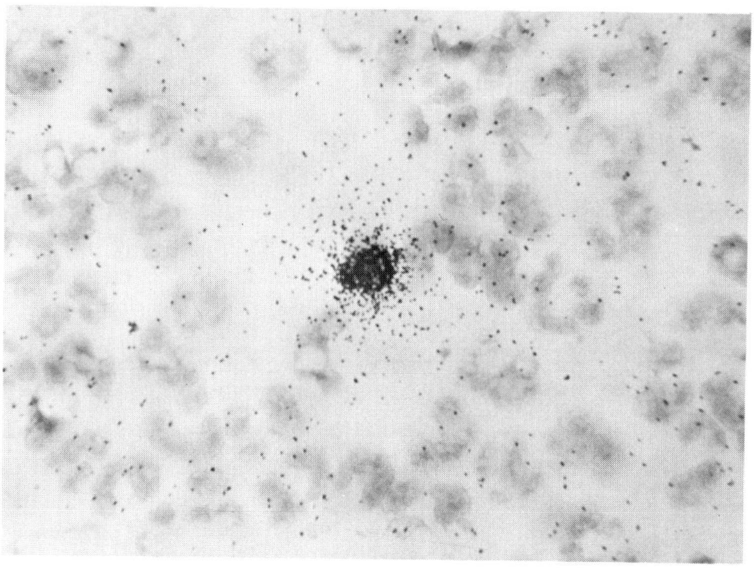

FIGURE 1. *In situ* hybridization of polymorphonuclear leukocytes (PMNs) using ^{35}S-labeled mix of cloned HCMV fragments. The PMNs were separated from fresh heparinized blood specimens by ficoll-hypaque gradient and lysis of red blood cells, and cytocentrifuged onto pretreated microscope slides. Hybridization was performed as previously described.[20] Briefly, the slides were incubated sequentially in 0.2 *N* HCl, 2 × SSC at 70°C, and digested with proteinase K in a Tris-$MgCl_2$-$CaCl_2$ buffer. The slides were dehydrated through graded alcohols. Hybridization was for 60 h. The slides were washed in a 50% formamide solution for 3 h, dipped in Kodak NTB2 nuclear track emulsion, and developed after 5 days. Slides were counter-stained with Giemsa. Figure A shows the absence of hybridization in the PMNs obtained from a HCMV seronegative subject. Figure B shows hybridization of the HCMV probe to PMNs obtained from an individual actively infected with HCMV.

with HCMV (Plate 1).* In these studies, we determined which brain cells obtained from a patient with encephalitis were infected with HCMV. Specimens were first hybridized with CMV-specific DNA probe nick translated with biotin and visualized with avidin-alkaline phosphatase with a Fast Blue BB chromagen. Glial fibrillary acid protein (GFAP) specific for glial cells was then detected using a polyclonal rabbit antiserum showing that the glial cells contained HCMV nucleic acid.

IV. USE OF HCMV PROBES FOR EVALUATION OF ANTIVIRAL AGENTS

We have used HCMV probes to assess the activity of antivirals against HCMV in vitro and in clinical specimens obtained from patients receiving viral-specific treatment. For in vitro studies, human fibroblast cells infected with clinical isolates are treated with increasing concentrations of antiviral agent(s). Following growth of the virus for 5 days, total DNA is extracted. Usually 1 μg of DNA is then applied to nitrocellulose filters for each concentration of antiviral and hybridized to a ^{32}P-labeled probe (fragment D). The amount of inhibition can be determined by the intensity of hybridization on autoradiography visually by using a densitometer or by comparing Cerenkov counts.

The effects of antiviral treatment on HCMV excretion can be determined in a patient by extracting DNA from sequential clinical specimens. We have found nucleic acid hybridization to be particularly valuable in following the inhibition of viral excretion in urine since many antivirals are concentrated in urine and can give artificially low or negative cultures. However, since nucleic acid hybridization detects viral DNA regardless of infectivity, the relative amount of virus in urine can be accurately assessed using dot/slot-blot hybridization.

V. USE OF DNA PROBES TO STUDY THE EPIDEMIOLOGY OF HCMV INFECTION

Restriction endonuclease digestion analysis has been widely used to determine the molecular relatedness of HCMV strains and to establish patterns of viral transmission. These molecular analyses are based on the marked genomic heterogeneity among random HCMV isolates. Epidemiologically related HCMV strains will exhibit the identical restriction enzyme digestion pattern, whereas different strains will have demonstrably different DNA fragments. The widespread use of restriction endonuclease digestion analysis is hampered, however, by the time required (which may be 6 months or longer) to grow the viral quantities necessary to prepare sufficient viral DNA to perform the digestions desired. In our studies of the polymorphism of the HCMV genome, we have found that heterogeneity is most frequent in fragments closest to the long and the short inverted repeat sequences of the joint regions. We found complete heterogeneity for all HCMV strains when hybridized to L-S junction fragments, designated F (*Eco*RI — WL) or H (*Eco*RI — WN). Hybridization of one of the L-S junction fragments to *Eco*RI-cleaved DNA from clinical isolates, transferred to nitrocellulose, can determine their molecular relatedness (Figure 2).[30] This method of comparing HCMV strains significantly decreases the amount of tissue culture required for performing molecular comparisons and should help to broaden the application of molecular epidemiology to study HCMV transmission.

VI. CONCLUSIONS

Recombinant DNA technology has provided necessary tools with which to elucidate many of the important questions relating to HCMV infection. We and others have applied DNA probes to study the epidemiology, pathogenesis, diagnosis, and treatment of HCMV. These studies

* Plate 1 appears following page 138.

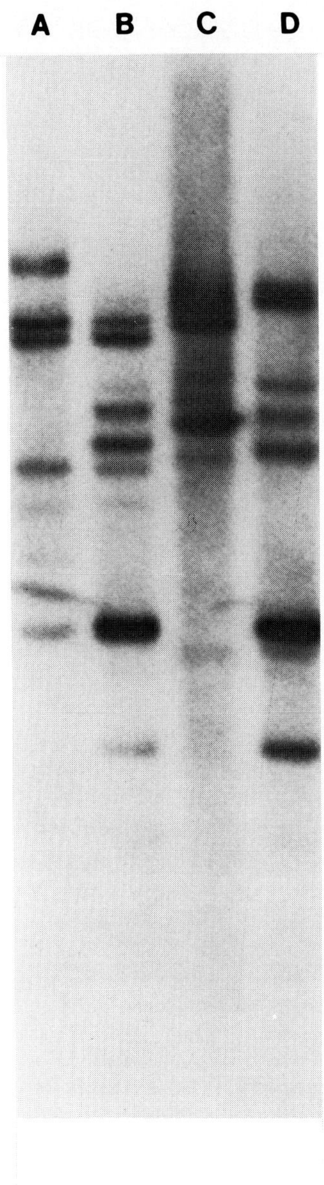

FIGURE 2. Autoradiograph of ^{32}P-labeled cloned HCMV DNA L-S junction fragment H hybridized to Southern blot of *Eco*RI-cleaved DNA from four HCMV isolates (A to D) cultured from urine of pediatric patients hospitalized on the same unit (isolates obtained courtesy of Dr. G. Demmler, Baylor University). All four isolates show a different pattern of hybridization to fragment H, indicating that there was no transmission of HCMV among these children.

comprise only the beginning of applying nucleic acid hybridization to clinical studies involving HCMV. Several technologic improvements are required before these procedures can be moved from the realm of the research laboratory into clinical laboratories. Extraction of nucleic acid for hybridization remains work intensive, and rapid methods for clean sample preparation are required. Additionally, development of sensitive and specific nonisotopic detection systems are needed to eliminate not only the potential hazards of radioactivity, but also to prolong the shelf life and decrease the time necessary for test performance. Many groups are currently working on these problems. It is likely that within the next 5 years, sufficient progress will have been made such that applications of these techniques will be common place for clinical viral diagnostic laboratories.

ACKNOWLEDGMENTS

This research was supported by National Institutes of Health grants CA35048 and HL/AI32471 (S.A.S) and CA34729 and AI20672 (D.H.S.), the Universitywide Task Force on AIDS, and Integrated Genetics.

REFERENCES

1. **Reynolds, D. W., Stagno, S., Stubbs, K. G., Dahle, A. J., Livingston, M. M., Saxon, S. S., and Alford, C. A.,** Inapparent congenital cytomegalovirus infection with elevated cord IgM levels: causal relation with auditory and mental deficiency, *N. Engl. J. Med.,* 290, 291, 1974.
2. **Starr, J. G., Bart, R. D., and Gold, E.,** Inapparent congenital cytomegalovirus infection: clinical and epidemiologic characteristics in early infancy, *N. Engl. J. Med.,* 282, 1075, 1970.
3. **Stagno, S., Reynolds, D. W., Huang, E.-S., Thames, S. D., Smith, R. J., and Alford, C. A.,** Congenital cytomegalovirus infection: occurrence in an immune population, *N. Engl. J. Med.,* 196, 1254, 1977.
4. **Marchevsky, A., Rosen, M. J., Chrystal, G., and Kleinerman, J.,** Pulmonary complications of the acquired immunodeficiency syndrome: a clinicopathologic study of 70 cases, *Hum. Pathol.,* 16, 659, 1985.
5. **Meyers, J. D., Spencer, H. C., Watts, J. C., Gregg, M. B., Steward, J. A., Troupin, R. H., and Thomas, E. D.,** Cytomegalovirus pneumonia after human marrow transplantation, *Ann. Intern. Med.,* 82, 181, 1975.
6. **Quinnan, G. V., Masur, H., Rook, A. H., Armstrong, G., Frederick, W. R., Epstein, J., Manischewitz, J. F., Macher, A. M., Jackson, L., Ames, J., Smith, H. A., Parker, M., Pearson, G. R., Parrillo, J., Mitchell, C., and Straus, S. E.,** Herpesvirus infections in the acquired immune deficiency syndrome, *JAMA,* 252, 72, 1984.
7. **Richardson, W. P., Colvin, R. B., Cheeseman, S. H., Tolkoff-Rubin, N. E., Herrin, J. T., Cosimi, A. B., Collins, A. B., Hirsch, M. S., McClusky, R. T., Russell, P. S., and Rubin, R. H.,** Glomerulopathy associated with cytomegalovirus viremia in renal allografts, *N. Engl. J. Med.,* 305, 56, 1981.
8. **Spector, D. H. and Spector, S. A.,** The oncogenic potential of human cytomegalovirus, in *Progress in Medical Virology,* Melnick, J. L., Ed., S. Karger, Basel, 1984, 45.
9. **Geelen, J. L. M. C., Walig, C., Wertheim, P., and van der Nordaa, J.,** Human cytomegalovirus DNA. I. Molecular weight and infectivity, *J. Virol.,* 26, 813, 1978.
10. **Tamashiro, J. C., Hock, L. J., and Spector, D. H.,** Construction of a cloned library of the EcoRI fragments from the EcoRI fragments from the human cytomegalovirus genome (strain AD169), *J. Virol.,* 42, 547, 1982.
11. **LeFemina, R. L. and Hayward, G. S.,** Structural organization of the DNA molecules from human cytomegalovirus, in *Animal Virus Genetics,* ICN-UCLA Symp. Mol. Cell. Biol., Fields, B. N., Jaenisch, A., and Fox, C. F., Eds., Academic Press, New York, 1980, 39.
12. **Spector, D. H., Hock, L., and Tamashiro, J. C.,** Cleavage maps for human cytomegalovirus DNA strain AD169 for restriction endonucleases EcoRI, Bgl II and Hind III, *J. Virol.,* 42, 552, 1982.
13. **Kilpatrick, B. A., Huang, E.-S., and Pagano, J. S.,** Analysis of cytomegalovirus genomes with restriction endonucleases Hind III and EcoRI., *J. Virol.,* 18, 1095, 1976.
14. **Huang, E.-S., Alford, C. A., Reynolds, D. W., Stagno, S., and Pass, R. F.,** Molecular epidemiology of cytomegalovirus infections in women and their infants, *N. Engl. J. Med.,* 303, 958, 1980.
15. **Spector, S. A.,** Transmission of cytomegalovirus among infants in hospital documented by restriction-endonuclease-digestion analyses, *Lancet,* 1, 379, 1982.
16. **Spector, S. A., Hirata, K. K., and Neuman, T. R.,** Identification of multiple cytomegalovirus strains in homosexual men with acquired immunodeficiency syndrome, *J. Infect. Dis.,* 150, 953, 1984.
17. **Adler, S. P.,** Molecular epidemiology of cytomegalovirus: evidence for viral transmission to parents from children infected at a day care center, *Pediatr. Infect. Dis.,* 5, 315, 1986.
18. **McDonough, S. H. and Spector, D. H.,** Transcription in human fibroblasts permissively infected by human cytomegalovirus strain AD169, *Virology,* 125, 31, 1983.
19. **Stenberg, R. M. and Stinski, M. F.,** Autoregulation of the human cytomegalovirus major immediate early gene, *J. Virol.,* 56, 676, 1985.

20. **Spector, S. A. and Spector, D. H.,** The use of DNA probes in studies of human cytomegalovirus, *Clin. Chem.*, 31, 1514, 1985.
21. **Spector, S. A., Rua, J. A., Spector, D. H., and McMillan, R.,** Detection of human cytomegalovirus in clinical specimens by DNA-DNA hybridization, *J. Infect. Dis.*, 150, 121, 1984.
22. **Chou, S. and Merigan, T.,** Rapid detection and quantification of human cytomegalovirus in urine through DNA hybridization, *N. Engl. J. Med.*, 308, 921, 1983.
23. **Buffone, G. J., Schimbor, C. M., Demmler, G. J., Wilson, D. R., and Darlington, J.,** Detection of cytomegalovirus in urine by nonisotopic DNA hybridization, *J. Infect. Dis.*, 154, 163, 1986.
24. **Spector, S. A.,** Cytomegalovirus: clinical applications of molecular biology, *Medical Virology*, de la Maza, L. M. and Peterson, E. M., Ed., Lawrence Erlbaum, Hillsdale, N.J., 1986, 11.
25. **Schrier, R. D., Nelson, J. A., and Oldstone, M. B. A.,** Detection of human cytomegalovirus in peripheral blood lymphocytes in a natural infection, *Science*, 230, 1048, 1985.
26. **Myerson, D., Hackman, R. C., Nelson, J. A., Ward, D. C., McDougall, J. K.,** Widespread presence of histologically occult cytomegalovirus, *Hum. Pathol.*, 153, 403, 1984.
27. **Myerson, D., Hackman, R. C., and Meyers, J. D.,** Diagnosis of cytomegalovirus pneumonia by in situ hybridization, *J. Infect. Dis.*, 150, 272, 1984.
28. **Spector, D. H., Shaw, S. H., Hock, L. J., Abrams, D., Mitsuyasu, R. T., and Gottlieb, M. S.,** Association of human cytomegalovirus with Kaposi's sarcoma in acquired immunodeficiency syndrome, in *Acquired Immunodeficiency Syndrome*, Alan R. Liss, New York, 1984, 109.
29. **Wiley, C. A., Schrier, R. D., Denaro, F. J., Nelson, J. A., Lampert, P. W., and Oldstone, M. B. A.,** Localization of cytomegalovirus proteins and genome during fulminant central nervous system infection in an AIDS patient, *J. Neuropathol. Exp. Neurol.*, 45, 127, 1986.
30. **Spector, S. A., Neuman, T. R., and Hirata, K. K.,** Rapid determination of molecular relatedness of human cytomegalovirus isolates, *J. Infect. Dis.*, 150, 953, 1984.

Chapter 12

DIAGNOSIS OF HUMAN PAPILLOMAVIRUS AND HUMAN CYTOMEGALOVIRUS BY HYBRIDIZATION HISTOCHEMISTRY

Anna Marie Beckmann and David Myerson

TABLE OF CONTENTS

I. INTRODUCTION

Hybridization histochemistry, or *in situ* hybridization, is a method that has been used as a sensitive and specific means for demonstration of viral nucleic acids in human tissues (Chapter 3, this volume). Until recently, routine use of *in situ* hybridization has generally been restricted to the research laboratory because of the need for radioactive probes and the protracted time period necessary for autoradiographic detection of hybridized probe. The invention of nonradioactive probes using biotinylated or other labeled nucleotides[1] has enabled *in situ* hybridization with colorimetric detection to be performed on routinely fixed, paraffin-embedded tissues.[2] Using this method of hybridization histochemistry, results can be obtained within 24 h.

Many human viral diseases have been studied by *in situ* hybridization, using either radioactive probes[3-6] or nonisotopic probes combined with immunohistochemistry.[7-21] Several of these studies have used tissues embedded in paraffin with resultant good preservation of tissue morphology as well as hybridization signals.[2,8-11,15-18,21] Since hybridization histochemistry can be used on routinely processed and stored tissue specimens, this method is applicable for use in a pathology laboratory and is within the means of any laboratory that employs standard immunocytochemical techniques. This chapter describes our experience and results using biotin-labeled DNA probes in *in situ* hybridization for detection of human papillomavirus (HPV) and human cytomegalovirus (HCMV) in routinely handled tissue specimens.

A. Human Papillomavirus (HPV)

HPVs are small DNA viruses which are epitheliotropic in the human host. HPVs cause a variety of benign proliferative lesions of the cutaneous and mucosal epithelia ranging from common and plantar warts of the skin to condylomata acuminata (genital warts) of the anogenital skin and mucosa. To date, more than 40 different HPV types have been described which can be differentiated on the basis of less than 50% cross-hybridization among the DNA genomes.[22] This means of classifying HPVs has some biologic as well as biochemical basis; in general, specific HPV types are associated with certain histologic appearances and anatomic locations of papillomas. Our interest is in HPV as a sexually transmitted disease (STD) and the role of the virus in the development or progression of anogenital cancers.

Infection of the anogenital region with HPV is an increasingly common occurrence.[23] Until recently, HPV has been a STD of little clinical concern and research effort. This lack of concern was due to the belief that genital warts had a benign course, as well as to the inability of clinicians to accurately recognize and diagnose clinically inapparent infections. Recently, a growing body of evidence has consistently associated the presence of HPVs with carcinomas of the genital tract in both men and women.[24]

HPV types 6 and 11 are usually found in histologically benign lesions of the anogenital region such as condylomas,[25] and in the so-called "flat" or nonpapilliferous warts of the cervix and vagina, also referred to as subclinical papillomavirus infection.[26,27] At this time, the HPVs that are of most clinical concern are types 16, 18, and 31, which have been found in a significant proportion of cancers of the cervix, vulva, penis, and anus.[24] HPV-16 has also been demonstrated in precancerous dysplastic lesions of the cervix, vulva, and perianal skin.[28-32a] Because of this demonstrated association of HPV infection with anogenital cancers, precise diagnosis of HPV infection becomes extremely important when considering the appropriate clinical management of the patient.

B. Human Cytomegalovirus (HCMV)

The use of hybridization analyses to study HCMV infections in the human host is discussed in detail in Chapter 11 of this volume. Our interest is in the use of biotinylated HCMV probes for diagnosis of HCMV infection in bone marrow transplant patients and for investigations of the pathogenesis of HCMV infection in these patients.

Infection with HCMV is a serious problem in such immunocompromised hosts as patients with acquired immunodeficiency syndrome[3,34] and those receiving organ[35] or marrow[36] transplants. HCMV infection is a significant cause of mortality in patients receiving allogeneic marrow grafts; as many as 20% of these patients eventually succumb to HCMV pneumonia.[36] A rapid, selective, and sensitive means for diagnosis of HCMV pneumonia is extremely important in the management of bone marrow transplant patients. New antiviral drugs for the treatment of HCMV infection are now being introduced[37] and are different from those antiviral agents used to treat pneumonias caused by herpes simplex (HSV) and varicella-zoster (VZV) viruses.

Open lung biopsy is often used for diagnosis of marrow transplant patients with acute pulmonary disease. Traditionally, isolation of the virus in cell culture from tissue homogenates has been the most sensitive means for diagnosis of HCMV infection. However, HCMV is a very labile virus that can be rendered noninfectious by improper handling and HCMV-specific cytopathic effects may require up to 6 weeks incubation time to be apparent in infected cell cultures.[38] Infection with HCMV is most usually diagnosed by routine histologic examination of hematoxylin and eosin (H & E) stained sections for the presence of distinctive nuclear and cytoplasmic inclusions, and also by immunochemistry for viral antigens on frozen sections.[39,40] Hybridization histochemistry has demonstrated the presence of HCMV in cells that do not appear infected by standard histologic criteria.[16] Therefore, we have been evaluating the use of hybridization histochemistry for the rapid diagnosis of HCMV infection.

II. HYBRIDIZATION METHODS

A. Preparation of Hybridization Probes

Molecularly cloned viral DNAs purified by gradient centrifugation can be labeled by nick translation with the thymidine analog biotinyl-E-aminocapronyl-deoxyuridine triphosphate[1] (bio-11-dUTP, Bethesda Research Laboratories, Gaithersburg, Md., Enzo Bicohem, Inc., New York City). Bio-11-deoxyruridine base pairs like thymidine, and the biotin moiety, are used for immunohistochemical detection of hybridized DNA sequences. The nick-translation reaction requires a two- to fourfold molar excess of unlabeled triphosphates and bio-11-dUTP to the nucleotide bases in the DNA to be labeled.[2] DNase I (CalBiochem-Behring, San Diego, Calif.) is used to introduce nicks into the DNA, and synthesis of new DNA is accomplished with *Escherichia coli* DNA polymerase I (BRL). A small amount of ^{3}H-dATP (5 μCi) is added to the reaction to monitor the incorporation of bio-11-dUTP molecules into the newly synthesized DNA strands. Biotin-labeled DNA is separated from unincorporated nucleotides by centrifugation through a Sephadex G-50 (Pharmacia, Piscataway, N.J.) column. The commercial availability of biotinylated probes is discussed in Section C.

1. HPV Hybridization Probes

The entire DNA genomes of plasmids containing HPV-1a,[41] -6,[42] -11,[43] -16,[31] -18,[44] and -31[45] were biotinylated for use in these studies. The plasmids were a gift from Drs. L. Gissmann and H. zur Hausen (HPV-1, -6, -11, -16, and 18), and from Dr. A. Lorincz (HPV-31). These plasmids contain the entire HPV genome inserted in pBR322. The pBR322 sequences are biotinylated in addition to the HPV DNA insert. However, any potential cross-hybridization of pBR sequences with bacterial DNA has not been a problem because bacteria are readily recognizable by microscopic inspection and because the HPV-specific hybridization product is always intranuclear.

2. HCMV Hybridization Probes

A combination of seven HCMV cosmid clones were biotinylated for use in these studies. These clones represent the entire HCMV genome[46] and were a gift from Dr. B. Fleckenstein.

Each cosmid was separately biotinylated and then combined to form the hybridization probe. In some studies, a commercially available biotinylated probe was used (Enzo Biochem, Inc.). This probe represents 18% of the HCMV genome.

B. *In Situ* Hybridization

1. Preparation of Tissue Specimens

These studies used routinely obtained tissues that had been fixed in neutral buffered formalin and embedded in paraffin. Tissue sections, approximately 3- to 5-µm thick, were cut onto acid-cleaned poly-D-lysine coated (mol wt 50,000; Sigma Chemical Company, St. Louis, Mo.) microscope slides.[47] Sections were dried for 24 to 48 h at 65°C.

Tissue sections were deparaffinized in xylene and then rinsed in absolute ethanol. Endogenous peroxidase activity was blocked by incubation in absolute methanol containing 1% hydrogen peroxide for 30 min at room temperature. Sections were hydrated through graded alcohols, treated with Pronase B (CalBiochem-Behring) (5 mg/ml, 5 min, room temperature) and postfixed with paraformaldehyde (4% in phosphate-buffered saline, 5 min, room temperature).[2] Tissue sections were then dehydrated through graded alcohols and air dried. Omission of the final dehydration steps and application of hybridization mixtures to "wet" tissue sections can improve the sensitivity of the *in situ* procedure for detection of HCMV[47-49] and HPV.[49a]

2. Hybridization

Hybridization mixtures consisted of (final concentration) 2 × SSC (SSC = 0.3 *M* NaCl, 0.03 *M* NaCitrate pH 7.0), 6.7% dextran sulfate (weight/volume), 400 µg/ml sheared herring sperm DNA, 2 µg/ml biotin-labeled DNA, and either 10% or 50% (volume/volume) deionized formamide. For HPV hybridizations, formamide concentrations of 10% were used for nonstringent hybridization approximating an equivalent of Tm – 42°C as determined from hybridization conditions described for detection of homology among heterologous polyomavirus genomes.[50] For stringent hybridizations, used to determine the specific HPV type present in a lesion and for all HCMV hybridizations, a final concentration of 50% formamide was used to approximate T_m – 17°C.

3. Detection of Hybridized Biotin-Labeled Probe

Following hybridization, tissue sections were washed in 2 × SSC (40 min, room temperature) and DNA:DNA hybrids were visualized by sequential immunocytochemical reactions. Nonspecific binding of antibody molecules was blocked by treatment of sections with 10% normal rabbit or goat serum/4% bovine serum albumin at 37°C for 10 min.[2] The choice of the blocking serum was dependent on the detection system chosen (see below). The biotinylated probe was detected by sequential treatment with avidin DH (Vector Laboratories, Burlingame, Calif.), goat anti-avidin (Vector Laboratories), biotinylated rabbit anti-goat, and avidin-biotinylated peroxidase complex (ABC) (Vector Laboratories). The two antibody steps may be replaced by a single step using biotinylated goat antiavidin (Vector Laboratories). Peroxidase moieties were visualized by reaction with nickel-complexed diaminobenzidine (DAB) containing 0.01% hydrogen peroxide[51] for 5 min at room temperature. All incubations with antibodies were performed at 37°C. Alternatively, biotinylated DNA probes were detected by sequential reactions with goat antibiotin (Vector Laboratories), biotinylated rabbit anti-goat, and ABC, with the same final step. Either detection scheme gives comparable results, although slightly more nonspecific background is noted with the use of avidin DH in the initial detection scheme due to slight nonspecific binding of avidin in the primary step. The tissue sections were lightly counterstained with Gill's hematoxylin or H & E.

All of our procedures are further discussed in detail by Myerson.[52] Details of similar procedures are also available.[2,49]

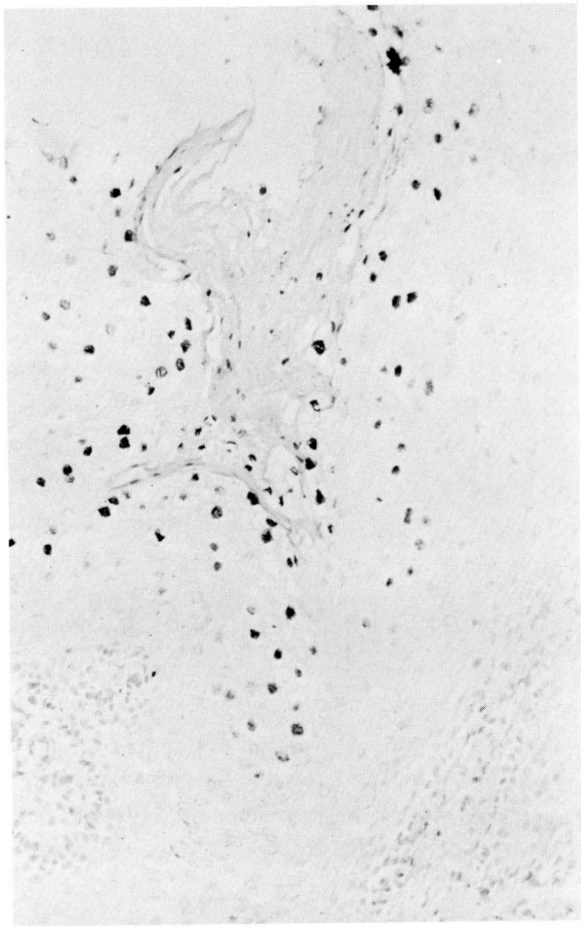

FIGURE 1. Vulvar exophytic condyloma. Hybridization histochemistry was performed with an HPV-6 probe. The hybridization is localized to the nuclei of the large number of epithelial cells that are infected.

C. Availability of Biotin-Labeled DNA Probes

The plasmid DNAs used as probes in these studies were generated for research purposes and were kindly provided by Drs. L. Gissmann and H. zur Hausen (HPV), and Dr. B. Fleckenstein (HCMV). There are some commercially available biotinylated viral DNA probes (HSV-1, -2, CMV, EBV, HBV, JCV, BKV, adenovirus type 5, HPV-1, and SV-40, — Enzo Biochem).

III. RESULTS OF HYBRIDIZATION STUDIES

A. Interpretation of Hybridized Tissue Sections

As with any interpretation of stained histologic sections, criteria for making a diagnosis must be developed. With HPV, a clearly hybridized nucleus is required for considering a section as positive. In the case of HCMV, either a hybridizing nucleus or multiple hybridizing cytoplasmic inclusions are necessary for a positive diagnosis. Representative examples of positive HPV and HCMV tissue sections are seen in Figures 1 and 2.

The methods described for *in situ* hybridization are usually free of nonspecific (background) staining. However, several problems can arise. First, some cells, like macrophages, contain cytoplasmic pigments. These pigments can usually be differentiated from the nickel-DAB by

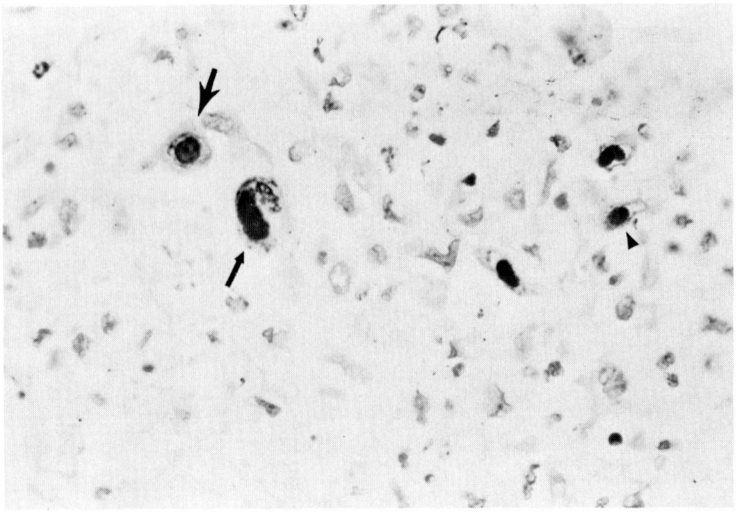

FIGURE 2. HCMV interstitial pneumonia. This lung contains cytomegalic cells in which hybridization with a HCMV probe is seen in both the nucleus (large arrow) and/or cytoplasm (small arrow). Small "occult infected" cells are also seen that hybridize, but are not morphologically cytomegalic (arrowhead).

high-power microscopy. However, nuclear hybridization is the preferred criterion for HPV and HCMV. Second, and more problematic, is nonspecific binding of the biotinylated probe, avidin, or antibodies to various tissue components. An important control to rule out nonspecific probe binding is the use of an unrelated biotinylated probe. Some tissue components that nonspecifically bind biotin-labeled probes are argentaffin cells in the gut crypts, neutral mucin, and the corpora amylacia. Endogenous biotin is not a problem in formalin-fixed, paraffin-embedded tissue and could be detected using controls without biotinylated probes.

B. Sensitivity

The question of the limits of sensitivity of hybridization histochemistry with biotin-labeled probes can be approached in two ways: first, by direct determination of the number of viral DNA molecules detectable by *in situ* hybridization; and second, by comparison of this method to a "gold-standard" means of diagnosis of a viral infectious agent. These two approaches are discussed below as separately evaluated for HPV and HCMV.

1. HPV

Direct determination of the sensitivity of *in situ* hybridization for detection of HPV is difficult to quantify, as the virus does not productively infect either human or animal cells in vitro. In addition, when evaluating tissue sections, another problem is that the distribution of virus DNA throughout an HPV-induced lesion is variable (see Section D). Southern hybridization of a genital condyloma, for example, allows determination of an average virus genome copy number per cell genome equivalent (e.g., 5 copies in 10 cell equivalents) but does not distinguish between 1 cell in 10 containing 5 copies of the HPV genome or 1 cell in 1000 containing 500 copies. We have attempted an estimation of the minimum number of HPV DNA molecules detectable by *in situ* hybridization in formalin-fixed, paraffin-embedded tissues in the following experiments.

A surgically removed, large vulvar condyloma was immediately divided in half; one half was snap frozen in liquid N_2 for Southern hybridization analysis, and the other half was fixed in formalin and embedded in paraffin for hybridization histochemistry. Southern hybridization found the lesion to contain 13 copes of the HPV-6 genome per cell genome equivalent. *In situ* hybridization with a HPV-6 probe in stringent conditions was performed on the paraffin-embedded tissue sections. Computer-generated microscopic fields were examined to insure randomness, and the resulting count showed that 1.7% of the cells hybridized. Assuming the most uneven or unfavorable distribution of HPV DNA (all of the DNA in the cells with a detectable hybridization product and no DNA in cells that did not hybridize), the sensitivity was 13/0.017 or 760 genome copies per cell.[52] When a correction for staining intensity was accounted for, the sensitivity was found to be about 370 copies per cell.[52] Crum et al.[11] have reported a similar sensitivity in cultured cells infected in vitro with SV-40, fixed in Carnoy's, and analyzed by *in situ* hybridization using alkaline phosphatase as the enzyme system for detection of hybridization. These investigators determined the sensitivity of *in situ* hybridization to be 800 copies of the SV-40 genome per cell genome equivalent.

2. HCMV

The clinical sensitivity of hybridization histochemistry for detection of HCMV DNA was determined by comparison of results obtained by this method with those by viral culture. Diagnosis of HCMV by each means was assayed in a series of 29 specimens obtained by open lung biopsy.[15] *In situ* hybridization was about as sensitive as immunofluorescent staining of frozen sections using monoclonal antibody 6C5 (Genetic Systems, Seattle, Wash.). The one case that was positive by immunofluorescence, but negative by culture, was later shown, by hybridization histochemistry, to contain two positive cells in a serial section. This confirms the suggestion that the sensitivity of *in situ* hybridization is readily affected by the distribution of infected cells throughout the entire specimen, as well as the capability of the method to detect infected cells in a specific tissue section.[15]

In collaboration with R. C. Hackman, C. Gleaves, G. B. McDonald, and J. D. Meyers, sequential upper gastrointestinal endoscopy biopsy studies were examined using several different antibodies, as well as by culture, centrifugation culture, and hybridization histochemistry.[53] Centrifugation culture is a recently described method that involves infecting a monolayer by centrifugation of tissue homogenates onto susceptible cells, and subsequent staining for HCMV immediate early antigen.[53] Centrifugation culture proved the most sensitive, but immunohistochemistry and hybridization histochemistry provided positive results in some cases that were negative by the culture techniques. The largest variation in results was accounted for by the differing pieces of tissue analyzed. These findings with gut tissue support the observations of an uneven HCMV distribution in the lung.[15] These results suggest that for the best diagnostic sensitivity, multiple sections, which encompass the largest possible tissue width, should be examined. This can be accomplished by putting up to 20 closely spaced sections of small biopsies on each slide.

C. Specificity

The specificity of *in situ* hybridization for HCMV and HPV, to the exclusion of other viruses, has been demonstrated by us[8,15,16] and by others,[11,48] by the use of the following control experiments. The specificity of biotinylated HCMV probes was analyzed by hybridizing these probes to sections from lungs infected with HSV, VZV, and adenovirus. No staining was present in these control sections. Nonspecific reactivity in a HCMV-positive section was assessed by hybridization with biotinylated probes of unrelated viruses such as HSV and VZV. Some sequence homology exists between the HCMV and the human genome that is detectable with Southern hybridization.[54] This cross-reactivity of the two genomes is not seen with *in situ* hybridization, since tissues not infected with HCMV do not stain with a biotinylated HCMV probe.

The specificity of hybridization histochemistry using biotinylated HPV DNA probes was analyzed in a similar manner to that described for HCMV. Control tissues from a HSV-2 induced lesion were hybridized with a HPV probe, and condylomas containing HPV-6 were hybridized with an HSV-2 probe.[55] No positive staining was seen in any of these tissues.

Table 1
DETERMINATION OF SPECIFIC HPV GENOTYPE BY *IN SITU* HYBRIDIZATION

	HPV probe (type)									
	1		6		11		16		18	
	NS[a]	S	NS	S	NS	S	NS	S	NS	S
Plantar wart	4+[b]	4+	–[c]	–	–	ND[d]	–	ND	–	ND
Vulvar condyloma	–	ND	4+	4+	2+	–	1+	–	1+	–
Laryngeal papilloma	–	ND	2+	–	3+	3+	ND	ND	ND	ND
Anal CIS[e]	ND	ND	1+	–	1+	–	4+	4+	2+	—
Anal verrucous carcinoma[e]	ND	ND	1+	–	ND	ND	2+	–	3+	3+

[a] NS, nonstringent; S, stringent: hybridization conditions.
[b] The degree of nuclear staining was scored from 1+ (least intense) to 4+ (most intense).
[c] –, absence of detectable DNA.
[d] ND, not done.
[e] Determination of HPV genotype was confirmed by Southern hybridization analysis.

We extended our studies of the specificity of *in situ* hybridization with biotinylated HPV DNA probes to determined if the particular HPV genotype present in a tissue specimen could be identified.[8] By Southern hybridization analysis, distinct HPV genotypes exhibit little cross-hybridization under stringent conditions (T_m – 10°C), but under conditions of reduced stringency (T_m – 42°C), cross-hybridization among heterologous genotypes is detectable.[56] We have analyzed a series of HPV-induced lesions using both homologous and heterologous probes in nonstringent and stringent conditions to evaluate the sensitivty and specificity of hybridization histochemistry.[8] A representative example of these experiments is shown in Table 1.

Anogenital condylomas and malignant lesions which contained HPV-6 or -16 DNA, as determined by Southern hybridization analyses, were tested. Also examined were plantar and laryngeal warts which should contain HPV-1 and -6 or -11 DNA, respectively. In nonstringent hybridization conditions, cross-reactivity among heterologous but related HPV-6, -11, -16, and -18 DNAs was detected. Unlike Southern hybridization, detectable hybridization between HPV-1 and -6 or -11 was not found. In more stringent conditions (T_m – 17°C), only the homologous virus sequences hybridized. The demonstration of the presence of a specific-HPV genotype by stringent *in situ* hybridization agrees with the results obtained with the use of Southern filter hybridizations that have established a correlation between the infecting HPV type and the anatomic location and pathologic characterization of the lesion.

D. Studies of the Pathogenesis of HPV and HCMV Infections

We are studying the pathobiology of HPV and HCMV infection in humans using biotinylated DNA probes for hybridization histochemistry to localize virus genomes in tissue specimens.[8,15,16,57,58] These investigations have used routinely processed, fixed, and embedded tissues obtained from pathology files at the Fred Hutchinson Cancer Research Center, Seattle, Wash. and the University of Washington Hospital, Seattle, Wash. and other hospitals in western Washington state. Highlights of our results are summarized below.

1. HPV

We have used *in situ* hybridization to examine over 500 human tissue specimens, including condylomas of the vulva, perianal skin, and penis, and premalignant and malignant lesions of the cervix, vulva, and anus. These tissues were obtained through collaborative studies with Drs.

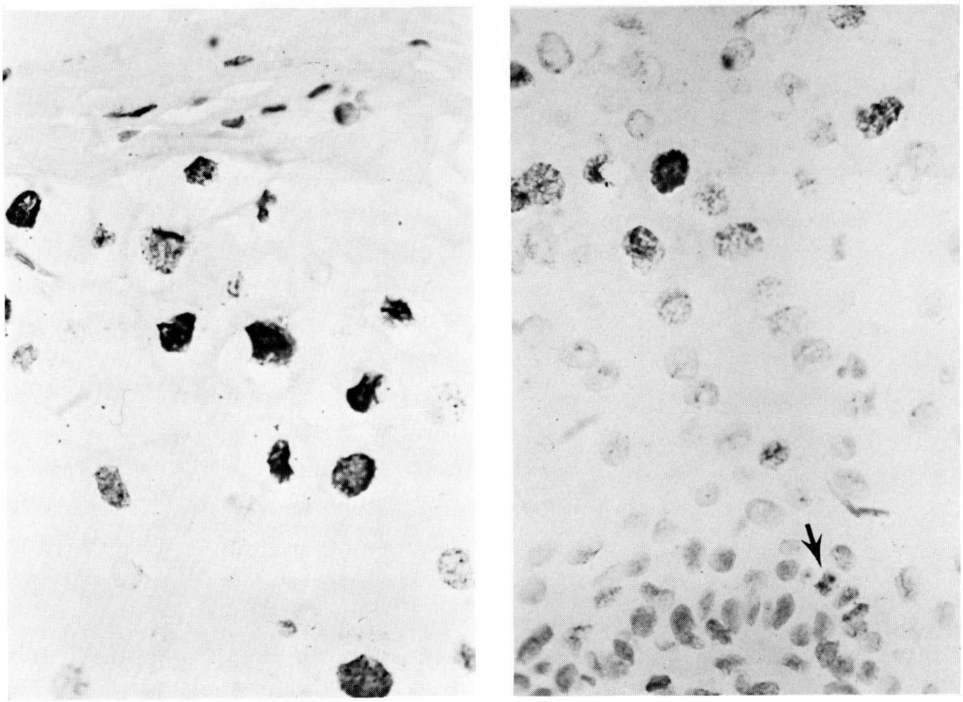

A B

FIGURE 3. Perianal exophytic condyloma hybridized with an HPV-6 probe. (A) Large amounts of viral DNA are seen in koilocytotic and parakeratotic cells at the surface of the lesion. (B) A slight hybridization signal is seen in parabasal cells (arrow).

N. Kiviat and K. Holmes (Harborview Medical Center STD Clinic, Seattle, Wash.), and with Dr. J. Daling (Fred Hutchinson Cancer Research Center, Seattle, Wash.).

In benign lesions of the anogenital tract, such as cervical, vulvar, penile, and perianal condylomas, *in situ* hybridization has shown that the intensity of the hybridization product, and correspondingly, the amount of viral DNA present, increased from cells deep in the epithelium to the more mature keratinocytes, and to the obvious koilocytotic and parakeratotic cells at the surface of the lesion (Figure 3A). Earlier observations by Orth and colleagues,[59,60] using radioactive probes for *in situ* analyses of HPV-induced skin lesions, showed a similar relationship in those lesions between HPV DNA replication and expression and the differentiation of epithelial cells.

A wart presumably arises by excessive replication and aberrant differentiation of an infected basal cell. We have also found small, but detectable, amounts of HPV DNA in parabasal cells (Figure 3B) supporting this concept. The intensity of nuclear straining was consistent among cells found in the same layer of the epithelium. Production of virus capsid antigens is sparse in genital warts, and many lesions have been examined that do not contain detectable viral late proteins.[61,62] These results suggest that the development of a genital wart may be the result of the presence and expression of nonstructural viral proteins rather than of a productive or lytic infection.

Stringent *in situ* hybridizations demonstrated that the HPV-6 genotype was the most prevalent virus in patients with histologically benign lesions. However, in five patients with anal condylomas who also had coexisting squamous cell carcinomas of the anal skin, two were found to be infected with HPV-16.

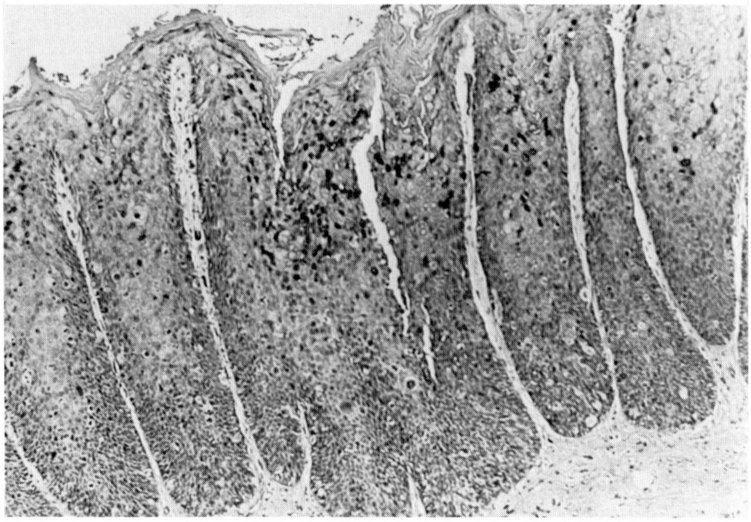

FIGURE 4. Carcinoma *in situ* of the vulva hybridized with an HPV-16 probe. Both dysplastic and hybridized cells are seen throughout the width of the epithelium. The intensity of hybridization is variable in the same levels of the epithelium.

In premalignant and malignant anogenital tumors, the pattern of HPV DNA distribution and the virus genotypes found in the lesions was different than that for benign condylomas. The tumors we have examined include dysplastic lesions, carcinoma *in situ* (CIS), and invasive squamous cell carcinomas of the cervix, vulva, and anus. We have also studied adenocarcinomas and cloacogenic carcinomas of the anus. In those tissue sections that contained HPV DNA, an uneven pattern of distribution of viral DNA through the same layers of the epithelium was seen. Most often, the majority of HPV-infected cells were found at the surface of the tissue section in cells exhibiting some morphologic changes indicating maturation. In many cases, viral DNA was found in regions of markedly abnormal epithelium, where atypical mitotic figures were seen, along with cells showing hyperchromatic nuclei with nuclear pleomorphism and increased nuclear/cytoplasmic ratios.[28] Figures 4 and 5 show representative examples of premalignant and malignant lesions of the vulva and anus, respectively. On the tissue level, HPV DNA was localized only in cells of squamous origin. No HPV DNA was found in the nonsquamous tissue of anal adenocarcinomas or cloacogenic carcinomas.

The identity of HPV genotypes present in these premalignant and malignant lesions also differed from that for benign condylomas. HPV-16 was the most common virus found in these tissue specimens. However, several anal cancers were found to contain HPV-6 DNA (all were from patients with coexisting condylomas), and one anal verrucous carcinoma contained HPV-18 DNA.[94]

2. HCMV

To investigate the pathogenesis of HCMV, tissues from four bone marrow transplant recipients were studied by hybridization histochemistry (Table 2). Complete sets of autopsy slides from each patient were studied. Two patients had died from disseminated HCMV, one died from interstitial pneumonia with histologic and culture evidence of HCMV, and one was seropositive for HCMV, but died from unrelated causes. In the two patients with disseminated HCMV (cases 1 and 2, Table 2), the virus was detected and localized in many normal-appearing cells, in addition to the obvious cytomegalic cells (see Figure 2). The occult infection was seen in endothelial cells, cardiac myocytes, spleen, and lymph node reticular cells, fibroblasts, and other cells. Mesenchymal cells, such as those readily infectable in tissue culture, were, in fact, infected by HCMV in vivo. Three patterns of infections were seen — necrotizing foci, diffuse

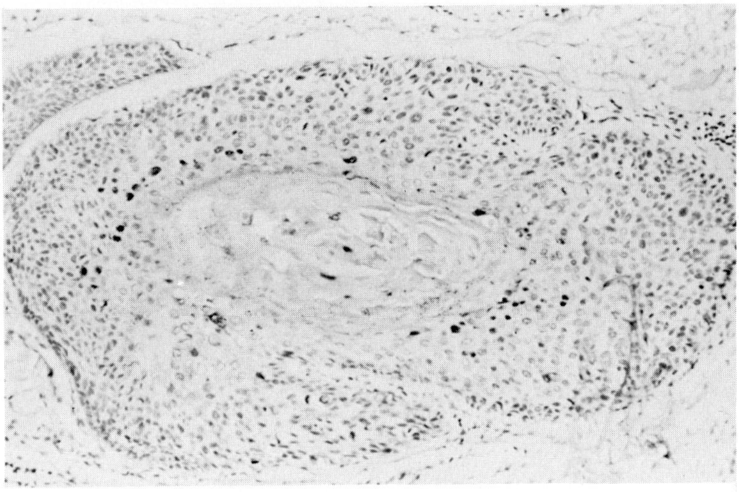

FIGURE 5. Dysplastic region associated with anal carcinoma hybridized with an HPV-16 probe. This region shows a markedly dysplastic epithelium with maturation and associated hybridization in the most superficial portions of the section. This is a typical example of what is found in anal cancers where dysplastic regions that contain HPV DNA are seen in close association with regions of frankly invasive cancer where no viral DNA is detectable by hybridization histochemistry.

foci, and scattered cells.[16,63-65] The findings support the hypothesis that necrotizing foci are initiated by hematogenous dissemination of HCMV-infected cells, producing a "miliary" focus. These results suggest that diffuse focus formation occurs by local viral spread via endothelial cell infection. HCMV infection of endothelial cells was confirmed by using antibodies to Factor VIII in adjacent sections.[66]

These findings indicate that, in disseminated disease, occult infection with HCMV accounts for a major portion of the HCMV-infected cells. The documentation of occult HCMV infection in the central nervous system[67] supports this interpretation. The HCMV seen in scattered cells in tissues such as the skin (arm), kidney, pituitary, and heart, probably represents persistent virus since viral RNA is not detected by the methods presented here. A persistent virus infection is more likely that a latent infection to contain detectable quantities of DNA. In the patient with HCMV pneumonia without disseminated HCMV (case 3), HCMV was present in low quantities in the lung (an average of 3.8 infected cells per section with a range of 0 to 14 cells in 10 sections). One pancreatic exocrine cell was also infected with HCMV (one section). The routine histology and culture performed on these tissues were negative for HCMV. The cytomegalic or near-cytomegalic appearance of these cells suggest the HCMV demonstrated is a persistent rather than a latent virus. The patient who was seropositive for HCMV had seroconverted post transplant (case 4), was very lightly infected in the lung (an average of one cell per section with a range of zero to three cells in seven sections) and in the pancreas (two cells in one section) (Figure 6).

IV. SIGNIFICANCE OF RESULTS

A. Comparison of *In Situ* Hybridization with Traditional Diagnostic Methods

Traditional laboratory methods for diagnosis of virus infections include isolation of the agent in cell culture, observation of specific virus-induced cytopathic changes or viral antigens or particles in infected tissues, and demonstration of circulating antiviral antibodies. Recently, the development of monoclonal antibodies for immunocytochemistry and DNA probes for hybridization analyses have been added to these traditional means for diagnosis. The use of the latter methodology for diagnosis of HPV and HCMV is discussed below.

Table 2
TISSUE DISTRIBUTION OF CMV IN FOUR CASES: IDENTIFICATION IN AUTOPSY SECTIONS

Case no.	Clinical	Location of CMV by hybridization histochemistry
1[a] (UPN 1459)	Disseminated CMV	Heart, lung, spleen, liver, pancreas, adrenal, kidney, uterus, fallopian tube, ovary, esophagus, stomach, jejunum, cecum, parotid, lymph node, breast, lip, skin
2[a] (UPN 1692)	Disseminated CMV	Heart, lung, spleen, liver, pancreas, adrenal, kidney, fallopian tube, ovary, esophagus, stomach, duodenum, cecum, colon, lymph node, pituitary
3 (UPN 888)	CMV interstitial pneumonia	Lung, pancreas
4 (UPN 1065)	CMV seropositive[b]	Lung, pancreas

[a] Previously reported.[16]
[b] Died from causes presumably unrelated to CMV.

1. HPV

Since HPV cannot, as yet, be propagated in cell culture, virus isolation for diagnosis of HPV infection is impossible. Also, the present lack of a system for production of virus antigens makes serologic tests not possible at this time. Systems for the production of viral late antigens are, however, being developed.[68] Traditionally, histology has been considered the gold standard for diagnosis of HPV infection. The criteria for making this diagnosis are the presence in tissue sections of koilocytosis, dyskeratosis, basal cell hyperplasia, and mild nuclear atypia.[69] However, there are many HPV-induced cervical lesions in which such changes are either less pronounced or inapparent, and some of these lesions contain significant nuclear atypia making a definitive histologic diagnosis more difficult. Also, Reid and colleagues[70] have shown the existence of a continuous morphologic spectrum between benign HPV infection and dysplastic lesions (cervical intraepithelial neoplasia; CIN). Several groups have demonstrated that high-grade lesions of the cervix may contain HPV DNA without histologic evidence of infection[24] and that HPV DNA can be demonstrated in normal epithelium adjacent to anogenital condylomata.[71] These findings further confound and limit the use of histology for diagnosis of HPV infection.

Our group and others have investigated the use of *in situ* hybridization as an adjunct method to histologic diagnosis. We have found that in typical anogenital condylomas, HPV DNA can always be visualized by hybridization histochemistry. In dysplastic lesions of the cervix classified as CIN I, II, or III, we have found that about half contain detectable HPV DNA.[93] Grussendorf-Conen[72] has also examined typical condylomas using tritium-labeled DNA probes and found all of those lesions contained HPV DNA. Crum et al.[11] used biotinylated DNA probes to demonstrate HPV DNA in 10 of 16 CIN lesions that also showed koilocytosis.

There are several possible explanations for the failure of hybridization histochemistry to detect HPV DNA sequences in cervical lesions that contain koilocytotic cells. First, koilocytes may not always be productively infected with HPV, although this seems unlikely. Second, these lesions might contain an HPV genotype unrelated to the currently available probes and, consequently, these sequences would not cross-hybridize. Lorincz et al.[73] have shown that about 60% of cervical lesions with abnormal cytologies contain an HPV type other than 6, 11, 16, 18, or 31. Third, HPV DNA replication in cervical dysplasias, in contrast to typical condylomas,

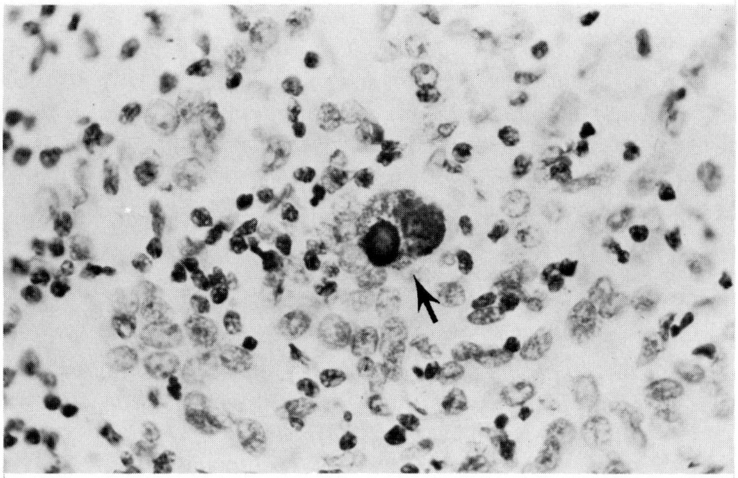

FIGURE 6. CMV in the pancreas. This pancreas section (case 4) contains a large hybridized cell most likely constituting early cytomegally. The intense hybridization and the probable cytomegally suggest that the virus has undergone considerable DNA replication in this cell.

may be restricted to low levels and, therefore, not detectable due to the limits of sensitivity of *in situ* hybridization with biotinylated DNA probes.

Cytologic examination of cervical smears (Papanicalaou smears) has also been used to diagnose HPV infection. In Pap smears, changes of koilocytosis, dyskeratosis, and atypical parabasal and multinucleated cells have all been described in association with HPV infection[26,74] with support from histologic, electron microscopic and immunocytochemical studies.[61,62,75,76] Cytology alone is not sensitive for the diagnosis of HPV infection, as studies comparing cytology with molecular hybridization for HPV DNA sequences have demonstrated the presence of HPV DNA in 2 to 29% of cervical specimens with no clinical or cytologic evidence of HPV infection.[73,77-80]

Gupta et al.[4] have successfully used ^{35}S-labeled DNA probes to detect HPV sequences in Pap smears showing dysplasia. We have also used biotinylated HPV DNA probes to examine Pap smears,[93] and have localized HPV DNA in cytologically abnormal cells.

Several techniques employing molecular hybridization for diagnosis of HPV infection have been used in a research laboratory setting. These include Southern filter or dot-blot hybridizations on DNA extracted from tissues,[24,77,78,80] filter *in situ* hybridization on exfoliated epithelial cells collected directly onto the filters,[79] *in situ* hybridization with radioactive probes on Pap smears[4] and tissue sections,[4,72,81,82] and hybridization histochemistry with biotinylated probes tissue sections.[8,11,57] Each of these methods has distinct advantages and disadvantages. Southern blot hybridization is the most sensitive and specific technique allowing detection of 0.1 copy of the HPV genome/cell genome equivalent and determination of subtypes within the genotype classification (i.e., distinguishing subtypes of HPV-6 and -11).[25] However, this method is time consuming, labor intensive, and requires radioactive probes. Dot-blot or filter *in situ* hybridizations are also sensitive and specific, somewhat less so than Southern hybridizations, have the same disadvantages, and additionally require a rigorously purified probe to avoid specific or nonspecific cross-reactivity. *In situ* hybridization allows localization of the virus genome to specific cell populations, use of routinely processed tissue specimens, and retrospective analysis of stored tissue blocks. Also, methods using biotinylated probes are more amenable to routine laboratory use than those methods employing radioactive probes. The use of radioactive probes allows detection of HPV in tissues determined, by Southern hybridizations, to have between 1 copy per cell with ^{3}H[81] and 50 to 100 copies per cell with ^{35}S.[82]

Several research groups have shown that filter hybridization methods are a sensitive means for diagnosis of HPV infection.[73,77-80] These studies have demonstrated that presence of HPV-6, -16, or -31 in 5 to 30% of cervical smears from women with no colposcopic or cytologic indication of any abnormalities. In patients with cervical dysplasias, the prevalence of HPV-16, -18, -31, or, as yet, unknown HPV types approaches 90% in some studies. These results indicate that molecular hybridization methods will prove to be the most sensitive for diagnosis of HPV infection when these techniques are successfully adapted for use in the clinical laboratory.

2. HCMV

The existence of a latent or persistent state for HCMV has been suspected from studies of renal transplant patients where the donor is the major source of primary HCMV infection in the recipient,[83-86] but neither histology or culture has demonstrated the presence of HCMV in nonviruric kidneys.[35,87] Also, HCMV can be transmitted by blood transfusions[88] when the virus is undetectable by culture. HCMV has been directly shown to exist in a latent state in leukocytes by *in situ* hybridization to the immediate early mRNA.[89] Immunohistochemistry has detected the protein associated with this mRNA[89] as well as other HCMV proteins.[90] Our studies have demonstrated widespread occult HCMV infection which probably represents persistent rather than latent virus infections.[16]

Isolation of HCMV in cell cultures is considered the gold standard for HCMV diagnosis. This method, along with centrifugation culture, can theoretically detect a single infectious virus particle. However, the use of culture often requires protracted time periods for detection of viral cytopathic effects, and the virus is fragile in transit. Routine histologic examination of infected tissues is easily and rapidly performed on sections and cytologic preparations, but histology only provides a definitive diagnosis in 75% of our lung biopsy cases confirmed by virus isolation.[39] Also, we have shown the presence of HCMV nucleic acids in many normal appearing cells in patients with disseminated HCMV disease.

The use of molecular hybridization techniques for detection of HCMV infection is discussed in Chapter 11 of this volume. In our experience, the use of biotinylated HCMV DNA probes on lung biopsies has given a sensitivity of greater than 90% as compared to culture. Similar findings using a ^{35}S-labeled HCMV probe have been reported.[91]

B. Applicability of Hybridization Histochemistry for Routine Diagnosis of HPV and HCMV Infections

Hybridization histochemistry with biotinylated probes has the potential to become a powerful tool for diagnosis for HPV and HCMV infections by the pathologist. This technique is applicable to routinely processed, formalin-fixed and paraffin-embedded tissues and allows excellent preservation of tissue architecture and cellular detail. Tissues that are obtained under nonsterile conditions, are contaminated, or are toxic to cell cultures can be utilized. The method does not require any special skills or facilities beyond those available in laboratories performing standard histochemical tests. The use of biotinylated probes (which are indefinitely stable) allows long-term storage of the probes, and permits reproducibility, as the same probe preparation can be used for an extended time period. The methodology used by us and others in a research laboratory setting is somewhat time consuming and labor intensive. However, methods have been described which permit rapid and simplified means for *in situ* hybridization in which tissues can be analyzed within 8 h from cutting of the sections.[47,48,92]

The need for routine molecular diagnostic methods is clear. For example, in the future, identification of a specific HPV in cytologically normal smears may significantly impact the follow-up care regimen for the patient. Rapid diagnosis of HCMV infection is important as the recent availability of drugs for treatment of HCMV infection makes therapeutic intervention possible. Hybridization histochemistry has the capability of becoming a valuable tool for molecular diagnosis of virus infections by the pathologist.

ACKNOWLEDGMENTS

We thank Dr. James McDougall for advice and support; Drs. Robert Hackman, Janet Daling, and Nancy Kiviat for communciation of unpublished results; David Sarju, Beth Miller, Mary Browne, and Charles Mahan for excellent technical assistance; Drs. Lutz Gissmann, Harald zur Hausen, Bernard Fleckenstein, and Denise Galloway for plasmid DNAs; and Regina Warmoth and Colleen Tuttle for preparation of this manuscript.

This work was supported in part by grants CA-15704, CA-18029, CA-29350, CA-32010, and CA-35881 from the National Cancer Institute and HL-36444 from the National Heart, Lung and Blood Institute, U.S. Department of Health and Human Services.

REFERENCES

1. **Langer, P. R., Waldrop, A. A., and Ward, D. C.,** Enzymatic synthesis of biotin-labeled polynucleotides: novel nucleic acid affinity probes, *Proc. Natl. Acad. Sci. U.S.A.*, 78, 6633, 1981.
2. **Brigati, D. J., Myerson, D., Leary, J. J., Spalholz, B., Travis, S. Z., Fong, C. K. Y., Hsiung, G. D., and Ward, D. C.,** Detection of viral genomes in cultured cells and paraffin-embedded tissue sections using biotin-labeled hybridization probes, *Virology*, 126, 32, 1983.
3. **Galloway, D. A., Fenoglio, C., Shevchuk, M., and McDougall, J. K.,** Detection of herpes simplex RNA in human sensory ganglia, *Virology*, 95, 205, 1979.
4. **Gupta, J., Gendelman, H. E., Naghashfar, Z., Gupta, P., Rosenshein, N., Sawada, E., Woodruff, J. D., and Shah, K.,** Specific identification of human papillomavirus type in cervical smears and paraffin sections by *in situ* hybridization with radioactive probes: a preliminary communication, *Int. J. Gynecol. Pathol.*, 4, 211, 1985.
5. **Haase, A. T.,** Analysis of viral infections by *in situ* hybridization, *J. Histochem. Cytochem.*, 34, 27, 1986.
6. **Harper, M. E., Marselle, L. M., Gallo, R. C., and Wong-Staal, F.,** Detection of lymphocytes expressing human T-lymphotropic virus type III in lymph nodes and peripheral blood from infected individuals by *in situ* hybridization, *Proc. Natl. Acad. Sci. U.S.A.*, 83, 772, 1986.
7. **Aksamit, A. J., Severe, J. L, and Major, E. O.,** Progressive multifocal leukoencephalopathy: JC virus detection by *in situ* hybridization compared with immunohistochemistry, *Neurology*, 36, 499, 1986.
8. **Beckmann, A. M., Myerson, D., Daling, J. R., Kiviat, N. B., and McDougall, J. K.,** Detection and localization of human papillomavirus DNA in human genital condylomas by *in situ* hybridization with biotinylated probes, *J. Med. Virol.*, 16, 265, 1983.
9. **Blum, H. E., Haase, A. T., and Vyas, G. N.,** Molecular pathogenesis of hepatitis B virus infection: simultaneous detection of viral DNA and antigens in paraffin-embedded liver sections, *Lancet*, 2, 771, 1984.
10. **Brambilla, C., Tackney, C., Hirschman, S. Z., Columbo, M., Dioguardi, M. K., Donati, M., and Paronetto, F.,** Varying nuclear staining intensity of hepatitis B virus DNA in human hepatocellular carcinoma, *Lab. Invest.*, 55, 475, 1986.
11. **Crum, C. P., Nagai, N., Levine, R. U., and Silverstein, S.,** *In situ* hybridization analysis of HPV-16 DNA sequences in early cervical neoplasia, *Am. J. Pathol.*, 123, 174, 1986.
12. **Forghani, B., Dupuis, K. W., and Schmidt, N. J.,** Rapid detection of herpes simplex virus DNA in human brain tissue by *in situ* hybridization, *J. Clin. Microbiol.*, 22, 656, 1985.
13. **Gibson, P. E., Gardner, S. D., and Field, A. M.,** Use of a molecular probe for detecting JCV DNA directly in human brain material, *J. Med. Virol.*, 18, 87, 1986.
14. **Gomez, S. A., Nascimento, J. P., Siqueira, M. M., Krawczuk, M. M., Pereira, H. G., and Russell, W. C.,** *In situ* hybridization with biotinylated DNA probes: a rapid diagnostic test for adenovirus upper respiratory infections, *J. Virol. Methods*, 12, 105, 1985.
15. **Myerson, D., Hackman, R. C., and Meyers, J. D.,** Diagnosis of cytomegaloviral pneumonia by *in situ* hybridization, *J. Infect. Dis.*, 15, 272, 1984.
16. **Myerson, D., Hackman, R. C., Nelson, J. A., Ward, C. C., and McDougall, J. K.,** Widespread presence of histologically occult cytomegalovirus, *Hum. Pathol.*, 15, 430, 1984.
17. **Negro, F., Berninger, M., Chiaberge, E., Gugliotta, P., Bassolati, G., Actis, G. C., Rizzetto, M., and Bonino, F.,** Detection of HBV-DNA by *in situ* hybridization using a biotin labelled probe, *J. Med. Virol.*, 15, 373, 1985.

18. **Rijntijes, P. J. M., Van Ditzhuijsen, J. M., Van Loon, A. M., Van Haelst, U. J. G. M., Bronkhorst, F. B., and Yap, S. H.,** Hepatitis B virus DNA detected in formalin-fixed liver specimens and its relation to serologic markers and histopathologic features in chronic liver disease, *Am. J. Pathol.*, 120, 411, 1985.
19. **Sixbey, J. W. and Pagano, J. S.,** Biotin-labeled DNA probes for detection of Epstein-Barr virus by *in situ* cytohybridizations, *Clin. Lab. Med.*, 5, 503, 1985.
20. **Sixbey, J. W., Nedrud, J. G., Raab-Traub, N., Hanes, R. A., and Pagano, J. S.,** Epstein-Barr virus replication in oropharyngeal epithelial cells, *N. Engl. J. Med.*, 310, 1225, 1984.
21. **Syrjanen, S., Syrjanen, K., Mantyjarvi, R., Parkkinen, S., Vayrynen, M., Saarikosi, S., and Castren, O.,** Human papillomavirus (HPV) DNA sequences demonstrated by *in situ* DNA hybridization in serial paraffin-embedded cervical biopsies, *Arch. Gynecol.*, 239, 39, 1986.
22. **Pfister, H., Kubke, J., Dietric, W., Iftner, T., and Fuch, P. G.,** Classification of the papillomaviruses-mapping the genome, in *Papillomaviruses*, Ciba Foundation Symposium, Vol. 120, Wiley, Chichester, England, 1986, 3.
23. Morbidity and Mortality Weekly Reports, Condyloma acuminatum-United States 1966—1981, *Morb. Mortal. Wkly. Rep.*, 32, 306, 1983.
24. **Gissmann, L. and Schneider, A.,** Human papillomavirus DNA in preneoplastic and neoplastic genital lesions, in *Viral Eiology of Cervical Cancer*, Peto, R. and zur Hausen, H., Eds., Cold Spring Harbor Laboratory, New York, 1986, 217.
25. **Gissmann, L., Wolnik, L., Ikenberg, H., Koldovsky, U., Schnurch, H. G., and zur Hausen, H.,** Human papillomavirus types 6 and 11 DNA sequences in genital and laryngeal papillomas and in some cervical cancers, *Proc. Natl. Acad. Sci. U.S.A.*, 80, 560, 1982.
26. **Meisels, A. and Fortin, R.,** Condylomatous lesions of the cervix and vagina. I. Cytologic patterns, *Acta Cytol.*, 20, 505, 1976.
27. **Reid, R., Laverty, C. R., Coppleson, M., Isarangkul, W., and Hills, E.,** Non-condylomatous cervical wart virus infection, *Obstet. Gynecol.*, 55, 476, 1980.
28. **Beckmann, A. M., Kiviat, N. B., Daling, J. R., Sherman, K. J., and McDougall, J. K.,** Human papillomavirus type 16 in multifocal neoplasias of the female genital tract, *Int. J. Gynecol. Pathol.*, 7, 39, 1988.
29. **Crum, C. P., Ikenberg, H., Richart, R. M., and Gissmann, L.,** Human papillomavirus type 16 and early cervical neoplasia, *N. Engl. J. Med.*, 310, 880, 1984.
30. **Crum, C. P., Mitao, M., Levine, R. U., and Silverstein, S.,** Cervical papillomaviruses segregate within morphologically distinct precancerous lesions, *J. Virol.*, 54, 675, 1985.
31. **Durst, M., Gissmann, L., Ikenberg, H., and zur Hausen, H.,** A papillomavirus DNA from a cervical carcinoma and its prevalence in cancer biopsy samples from different geographic regions, *Proc. Natl. Acad. Sci. U.S.A.*, 80, 3812, 1983.
32. **Ikenberg, H., Gissmann, L., Gross, G., Grussendorf-Conen, E.-I., and zur Hausen, H.,** Human papillomavirus type-16-related DNA in genital Bowen's disease and in bowenoid papulosis, *Int. J. Cancer*, 32, 563, 1983.
32a. **Beckmann, A. M., Daling, J. R., and McDougall, J. K.,** unpublished data.
33. **Drew, W. L., Miher, R. C., Ziegler, J. K., Gullett, J. H., Abrams, D. I., Cohant, M. A., Hung, E.-S., Groundwater, J. R., Volberding, P., and Mintz, L.,** Cytomegalovirus and Kaposi's sarcoma in young homosexual men, *Lancet*, 2, 125, 1982.
34. **McDougall, J. K., Buonaguro, F. M., Galloway, D. A., Nelson, J. A., and Smith, P. P.,** DNA virus and EM studies in AIDS, in *Acquired Immune Deficiency Syndrome*, Gottlieb, M. S. and Groopman, J. E., Eds., UCLA Symp. Mol. Cell. Biol., Vol. 16, Alan R. Liss, New York, 1984, 139.
35. **Ho, M.,** *Cytomegalovirus; Biology and Infection*, Plenum Medical Books, New York, 1982.
36. **Meyers, J. D., Flournoy, N., and Thomas, E. D.,** Nonbacterial pneumonia after allogenic marrow transplantation: a review of ten years' experience, *Rev. Infect. Dis.*, 4, 1119, 1982.
37. **Shepp, D. H., Dandiliker, P. S., deMiranda, P., Burnette, T. C., Cederberg, D. M., Kirk, L. E., and Meyers, J. D.,** Activity of 9-[2-hydroxy-1 (hydroxymethyl) ethoxymethyl] guanine in the treatment of cytomegalovirus pneumonia, *Ann. Intern. Med.*, 103, 368, 1985.
38. **Stagno, S., Pass, R. F., Reynolds, D. W., Alford, C. A.,** Diagnosis of cytomegalovirus infections, in *The Human Herpesviruses*, Nahmias, A. J., Dawdle, W. R., and Schinazi, R. F., Eds., Elsevier, New York, 363, 1981.
39. **Hackman, R. C., Myerson, D., Myers, J. D., Shulman, H. M., Sale, G. E., Goldstein, L. C., Rastetter, M., Flournoy, N., and Thomas, E. D.,** Rapid diagnosis of cytomegaloviral pneumonia by tissue immunofluorescence with a murine monoclonal antibody, *J. Infect. Dis.*, 151, 325, 1985.
40. **Volpi, A., Whitley, R. J., Ceballos, R., Stagno, S., and Pereira, L.,** Rapid diagnosis of pneumonia due to cytomegalovirus with specific monoclonal antibodies, *J. Infect. Dis.*, 147, 1119, 1983.
41. **Clad, A., Gissmann, L., Meier, B., Freesell, K., and Schwarz, E.,** Molecular cloning and partial nucleotide sequence of human papillomavirus type la DNA, *Virology*, 118, 254, 1982.
42. **deVilliers, E.-M., Gissmann, L., and zur Hausen, H.,** Molecular cloning of viral DNA from human genital warts, *J. Virol.*, 40, 932, 1981.

43. **Gissmann, L., Diehl, V., Schultz-Coulon, H.-J., and zur Hausen, H.,** Molecular cloning and characterization of human papillomavirus DNA derived from a laryngeal papilloma, *J. Virol.,* 44, 393, 1982.
44. **Boshart, M., Gissmann, L., Ikenberg, H., Kleinheinz, A., Scheurlen, W., and zur Hausen, H.,** A new type of papillomavirus DNA, its presence in genital cancer biopsies and in cell lines derived from cervical cancer, *EMBO J.,* 3, 1151, 1984.
45. **Lorincz, A. T., Lancaster, W. D., and Temple, G. F.,** Cloning and characterization of the DNA of a new human papillomavirus from a woman with dysplasia of the uterine cervix, *J. Virol.,* 58, 225, 1986.
46. **Fleckenstein, B., Muller, I., and Collins, J.,** Cloning of the complete human cytomegalovirus genome in cosmids, *Gene,* 18, 39, 1982.
47. **Przepiorka, D. and Myerson, D.,** A single step silver enhancement method permitting rapid diagnosis of cytomegalovirus infection in formalin-fixed, paraffin-embedded tissue sections by *in situ* hybridization and immunoperoxidase detection, *J. Histochem. Cytochem.,* 34, 1731, 1986.
48. **Unger, E. R., Budgeon, L. R., Myerson, D., and Brigati, D. J.,** Viral diagnosis by *in situ* hybridization: description of a rapid simplified colorimetric method, *Am. J. Surg. Pathol.,* 10, 1, 1986.
49. **Singer, R. H., Lawrence, J. B., and Villnave, C.,** Optimization of *in situ* hybridization using isotopic and non-isotopic detection methods, *BioTechniques,* 4, 230, 1986.
49a. **Sarju, D. A. and Beckmann, A. M.,** unpublished data.
50. **Howley, P. M., Israel, M. A., Law, M.-F., and Martin, M. A.,** A rapid method for detecting and mapping homology between heterologous DNAs. Evaluation of polyoma virus genomes, *J. Biol. Chem.,* 254, 4876, 1979.
51. **Hsu, S. M. and Soban, E.,** Color modification of diaminobenzidine (DAB) precipitation by metallic ions and its application for double immunohistochemistry, *J. Histochem. Cytochem.,* 30, 1079, 1982.
52. **Myerson, D.,** Hybridization histochemistry and viral diagnosis, in *Diagnostic Immunohistochemistry,* Colvin, R. B., Bhan, A. K., and McCluskey, R. T., Eds., Raven Press, New York, in press, 1988.
53. **Hackman, R. C., Gleaves, C. A., Myerson, D., McDonald, G. B., and Meyers, J. D.,** Immunohistology and *in situ* hybridization in the diagnosis of gastrointestinal infections, *Am. J. Surg. Pathol.,* 11, 154, 1987.
54. **Ruger, R., Bornkamm, G. W., and Fleckenstein, B.,** Human cytomegalovirus DNA sequences with homologies to the cellular genome, *J. Gen. Virol.,* 65, 1351, 1984.
55. **Galloway, D. A. and Swain, M. A.,** Cloning of herpes simplex virus type 2 fragments in a plasmid vector, *Gene,* 11, 253, 1980.
56. **Law, M.-F., Lancaster, W. D., and Howley, P. M.,** Conserved sequences among the genomes of papillomaviruses, *J. Virol.,* 32, 199, 1979.
57. **McDougall, J. K., Beckmann, A. M., and Kiviat, N. K.,** Methods for diagnosing papillomavirus infection, in *Papillomaviruses,* CIBA Found. Symp., Vol. 120, Wiley, Chichester, England, 1986, 86.
58. **McDougall, J. K., Myerson, D., and Beckmann, A. M.,** Detection of viral DNA and RNA by *in situ* hybridization, *J. Histochem. Cytochem.,* 34, 33, 1986.
59. **Orth, G., Breitburd, F., Favre, M., and Croissant, O.,** Papillomaviruses: possible role in human cancer, in *Origin of Human Cancer,* Hiatt, H. H., Watson, J. D., and Winsten, J. A., Eds., Cold Spring Harbor Laboratory Press, Cold Spring Harbor, N.Y., 1977, 1043.
60. **Orth, G., Favre, M., Breitburd, F., Croissant, O., Jablonska, S., Obalak, S., Jarzabek-Chorzelska, M., and Rzesa, G.,** Epidermodysplasia Verruciforms: a model for the role of papilloma viruses in human cancer, in *Viruses in Naturally Occurring Cancers,* Essex, M., Todaro, G., and zur Hausen, H., Eds., Cold Spring Harbor Laboratory Press, Cold Spring Harbor, N.Y., 1980, 259.
61. **Morin, C., Braun, L., Casas-Cordero, M., Shah, K. V., Roy, M., Fortier, M., and Meisels, A.,** Confirmation of the papillomavirus etiology of the condylomatous lesions of the cervix by the peroxidase anti-peroxidase technique, *J. Natl. Cancer Inst.,* 66, 831, 1981.
62. **Woodruff, J. D., Braun, L., Cavalieri, R., Gupta, P., Pass, F., and Shah, K. V.,** Immunologic identification of papillomavirus antigen in condyloma tissue from the female genital tract, *Obstet. Gynecol.,* 56, 727, 1980.
63. **Beschorner, W. E., Hutchins, G. M., Saral, R., Tutschka, P. J., and Santos, G. W.,** Cytomegalovirus pneumonia in bone marrow transplant recipients: miliary and diffuse patterns, *Am. Rev. Respir. Dis.,* 122, 107, 1980.
64. **Craighead, J. E.,** Cytomegalovirus pulmonary disease, *Pathobiol. Annu.,* 5, 197, 1975.
65. **Nash, G.,** Pathologic features of the lung in the immunocompromised host, *Hum. Pathol.,* 13, 841, 1982.
66. **Sehested, M. and Hou-Jensen, K.,** Factor VIII related antigen as an endothelial cell marker in benign and malignant diseases, *Virchows Arch. A,* 391, 217, 1981.
67. **Wiley, C. A., Schrier, R. D., Denaro, F. J., Nelson, J. A., Lampert, P. W., and Oldstone, M. B. A.,** Localization of cytomegalovirus proteins and genome during fulminant central nervous system infection in an AIDS patients, *J. Neuropathol. Exp. Neurol.,* 45, 127, 1986.
68. **Firzlaff, J. M., Hsia, C. N. L., Halbert, P. H. L., Jenison, S. A., and Galloway, D. A.,** Polyclonal antibodies to human papillomavirus type 6b and type 16 bacterially derived fusion proteins, in *Cancer Cells 5 — Papillomaviruses,* Steinberg, B. M., Brandsma, J. L., and Taichman, L. B., Eds., Cold Spring Harbor Laboratory Press, Cold Spring Harbor, 1987, 105.

69. **Meisels, A., Morin, C., and Casas-Cordero, M.,** Human papillomavirus infection of the uterine cervix, *Int. J. Gynecol. Pathol.,* 1, 75, 1982.
70. **Reid, R., Crum, C. P., Herschman, B. R., Fu, Y. S., Braun, L., Shah, K. V., Agronow, S. J., and Stanhope, C. R.,** Genital warts and cervical cancer. III. Subclinical papillomaviral infection and cervical neoplasia are linked by a spectrum of continuous morphologic and biologic change, *Cancer,* 53, 943, 1984.
71. **Ferenczy, A., Mitao, M., Nagai, N., Silverstein, S. J., and Crum, C. P.,** Latent papillomavirus and recurring genital warts, *N. Engl. J. Med.,* 313, 784, 1985.
72. **Grussendorf-Conen, E.-I.,** *In situ* hybridization with papillomavirus DNA genital lesions, in *Viral Etiology of Cervical Caner,* Peto, R. and zur Hausen, H., Eds., Cold Spring Harbor Laboratory Press, Cold Spring Harbor, N.Y., 1986, 239.
73. **Lorincz, A. T., Lancaster, W. D., Kurman, R. J., Jenson, A. B., Temple, G. F.,** Characterization of human papillomaviruses in cervical neoplasias and their detection in routine clinical screening, in *Viral Etiology of Cervical Cancer,* Peto, R. and zur Hausen, H., Eds., Cold Spring Harbor Laboratory Press, Cold Spring Harbor, N.Y., 1986, 225.
74. **Purola, E. and Savia, E.,** Cytology of gynecologic condyloma acuminatum, *Acta Cytol.,* 21, 26, 1977.
75. **Hills, E. and Laverty, C. R.,** Electron microscopic detection of papillomavirus particles in selected koilocytotic cells in routine cervical smears, *Acta Cytol.,* 23, 53, 1979.
76. **Kurman, R. J., Shah, K. H., Lancaster, W. D., and Jenson, A. B.,** Immunoperoxidase localization of papillomavirus antigens in cervical dysplasia and vulvar condylomas, *Am. J. Obstet. Gynecol.,* 140, 931, 1981.
77. **Burk, R. D., Kadish, A. S., Calderin, S., and Romney, S. L.,** Human papillomavirus infection of the cervix detected by cervicovaginal lavage and molecular hybridization: correlation with biopsy results and Papanicolaou smear, *Am. J. Obstet. Gynecol.,* 154, 982, 1986.
78. **Schneider, A., Kraus, H., Schuhmann, K., and Gissmann, L.,** Papillomavirus infection of the lower genital tract: detection of viral DNA in gynecological swabs, *Int. J Cancer,* 35, 443, 1985.
79. **Wagner, D., Ikenberg, H., Boehm, N., and Gissmann, L.,** Identification of human papillomavirus in cervical swabs by deoxyribonucleic acid *in situ* hybridization, *Obstet. Gynecol.,* 64, 767, 1985.
80. **Wickenden, C., Steele, A., Malcolm, A. D. B., and Coleman, D. V.,** Screening for wart virus infection in normal and abnormal cervices by DNA hybridization of cervical scrapes, *Lancet,* 1, 65, 1985.
81. **Ostrow, R., Manias, D., Clark, B., Okagaki, T., Twiggs, L., and Faras, A.,** Detection of papillomavirus-specific sequences in human genital tumors by *in situ* hybridization, in *Viral Etiology of Cervical Cancer,* Peto, R. and zur Hausen, H., Eds., Cold Spring Harbor Laboratory Press, Cold Spring Harbor, N.Y., 1986, 253.
82. **Gupta, J., Schneider, A., and Shah, K. V.,** Detection of papillomavirus antigen and DNA in cells and tissues, in *Viral Etiology of Cervical Cancer,* Peto, R. and zur Hausen, H., Eds., Cold Spring Harbor Laboratory Press, Cold Spring Harbor, N.Y., 1986, 247.
83. **Ho, M., Suwansirikue, S., Dowling, J. N., Youngblood, L. A., and Armstrong, J. A.,** The transplanted kidney as a source of cytomegalovirus infection, *N. Engl. J. Med.,* 293, 1109, 1975.
84. **Betts, R. F., Freeman, R. B., Douglas, R. G., Jr., Talley, T. E., and Rundell, B.,** Transmission of cytomegalovirus infection with renal allograft, *Kidney Int.,* 8, 385, 1975.
85. **Tobin, J. O., Warrell, M. J., and Morris, P. J.,** Cytomegalovirus infection and renal transplantation, *Lancet,* 1, 926, 1979.
86. **Marker, S. C., Howard, R. J., Simmons, R. L., Kalis, J. M., Connely, D. P., Najarian, J. S., and Balfour, H. H., Jr.,** Cytomegalovirus infection: a quantitative study of 320 consecutive renal transplants, *Surgery,* 89, 660, 1981.
87. **Naraqi, S., Jackson, G. G., Jonasson, O., and Rubenis, M.,** Search for latent cytomegalovirus in renal allografts, *Infect. Immun.,* 19, 699, 1978.
88. **Lang, D. J., Ebert, P. A., Rodgers, B. M., Boggess, H. P., and Rixse, R. S.,** Reduction of postperfusion cytomegalovirus infections following the use of leukocyte depleted blood, *Transfusion,* 17, 391, 1977.
89. **Schrier, R. D., Nelson, J. A., and Oldstone, M. B. A.,** Detection of human cytomegalovirus in peripheral blood lymphocytes in a natural infection, *Science,* 230, 1048, 1985.
90. **Einhorn, L. and Ost, A.,** Cytomegalovirus infection of human blood cells, *J. Infect. Dis.,* 149, 207, 1984.
91. **Maters, A. W.,** Detection of cytomegalovirus in lung tissue by *in situ* hybridization: comparison with cytopathologic antigen detection and tissue culture results, in *Abstr. Annu. Meet., Am. Soc. Microbiol.,* Am. Soc. Microbiol., Washington, D.C., 1987, 309.
92. **Todd, J. A., Jou, H. L., and Yang, H. L.,** A rapid assay for the detection of human papillomavirus in cervical biopsies, in *Abstr. Annu. Meet. Am. Sco. Microbiol.,* Am. Soc. Microbiol., Washington, D.C., 1987, 309.
93. **Beckmann, A. M., Kiviat, N. B., and McDougall, J. K.,** unpublished observations, 1987.
94. **Beckmann, A. M., Daling, J. R., and McDougall, J. K.,** Human Papillomavirus Infection and Anal Cancer, submitted, 1988.

Chapter 13

BLOOD-BORNE VIRAL INFECTIONS

Hepatitis B Virus and Human Immunodeficiency Virus

Paul P. Ulrich, Hubert E. Blum, Michael P. Busch, and Girish N. Vyas

TABLE OF CONTENTS

I. INTRODUCTION

The recent advances in molecular biology, including recombinant DNA technology, have provided novel opportunities in the diagnostic and pathogenetic investigation of viral infections. The technique of nucleic acid hybridization permits detection of small viral genes within the abundant number of nucleotides of a eukaryotic cell genome estimated to be 10^{10}.[1] Hybridization is the formation of a nucleic acid duplex between two complementary nucleotide sequences, i.e., the target nucleotide sequence and the labeled probe. Both radiolabeled and nonisotopically labeled single-stranded (ss) and double-stranded (ds) DNA as well as RNA probes have been used.

The detection of viral agents present in tissues, cells, or body fluids is usually performed after chemical extraction of nucleic acids that are assayed in dot-blot hybridization (DBH) and transfer-blot hybridization (TBH) analyses. The TBH for analysis of DNA and RNA are termed Southern and Northern blot (SBH and NBH), respectively. *In situ* hybridization (ISH) analyses are performed using intact cells or tissue sections. The three distinct analytic approaches and their respective applications are illustrated in Table 1. In principle, each of the three approaches to nucleic acid hybridization illustrated in Table 1 yields comparable qualitative as well as quantitative information about the DNA/RNA.

We have used these three standard methods to analyze blood-borne viral infections, specifically hepatitis B virus (HBV) and human immunodeficiency virus (HIV).

II. HEPATITIS B VIRUS

A. Clinical Perspective

Both sporadic and epidemic forms of jaundice have been historically noted, but only in 1839 was it recognized to be a result of hepatocellular necrosis. The viral etiology of hepatocellular necrosis was established by a series of experimental inoculations of human sera from two epidemiologically distinct hepatitis cases (MS-1 and MS-2) into human subjects, marmosets, and chimpanzees, which led to a distinction of two etiologic agents: hepatitis A virus (HAV) and HBV.[2] Definitive laboratory diagnosis of clinically indistinguishable infections with HAV and HBV is possible with the serologic assays for IgM anti-HAV and a variety of immunoassays for HBV, including hepatitis B surface antigen and antibodies (HBsAg/anti-HBs), hepatitis e antigen and antibodies (HBeAg/anti-HBe), and hepatitis B core antibodies (anti-HBc). In patients with no clinical explanation for hepatitis or no laboratory evidence of HAV and HBV infection, the diagnosis by exclusion is non-A, non-B (NANB) hepatitis. In fact, NANB hepatitis is the most major cause of transfusion-associated hepatitis, and no etiological agents or serologic markers of infection are defined.[3]

B. Rationale for the Use of Probes to Study HBV

During the last two decades, more than 20,000 papers and 32 books have been published on viral hepatitis. This plethora of information and consistent failure to propagate HBV in vitro led to cloning HBV DNA from human plasma and determination of complete nucleotide sequence of HBV DNA.[4]

The cloned HBV DNA provides a new molecular probe for direct assay of viral nucleic acids in the serum and the tissues of humans and animal models infected by hepatotropic DNA viruses termed hepadna viruses.[5] The major impetus for the application of nucleic acid hybridization technology stems from the fact that an estimated 200 million people in the world are chronic carriers of HBV, and at least a million of them live in the U.S. It is estimated that 25% of these carriers have the risk of developing chronic liver disease, while others are apparently healthy, but serve as an epidemiologic reservoir for transmission of HBV infection through blood or sexual contacts. In addition, the chronic carriers are at a 200-fold increased risk of developing hepatocellular carcinoma.[6]

Table 1
HYBRIDIZATION TECHNIQUES: DETECTION AND CHARACTERIZATION OF HBV DNA/RNA BY DOT-BLOT, TRANSFER-BLOT, AND *IN SITU* HYBRIDIZATION

	Dot blot	Transfer blot (Southern/Northern)	*In situ* hybridization
DNA	+	+	+
RNA	+	+	+
Quantitation	+	+	+
Sensitivity (DNA)	0.1—0.5 pg	0.5—1 pg (per molecular species)	single copy (per cell)
Characterization			
RNA size classes	–	+	–
DNA size classes	–	+	–
DNA conformation	–	+	–
DNA integration	–	+	–
Restriction analysis	–	+	–
Replicative forms (ssDNA)	+	+	+
Single-cell analysis	–	–	+
Tissue pattern of viral DNA, RNA transcripts	–	–	+
Detection of rare/focal nucleic acid sequences	–	–	+
Combination of DNA/RNA hybridization with immunohistochemistry	–	–	+

When the cloned HBV DNA was used as a molecular probe for a direct demonstration of HBV DNA in patients with acute and chronic viral hepatitis, it was observed that the virus was not only found in the patients with immunological markers of HBV infection but also detected in the serum and tissues of certain NANB hepatitis patients.[7] Thus, occult HBV infection in the absence of standard serologic markers in the serum would suggest that a proportion of the so-called NANB hepatitis may indeed be due to viral agents that are antigenically distinct from the HBV, but sharing nucleic acid homology.[8] Therefore, we term them HBV-like agents.

Herein, we present a contemporary assessment of the use of cloned HBV DNA probes for studies of both the pathogenesis of persistant infection and the mechanism of hepatocellular injury in HBV infections.

C. Methods and Results of HBV Probe Studies

The complete HBV genome of 3200 bp (HBsAg subtype adw_2), introduced at the unique *Eco*RI cleavage site into pBH20, was propagated in Escherichia coli.[9] Recombinant plasmid was recovered by density gradient centrifugation and HBV DNA was purified from the plasmid by cleaving with *Eco*RI following separation by preparative agarose gel electrophoresis. The HBV DNA was uniformly labeled by nick translation with ^{32}P-dNTPs for DBH and TBH.[10] For ISH, nick-translated ^{3}H-labeled or ^{35}S-labeled probes with an average length of 300 bp have been used.[11] For the synthesis of ss probes, the purified HBV DNA was religated and cut with *Hin*cII to yield five fragments.[12] The fragments were ligated subsequently into M13 phage vector and propagated in E. coli. Using the universal primer, radiolabeled dNTPs, and Klenow fragment, the complementary strand to be used as hybridization probe was synthesized. The resulting products are as, polarity- and region-specific, subgenomic DNA probes.

To reveal a high fidelity of base pairing, hybridization and posthybridization washings are usually carried out under stringent conditions. Sometimes less stringent conditions are desired,

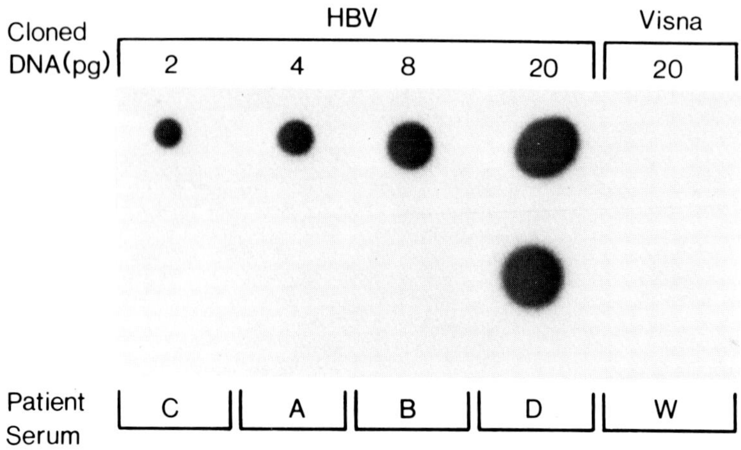

FIGURE 1. Autoradiogram showing HBV DNA dot-blot hybridization with various amounts of cloned HBV DNA and unrelated Visna cDNA (upper row); nucleic acids extracted from normal control serum (C), and three HBsAg/anti-HBe-positive sera (A, B, and W) and a HBsAg/HBeAg-positive serum (D) (lower row).

i.e., for the detection of viruses sharing partial homologies in the nucleic acid sequence. The dot-blot hybridization assay is specific for the detection of HBV DNA, since an unrelated radiolabeled probe (e.g., pBR322) or DNA from an unrelated virus (e.g., Visna cDNA) do not result in a detectable hybrid formation as illustrated in Figure 1. This assay readily identifies HBV DNA in the serum from a carrier serologically positive for HBsAg, HBeAg, and anti-HBc, and indicates active viral replication in the liver and high infectivity of the serum from patient D (Figure 1). Three patients (A, B, and W, Figure 1) serologically positive for HBsAg, anti-HBc, and for antibodies against HBeAg, are negative for serum HBV DNA. However, negative results do not exclude the occurrence of HBV in serum or active viral replication in the patient's liver because the minimal HBV DNA detectable in DBH is 0.1 pg, which is about 30,000 copies of HBV DNA. On the other hand, studies have now established that dot-blot analyses are more sensitive for the detection of HBV in the serum than radioimmunoassay for HBeAg and tests for DNA polymerase activity.[13] Additionally, HBV DNA has been detected by dot-blot analyses of HBsAg-negative sera with and without antibodies to HBsAg or HBcAg.[13] These findings provide remarkable evidence that HBsAg-positive/HBeAg-negative patients and HBsAg-negative individuals with or without antibodies to HBV antigens (the so-called non-A, non-B hepatitis) may have active HBV replication in their livers and have to be considered potentially infectious. Dot-blot hybridization can also be applied for the detection of viral DNA or RNA in tissue extracts or subcellular fractions. Quantitation of HBV DNA is possible by densitometry or liquid scintillation counting, relative to standards of known amounts of viral nucleic acids.

While early methods required extraction of the HBV DNA from serum followed by transfer to a membrane, more recent techniques involved direct application of alkali-denatured serum samples to nitrocellulose filters.[23,24] The so-called spotting hybridization method is a marked improvement in terms of simplicity and sensitivity, which is about three times higher than that for the conventional DBH after DNA extraction. During the phenol extraction and ethanol precipitation procedure, about two thirds of HBV DNA in the serum is lost.[23] Therefore, the direct spotting method is more suitable for the quantitation of HBV DNA in serum. Most recently, Abbott Laboratories have developed a liquid-phase hybridization assay for the detection of HBV DNA in serum. After denaturation of the serum and hybridization with a ss ^{125}I-labeled DNA probe of 350 bases, the mixture is loaded onto a column. In the system, ^{125}I-labeled hybrid molecules are eluted, whereas unhybridized ^{125}I-HBV probe is retained by the column. The eluate is collected in one step and read in a gamma counter. We investigated 20

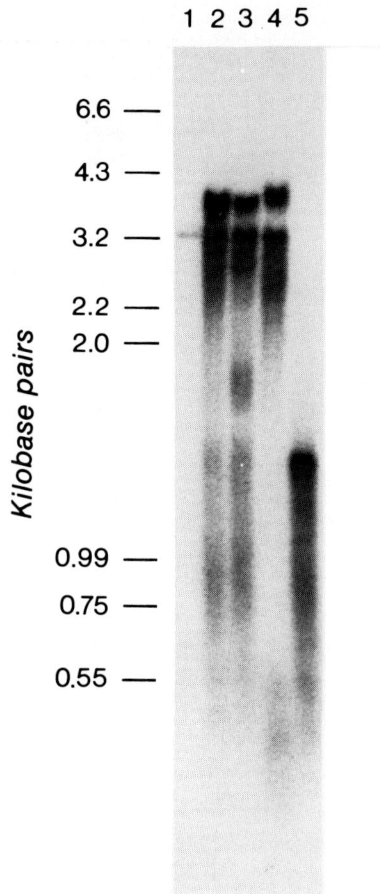

FIGURE 2. Southern blot hybridization of liver DNA from a patient with HBsAg/HBeAg-positive chronic active hepatitis showing exclusively free viral DNA with relaxed circular forms (3800 bp), complemented linear forms (3200 bp), and nascent viral DNA molecules (<3200 bp). Lane 1: 5 pg cloned HBV DNA as molecular weight standard. Lane 2: 20 mg uncut cellular DNA. Lane 3: 20 mg cellular DNA digested with *Eco*RI. Lane 4: 20 mg cellular DNA digested with nuclease S1, demonstrating that the small molecular weight viral DNA species are predominantly single stranded. Lane 5: 20 mg heat-denatured cellular DNA, demonstrating that the single-stranded full-length HBV genome migrates into the 1.4-kb position.

serum samples of HBsAg carriers. HBV DNA was detected in 10 out of 10 HBeAg-positive sera. Borderline results were revealed in 3 HBeAg-negative sera, and the remaining 7 were negative for HBV DNA.

Southern blot analyses of DNA extracted from the liver tissue have demonstrated that HBV DNA in hepatocytes can exist in a free, extrachromosomal, actively replicating form, as well as integrated into the cellular genome. The hybridization pattern in a patient with active viral replication is characterized by the presence of a large amount of low molecular weight (<3200 bp), nascent DNA molecules (Figure 2). These species were shown to be predominantly ss and of minus strand polarity, demonstrating the asymmetric replication of the HBV genome.[12] In contrast, PLC/PRF/5 cells (Alexander cells) derived from a human hepatocellular carcinoma have HBV DNA predominantly integrated into the host genome. Following restriction endonuclease digestion of the cellular DNA (Figure 3), hybridizing DNA molecules larger than 3200 bp are produced as a result of the flanking sequences from the host genome. Free viral DNA has been generally found in the liver of HBeAg-positive patients, while integrated forms are

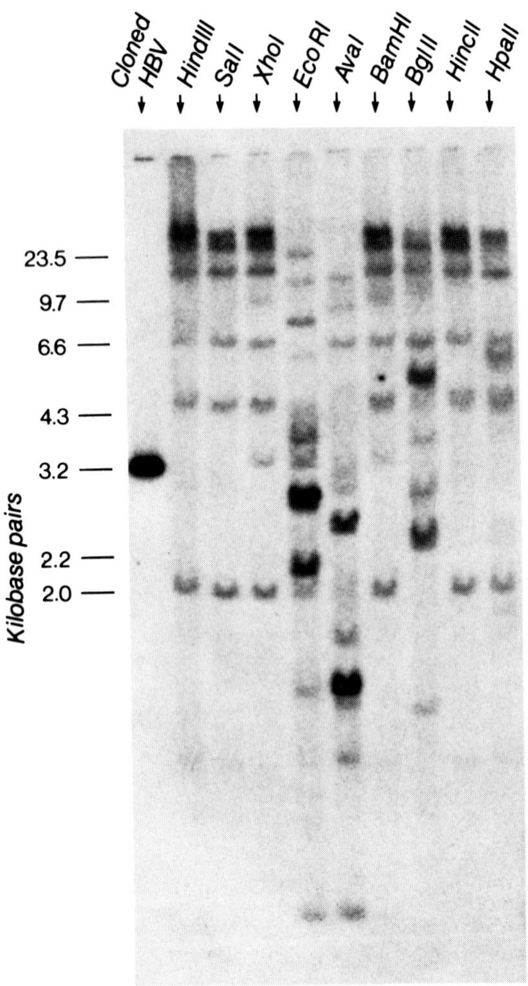

FIGURE 3. Southern blot hybridization of 20 mg of DNA from PLC/PRF/5 cells (Alexander cells) after digestion with restriction enzymes indicated on top showing predominantly integrated HBV DNA of molecular weight larger than 3200 bp due to flanking host cellular sequences.

predominantly associated with HBeAg-negative chronic liver disease and hepatocellular carcinoma1.[14,15] However, it is important to note that hepatocellular DNA, without endonuclease digestion, has shown free episomal or extrachromosomal replicative forms of molecular weight higher than 3200 bp in several HBsAg-negative patients with histologically proven chronic liver disease.[7] Thus, free viral DNA with molecular weight higher than 3200 bp may be confused with integrated forms, unless hepatocellular DNA is concurrently tested without hydrolysis as well as after hydrolysis with different restriction endonucleases, e.g., *Eco*RI, *Hind*III, and *Hinc*II. These findings imply that in certain cases of serologically defined NANB hepatitis, the chronic liver disease may be due to infection by HBV or HBV-related agents.

NBH is for RNA the equivalent of the SBH procedure for DNA.[16] For example, the HBV transcripts produced in the PLC/PRF/5 cell line (Alexander cells) have been analyzed by Rutter et al.[17] In order to test qualitatively the proportion of mRNAs derived from the integrated HBV sequence, the host sequences of the longest cDNA clone (pAL2) were isolated, subcloned, and used as a probe for NBH with Alexander-cell mRNA. This probe detects heterogeneous

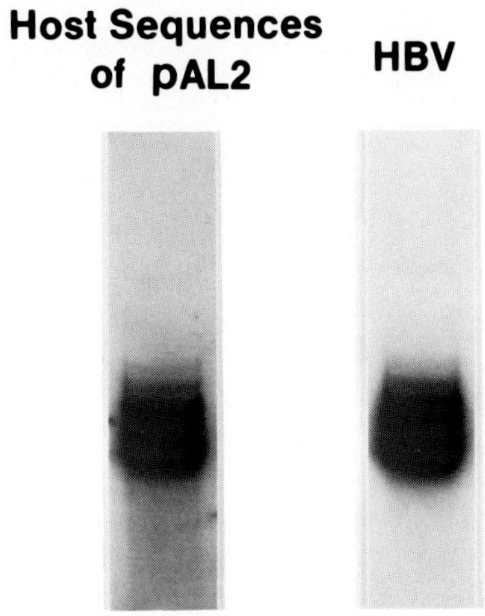

FIGURE 4. Northern blot analysis of Alexander cell mRNAs. Ten mg of PLC/PRF/5 poly A RNA were run on each lane. In the right panel, the entire cloned HBV DNA was used as the probe. In the left panel, the subclonal host sequence of pAL2 was used.

population of RNA molecules, and the pattern was very similar to that obtained when the entire cloned HBV DNA was used as a probe (Figure 4).

The principle of the *in situ* hybridization (ISH) procedure is outlined in Figure 5. Greatest sensitivity of ISH can be obtained using minimum probe concentrations, which saturate all the target nucleic acids. This probe concentration has to be predetermined by an experiment hybridizing different probe concentrations to the target sequence. As a guideline, Haase et al.[18] recommend for viral genomes in the range of 3000 to 10,000 bp, using 0.06 ng/μl for 1 to 10 copies per cell, 0.14 ng/μl for 20 to 100 copies per cell, and 2 to 4 ng/μl for 1000 and more copies per cell. We have used nick-translated DNA probes at a concentration of 0.1 ng/μl for HBV. Additionally, the average size of the probe is considered critical for the sensitivity of ISH.[19-21] In our hands, a probe size of about 300 bp results in the best signal to noise ratio. The ratio is decreased for smaller probes due to an increase in the background and for larger due to a decreased diffusion of probe into the cells. Under optimal conditions for ISH, HBV DNA is readily detected in infected hepatocytes. A higher focal tissue distribution with infected cells next to uninfected cells has been observed. The patients with active viral replication displayed a predominantly cytoplasmic hybridization pattern (Figure 6). These cytoplasmic viral nucleic acid species were shown to be mostly ss and of minus strand polarity, demonstrating, at the cellular level, the asymmetric replication of HBV DNA.[12] Due to the high sensitivity of the ISH assay and visualization of the infected cells, HBV DNA could be demonstrated by us not only in the hepatocytes, but also in the bile duct epithelium, in the endothelial cells, and the smooth muscle cells of blood vessel walls, as well as in the spleen.[11,12a] This assay has identified for the first time nonhepatocytes as target cells for HBV infection, thus extending the known host cell range for HBV. These findings are of particular significance for the understanding of the biology of HBV and consistent with several recent reports demonstrating HBV in nonhepatocytes (Table 2).

The combination of both immunochemistry and ISH provide the opportunity for simultaneous detection of viral genes and gene products at the single-cell level in the same cell preparation. Blum et al.[22] have used this combination for studies of HBV in human liver. At first, the viral antigens have to be detected by specific antibodies following their visualization with an enzymatic detection system, then the rigorous ISH procedure can be performed. After comple-

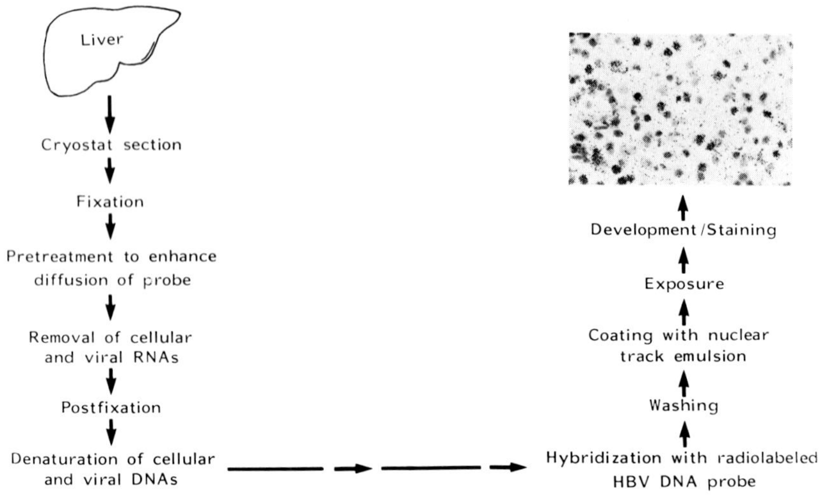

FIGURE 5. Schematic illustration of the *in situ* hybridization technique as applied to studying HBV in human liver.

tion of both assays, the final product will have cells in which silver grains indicate HBV genes and the blue substrate colors demonstrate expression of HBV antigens. In the liver from a patient with chronic active hepatitis serologically positive for HBsAg and HBeAg, actively replicating HBV DNA species were demonstrated predominantly in the cytoplasm. While HGsAg was found exclusively in hepatocyte cytoplasm, HBcAg was predominantly expressed in the liver cell nuclei and was not detected in most hepatocytes with high levels of viral replication. Conversely, most liver cells in which HBcAg was found did not contain replicating HBV. Because no pathological changes could be identified either in hepatocytes with high levels of HBV replication or expression of HBcAg, the liver cell damage in this patient with chronic hepatitis B was presumably induced by other mechanisms.

In summary, the nucleic acid hybridization technology is evolving and being applied in the detection of HBV in serum and cells of infected individuals. The three principal methodologies outlined in Table 1 have revealed the following: (1) direct evidence of HBV infectivity in serum of HBsAg-positive as well as some HBsAg-negative patients with hepatitis; (2) defined a unique mechanism of DNA virus replication through use of reverse transcriptase; (3) replication of HBV DNA in nonhepatocytes including lymphoreticular system; and (4) lack of HBcAg gene expression in immunopathogenesis of hepatocellular injury in chronic hepatitis.

III. HUMAN IMMUNODEFICIENCY VIRUS

The disease entity, acquired immune deficiency syndrome (AIDS), has in the last few years attracted international attention, and rightfully so. As of February 1988, over 50,000 cases had been reported in the U.S., of which more than 50% had been fatal. Except for cases occurring in persons of Haitian origin (4%), most have been associated with intimate personal contact among gay males (72%) or illicit intravenous drug use (17%). Although AIDS occurs most frequently in promiscuous gay males, both heterosexual and homosexual transmission are known to occur.

The suspected AIDS agent was identified in 1983/84 as a Type D retrovirus variously called human T-cell lymphotropic virus-III (HTLV-III) and lymphadenopathy associated virus (LAV).[25] The agent is a retrovirus belonging to the Lentivirinae group and has been recently called the human immunodeficiency virus (HIV).[26,27]

The transmission of HIV by blood transfusion is now widely recognized. About 3.0% of

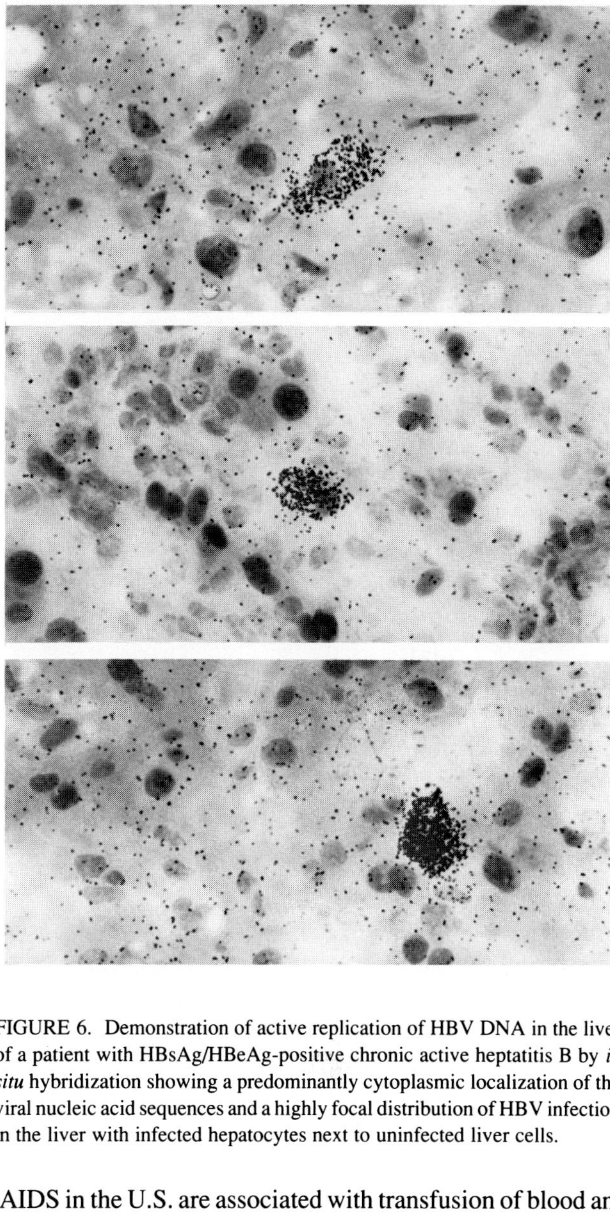

FIGURE 6. Demonstration of active replication of HBV DNA in the liver of a patient with HBsAg/HBeAg-positive chronic active heptatitis B by *in situ* hybridization showing a predominantly cytoplasmic localization of the viral nucleic acid sequences and a highly focal distribution of HBV infection in the liver with infected hepatocytes next to uninfected liver cells.

reported cases of AIDS in the U.S. are associated with transfusion of blood and blood products. As of April 1987, a total of 1057 cases were reported, with 732 among recipients of blood transfusions and 325 among hemophiliacs. AIDS cases are classified as transfusion associated, if the patient has received blood or blood components between 1978 and the onset of illness, and no other potential risk factors are identified. Of transfusion-associated cases, 90% were transfused before recognition of a connection between transfusion and AIDS and before steps were taken to eliminate blood donors from groups at risk for AIDS. Protection of the national blood and plasma supply from HIV contamination has been one of the highest public health priorities related to the AIDS epidemic.

Licensed tests utilizing the enzyme-linked immunosorbent assay (ELISA) technology to detect the HIV antibody have become available since March of 1985. These tests have high sensitivity and, when strongly reactive, high specificity in correctly identifying a person who has been exposed to HIV. Antibodies to HIV are found in 100% of AIDS patients and up to 85% of persons in the high-risk categories for developing or transmitting AIDS.[28] Because of epidemi-

Table 2
HBV DNA IN NONHEPATOCYTES

Tissue/cells	Ref.
Bile duct epithelium, endothelial cells, smooth muscle cells	11
Skin, pancreas, kidney	44
Kaposi's sarcoma	45
Spleen	12a
Bone marrow or blood cells	
Lymphoblastoid cells	42, 46, 47
White blood cells	48
Mononuclear cells	43, 49—52
Polymorphonuclear leukocytes	53

ologic association of anti-HIV and anti-HBc in homosexual men, HBV has been suspected to be a cofactor in AIDS.[29] All the 53 cases of transfusion-associated AIDS encountered in San Francisco were among the patients transfused before the institution of the HIV antibody testing. Antibody testing so far has not been able to classify infected persons as protected or unprotected. In addition, it is known that a virus-positive, antibody-negative time window does exist.[28] The incubation of AIDS is thought to range from 1 to 5 years and possibly as long as 15 years.[30] In suspected transfusion-associated AIDS cases, incubation periods up to 7 years have been reported. In some of our studies, it appears that transmission of HIV infection from HIV seropositive individuals is close to 100%.[30a] If so, this infectivity rate is indeed disconcerting, considering that an estimated 12,000 patients have been transfused with infected blood in the years before introduction of the HIV antibody test.[30b]

The presence of HIV in plasma or peripheral blood mononuclear cells (PBMC) is difficult to demonstrate by immunochemical or ISH analyses unless the specimens are cultured and monitored for their viral gene expression.[31,32]

HIV culture is a relatively new and, as yet, unstandardized technique with a great variation in the efficiency of HIV isolation. We have employed the culture method of D. Gallo et al.,[33] in the course of which fresh or frozen PBMC or plasma were cocultured with 3 days phytohem-magglutinin (PHA)-stimulated normal PBMC and assayed twice a week for HIV expression. The supernatant fluid was monitored for HIV by assaying for the virion-associated reverse transcriptase (RT). Additionally, a newly established enzyme-linked antigen capture assay (AC) for the production of mature virus in the supernatant was also employed. Further investigations will determine whether AC can replace the RT assay, which is technically complex, cumbersome, and expensive. For characterizing the early events in intracellular HIV replication and expression of the virus in cultures, we have used immunoperoxidase (IP) analysis with monoclonal antibodies against viral core proteins (p18) and ISH.[31] As a probe for ISH, we have used agarose gel-purified, nick-tranlated cDNAfragments of HIV that were cloned into the *Eco*RI site of the pUC19 vector.[34]

Other investigators have employed RNA probes for the detection of HIV in blood cells and tissues.[32,35] In some cases, RNA probes have shown higher sensitivity than equivalent nick-translated ds DNA probes in both ISH and NBH.[36,37] This is most probably due to the higher stability of the RNA:RNA hybrids compared to the RNA:DNA or DNA:DNA hybrids, and also due to the fact that reannealing of RNA probes does not occur. However, we[31] have optimized our hybridization protocol for the use of DNA probes and found the results for HIV similar to those of Harper et al.[32] We favor DNA probes because of their stability and convenience of not requiring further subcloning into SP6 vectors.

The ISH assay is a specific and sensitive tool for the localization and quantitation of viral

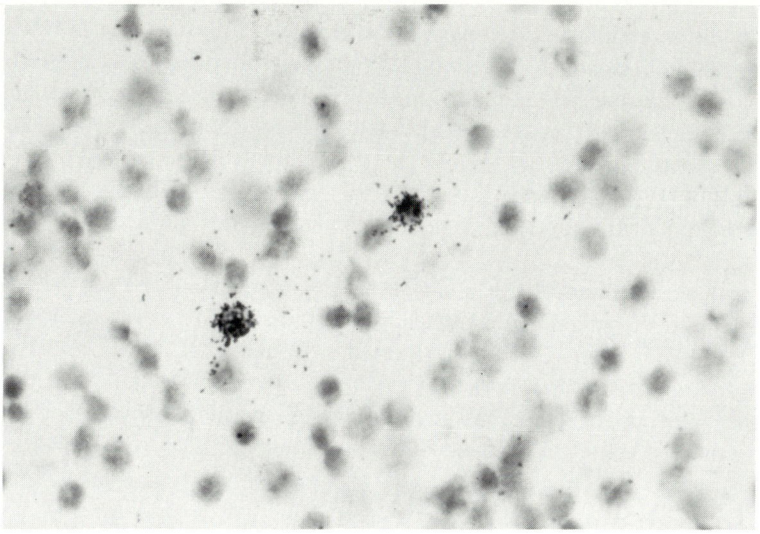

FIGURE 7. *In situ* hybridization of ^{35}S-labeled HIV probe with cultured PBMC from an anti-HIV-positive patient.

nucleic acids at the single-cell level (Figure 7). However, since ISH is a laborious and technically complex procedure, we have alternatively used an optimized immunocytochemical assay as the primary method for monitoring HIV in cultures. Viral core (p18) and envelope (gp41) proteins were detected with a mixture of monoclonal antibodies and visualized by the alkaline phosphatase antialkaline phosphatase (APAAP) technique. Using APAAP, we are generally able to detect viral specific antigens at day 7 and sometimes as early as day 3 in cultured PBMC. Samples that were APAAP-negative for 21 days in culture were shown to remain virus negative as confirmed by ISH and RT assay.

Uniform seroconversion in all of the recipients of anti-HIV-positive blood in San Francisco led us to investigate culturing of HIV from their PBMC to test for the incidence of viremia following a single parenteral exposure in an otherwise low-risk group.[38] We could isolate HIV from 15/18 (83%) of the representative anti-HIV-positive recipients of blood transfusions. Only two of the culture-positive patients had clinical symptoms of AIDS-related complex, while others were asymptomatic after 2 years of blood transfusion. To determine the inoculum of HIV-infected PBMC necessary for virus culture, limiting dilutions of PBMC from 10 asymtomatic anti-HIV-positive persons were cultured.[39] The dilutional culture studies revealed a higher proportion of HIV-infected cells ranging from 1:100 to 1:10,000 in comparison with exceedingly rare (less than 1:10,000) HIV-positive cells detected either by ISH or IP/APAAP analyses of uncultured cells.[31,32] Thus, a large proportion of in vivo-infected cells are not producing detectable levels of HIV nucleic acids and proteins without in vitro stimulation in the coculture procedure, suggesting that they are latently infected and require in vitro stimulation for the expression of viral gene products.

A procedure, termed polymerase chain reaction (PCR), was developed at Cetus Corporation in Emoryville, Calif. for in vitro DNA amplification.[40] The amplification procedure is based on repeated cycles of denaturation, oligonucleotide primer annealing, and primer extension by Klenow fragment of DNA polymerase, which results in an exponential increase in copies of the region flanked by the primers. After 20 cycles, the target sequence is amplified about 200,000-fold. We have used oligonucleotide primers for a conserved 110-bp sequence in the gag region of HIV. To increase the sensitivity of the procedure, the amplified gag sequence was detected by hybridization using a 40-base oligonucleotide probe, which was 5′-phosphorylated with ^{32}P-

ATP and polynucleotide kinase. By employing this procedure, we could identify HIV sequences in DNA extracted from mononuclear cells of fresh blood and bone marrow of anti-HIV-positive, asymptomatic individuals, thereby eliminating the lengthy process required for cell culturing.[41]

The AIDS epidemic and transfusion-associated AIDS in recipients of blood and hemophiliacs have stimulated extraordinary interest in study of molecular virology of HIV and the cofactorial role of HBV and cytomegalovirus (CMV). We are currently investigating the value of double-label ISH for assessment of simultaneous infection with HIV and either HBV or CMV in the pathogenesis of AIDS.

In SBH analysis of cultured PBMC, Laure et al.[42] reported not only HIV in the supernatants, but HBV DNA in the cellular DNA extracted from cultured PBMC. The presence of episomal HBV DNA in PBMC of all AIDS patients reported by Noonan et al.[43] confirms the observation of Laure et al.,[42] implicating HBV as a cofactor in etiology of AIDS.[43] However, these reports are at variance with our negative results in ISH analyses, and further work needs to be done to reconcile the current enigma of HIV and HBV coinfection in PBMC of AIDS patients.

IV. SUMMARY

Blood-borne viral agents are generally present in the circulating leukocytes as well as in the plasma. Such infectious agents are transmitted to recipients of blood transfusion. Nucleic acid hybridization has been recently used as a powerful tool to gain new insights into the biology of HIV, HBV, and CMV infections. However, such molecular analysis still remains a research tool that needs to refined further. The fact that kits for dot-blot analysis are commercially available and ready to be used in clinical diagnosis of HIV, HBV, and hepatitis D virus connotes the practical application of unprecedented technology entering the field of transfusion medicine.

ACKNOWLEDGMENT

Hubert E. Blum is recipient of the Heisenberg Award. This work was supported in part by NHLBI Grant POI HL-36489, Contract NOI-HL-86-2074, and NIAID Contract NOI-AI-32519.

REFERENCES

1. **Becker, W. M.,** *The World of the Cell,* Benjamin/Cummings, Menlo Park, Calif., 1986.
2. **Schmid, R.,** Keynote address: viral hepatitis: some historical perspectives, in *Viral Hepatitis and Liver Disease,* Vyas, G. N., Hoofnagle, J. H., and Dienstag, J. L., Eds., Grune & Stratton, Orlando, Fla., 1984, 1.
3. **Gerety, R. J., Tabor, E., Schaff, Z., Seto, B., and Coleman W. G.,** Non-A, non-B hepatitis agents, in *Viral Hepatitis and Liver Disease,* Vyas, G. N., Hoofnagle, J. H., and Dienstag, J. L., Eds., Grune & Stratton, Orlando, Fla., 1984, 23.
4. **Tiollais, P., Dejean, A., Brechot, C., Michel, M.-L., Sonigo, P., and Wain-Hobson, S.,** Structure of hepatitis B virus DNA, in *Viral Hepatitis and Liver Disease,* Vyas, G. N., Hoofnagle, J. H., and Dienstag, J. L., Eds., Grune & Stratton, Orlando, Fla., 1984, 49.
5. **Summers, J.,** Three recently-described animal virus models for human hepatitis B virus, *Hepatology,* 1, 179, 1981.
6. **Beasley, R. P. and Hwang, L.-Y.,** Epidemiology of hepatocellular carcinoma, in *Viral Hepatitis and Liver Diseases,* Vyas, G. N., Hoofnagle, J. H., and Dienstag, J. L., Eds., Grune & Stratton, Orlando, Fla., 1984, 209.
7. **Figus, A., Blum, H. E., Vyas, G. N., De Vergilis, S., Cao, A., Lippi, M., Lai, E., and Balestiere, A.,** Hepatitis B viral nucleotide sequences in non-A, non-B or hepatitis B virus-related chronic liver disease, *Hepatology,* 4, 364, 1984.
8. **Vyas, G. N. and Perkins, H. A.,** Non-B posttransfusion hepatitis associated with hepatitis B core antibiodies in donor blood, *N. Engl. J. Med.,* 306, 749, 1982.

9. **Cummings, J. W., Browne, J. K., Salser, W., Tyler, G. V., Snyder, R. L., Smolec, J. M., and Summers, J.,** Isolation, characterization and comparison of recombinant DNA derived from genomes of human hepatitis B virus and woodchuck hepatitis virus, *Proc. Natl. Acad. Sci., U.S.A.,* 77, 1842, 1980.
10. **Rigby, P. W. J., Dieckmann, M., Rhodes, C., and Berg, P.,** Labelling deoxyribonucleic acid to high specific activity in vitro by nick translation with DNA polymerase I, *J. Mol. Biol.,* 113, 237, 1977.
11. **Blum, H. E., Stowring, L., Figus, A., Montgomery, C. K., Haase, A. T., and Vyas, G. N.,** Detection of hepatitis B virus DNA in hepatocytes, bile duct epithelium, and vascular elements by in situ hybridization, *Proc. Natl. Acad. Sci., U.S.A.,* 80, 6685, 1983.
12. **Blum, H. E., Haase, A. T., Harris, J. D., Walker, D., and Vyas, G. N.,** Asymmetric replication of hepatitis B virus DNA in human liver: demonstration of cytoplasmic minus-strand DNA by blot analyses and in situ hybridization, *Virology,* 139, 87, 1984.

12a. **Blum, H. E. and Vyas, G. N.,** Hepatitus B virus in nonhepatocytes, *Lancet,* 2, 920, 1983.

13. **Scotto, J., Hadchouel, M., Hery, C., Yvart, J., Tiollais, P., and Brechot, C.,** Detection of hepatitis B virus DNA in serum by a simple spot hybridization technique: comparison with results for other viral markers, *Hepatology,* 3, 279, 1983.
14. **Brechot, C., Hadchouel, M., Scotto, J., Degos, F., Charnay, P., and Tiollais, P.,** Detection of hepatitis B virus DNA in liver and serum: a direct appraisal of the chronic carrier state, *Lancet,* 2, 765, 1981.
15. **Brechot, C., Hadchouel, M., Scotto, J., Fonck, M., Potet, F., Vyas, G. N., and Tiollais, P.,** State of hepatitis B virus DNA in hpeatocytes of patients with hepatitis B surface antigen-positive and -negative liver diseases, *Proc. Natl. Acad. Sci. U.S.A.,* 78, 3906, 1981.
16. **Maniatis, T., Fritsch, E. F., and Sambrook, J.,** *Molecular Cloning. A Laboratory Manual,* Cold Spring Harbor Laboratory, Cold Spring Harbor, N.Y., 1982.
17. **Rutter, W. J., Ziemer, M., Ou, J., Shaul, Y., Laub, O., Garcia, P., and Standring, D.,** Transcription units of HBV genes and expression of integrated viral sequences, in *Viral Hepatitis and Liver Disease,* Vyas, G. N., Hoofnagle, J. H., and Dienstag, J. L., Eds., Grune & Stratton, Orlando, Fla., 1984, 67.
18. **Haase, A., Brahic, M., Stowring, L., and Blum, H.,** Detection of viral nucleic acids by in situ hybridization, in *Methods in Viriolgy,* Vol. 7, Maromorosch, K. and Koprowski, H., Eds., Academic Press, New York, 1984, 189.
19. **Brahic, M. and Haase, A. T.,** Detection of viral sequences of low reiteration frequency by in situ hybridization, *Proc. Natl. Acad. Sci. U.S.A.,* 75, 6125, 1978.
20. **Angerer, L. M. and Angerer, R. C.,** Detection of poly A RNA in sea urchin eggs and embryos by quantitative in situ hybridization, *Nucleic Acids Res.,* 9, 2819, 1981.
21. **Moench, T. R., Gendelman, H. E., Clements, J. E., Narayan, O., and Griffin, D. E.,** Efficiency of in situ hybridization as a function of probe size and fixation technique, *J. Virol. Methods,* 11, 119, 1985.
22. **Blum, H. E., Haase, A. T., and Vyas, G. N.,** Molecular pathogenesis of hepatitis B virus infection: simultaneous detection of viral DNA and antigens in paraffin-embedded liver sections, *Lancet,* 2, 771, 1984.
23. **Shimizu, Y., Ida, S., Matsukura, T., and Yuasa, T.,** Determination of hepatitis B virus DNA in serum by molecular hybridization, *Microbiol. Immunol.,* 28, 1117, 1984.
24. **Morace, G., von der Helm, K., Jilg, W., and Deihardt, F.,** Detetion of hepatitis B virus DNA in serum by a rapid filtration-hybridization assay, *J. Virol. Methods,* 12, 235, 1985.
25. **Gallo, R. C. and Montagnier, L.,** The chronology of AIDS research, *Nature (London),* 326, 435, 1987.
26. **Haase, A. T.,** Pathogenesis of lentivirus infections, *Nature (London),* 322, 130, 1986.
27. **Coffin, J., Haase, A. T., Levy, J. A., et al.,** What to call the AIDS virus?, *Nature (London),* 321, 10, 1986.
28. **Allain, J. P., Laurian, Y. and Paul, D. A.,** Serologic markers in early stages of HIV infections in hemophiliacs, *Lancet,* 2, 1233, 1986.
29. **Anderson, R. E., Winkelstein, W., Blum, H. E., and Vyas, G. N.,** Hepatitis B virus infection in the acquired immune deficiency syndrome, in *Viral Hepatitis and Liver Disease,* Vyas, G. N., Hoofnagle, J. H., and Dienstag, J. L., Eds., Grune & Stratton, Orlando, Fla., 1984, 339.
30. **Rees, M.,** The sombre view of AIDS, *Nature (London),* 326, 343, 1987.

30a. **Donegan, E. and the Transfusion Safety Study Group,** Serum antibody positivity in recipients of anti-HIV positive blood, *Transfusion,* 26, 576, 1986.

30b. **Peterman, T. S., Lui, K., Lawrence, D. N., and Allen, J. R.,** Estimating the risk of transfusion-associated AIDS and HIV infection, *Transfusion,* 27, 371, 1987.

31. **Busch, M. P., Rajagopalan, M. S., Gantz, D. M., Fu, S., Steimer, K. S., and Vyas, G. N.,** In situ hybridization and immunochemistry for improved assessment of HIV cultures, *Am. J. Clin. Pathol.,* 88, 673, 1987.
32. **Harper, M. E., Marselle, L. M., Gallo, R. C., and Wong-Stall, F.,** Detection of lymphocytes expressing human T-lymphotropic virus type III in lymph nodes and peripheral blood from infected individuals by in situ hybridization, *Proc. Natl. Acad. Sci. U.S.A.,* 83, 772, 1986.
33. **Gallo, D., Kimpton, J. S., and Dailey, P. J.,** Comparitive studies on the use of fresh and frozen peripheral blood lymphocyte specimens for the isolation of HIV and the effects of cell lysis on isolation efficiency, *J. Clin. Microbiol.,* 25, 1291, 1987.

34. **Luciw, P. A., Potter, S. J., Steimer, K., and Dina, D.,** Molecular cloning of AIDS-associated retrovirus, *Nature (London),* 312, 760, 1984.
35. **Stoler, M. H., Eskin, T. A., Benn, S., Angerer, R. C., and Angerer, L. M.,** Human T-cell lymphotropic virus type III infection of the central nervous system, *JAMA,* 256, 2360, 1986.
36. **Cox, K. H., Deleon, D. V., Angerer, L. M., and Angerer, R. C.,** Detection of mRNAs in sea urchin embryos by in situ hybridization using asymmetric RNA probes, *Dev. Biol.,* 101, 485, 1984.
37. **Krieg, P. A. and Melton, D. A.,** Functional messenger RNAs are produced by SP6 in vitro transcription of cloned cDNAs, *Nucleic Acids Res.,* 12, 7057, 1984.
38. **Ulrich, P. P., Donegan, E., El-Beik, T., Shriver, K., and Vyas, G. N.,** Culture of human immunodeficiency virus from seropositive recipients of HIV-infected blood correlated with lymphocyte subsets, *Abstr. 3rd Int. Conf. Acquired Immunodeficiency Syndrome (AIDS),* Washington, D.C., 1987.
39. **Ulrich, P. P., El-Beik, T., Busch, M. P., Donegan, E., and Vyas, G. N.,** Human immunodeficiency virus cultured from limiting dilutions of the peripheral blood mononuclear cells of asymptomatic seropositive persons, *Abstr. 3rd Int. Conf. Acquired Immunodeficiency Syndrome (AIDS),* Washington, D.C., 1987.
40. **Saiki, R. K., Scharf, S., Faloona, F., Mullis, K. B., Horn, G. T., Ehrlich, H. A., and Arnheim, N.,** Enzymatic amplification of beta-globin genomic sequences and restriction site analysis for diagnosis of sickle cell anemia, *Science,* 230, 1350, 1985.
41. **Sninsky, J. J., Kwok, S., Mack, D., Ehrlich, G., Ulrich, P. P., Poiesz, B., et al.,** Identification of HIV sequence in DNA extracted directly from peripheral blood lymphocytes and bone marrow using in vitro DNA amplification, *Abstr. 3rd Int. Conf. Acquired Immunodeficiency Syndrome (AIDS),* Washington, D.C., 1987.
42. **Laure, F., Zagury, D., Saimont, A. G., Gallo, R. C., Hahn, B. H., and Brechot, C.,** Hepatitis B virus DNA sequences in lymphoid cells from patients with AIDS and AIDS-related complex, *Science,* 229, 561, 1985.
43. **Noonan, C. A., Yoffe, B., Mansell, P. W. A., Melnick, J. L., and Hollinger, F. B.,** Extrachromosomal sequences of hepatitis B virus DNA in peripheral blood mononuclear cells of AIDS patients, *Proc. Natl. Acad. Sci. U.S.A.,* 83, 5698, 1986.
44. **Dejean, A., Lugassy, C., Zafrani, S., Tiollais, P., and Brechot, C.,** Detection of hepatitis B virus DNA in pancreas, kidney and skin of two human carriers of the virus, *J. Gen. Virol.,* 65, 651, 1984.
45. **Siddiqui, A.,** Hepatitis B virus DNA in Kaposi sarcoma, *Proc. Natl. Acad. Sci. U.S.A.,* 80, 4861, 1983.
46. **Romet-Lemonne, J.-L., McLane, M. F., Elfassi, E., Haseltine, W. A., Azocar, J., and Essex, M.,** Hepatitis B virus infection in cultured human lymphoblastoid cells, *Science,* 221, 667, 1983.
47. **Elfassi, E., Romet-Lemone, J.-L., Essex, M., McLane, M. F., and Haseltine, W. A.,** Evidence of extrachromosomal forms of hepatitis B virus DNA in a bone marrow culture obtained from a patient recently infected with hepatitis B virus, *Proc. Natl. Acad. Sci. U.S.A.,* 81, 3526, 1984.
48. **Lie-Injo, L. E., Balasegaram, M., Lopez, C. G., and Herrera, A. R.,** Hepatitis B virus DNA in liver and white blood cells of patients with hepatoma, *DNA,* 2, 299, 1983.
49. **Pontisso, P., Poon, M. C., Tiollais, P., and Brechot, C.,** Detection of hepatitis B virus DNA in human blood mononuclear cells, *Br. Med. J.,* 288, 1563, 1984.
50. **Morichika, S., Hada, H., Arima, T., Togawa, K., Watanabe, M., and Nagashima, H.,** Hepatitis B virus DNA eplication in peripheral blood mononuclear cells, *Lancet,* 2, 1431, 1985.
51. **Pasquinelli, C., Laur, F., Chatenoud, L. Beaurin, G., Gazengel, C., Bismuth, H., Degos, F., Tiollais, P., Bach, J. F., and Brechot, C.,** Hepatitis B virus DNA in mononuclear blood cells, *J. Hepatol.,* 3, 95, 1986.
52. **Yoffe, B., Noonan, C. A., Melnick, J. L., and Hollinger, F. B.,** Hepatitis B virus DNA in mononuclear cells and analysis of cell subsets for the presence of replicative intermediates of viral DNA, *J. Infect. Dis.,* 153, 471, 1986.
53. **Hoar, D. I., Bowen, T., Matheson, D., and Poon, M. C.,** Hepatitis B virus DNA is encirched in polymorphonuclear leukocytes, *Blood,* 66, 1253, 1985.

Chapter 14

DETECTION OF VARICELLA-ZOSTER, HERPES SIMPLEX, AND ENTERIC ADENOVIRUSES USING NUCLEIC ACID HYBRIDIZATION

Kenneth D. Croen, Jeffrey M. Ostrove, and Stephen E. Straus

TABLE OF CONTENTS

I. INTRODUCTION

The three families of DNA viruses to be discussed in this chapter are all common human pathogens, and our understanding of each has benefited greatly by advances in the field of molecular biology. Hybridization technology has played a fundamental role in unraveling many questions about these pathogens including their genetics, mechanisms of replication and gene expression, structure and organization of the virions, and the state of latency. This powerful technology inevitably made its way into the clinical arena where it has already begun to facilitate diagnosis, define epidermiology, and clarify the pathogenesis of disease processes. We will address the molecular biology of each virus relevant to hybridization studies, and then discuss some of the research and clinical applications of each.

II. HYBRIDIZATION TECHNIQUES AND SAMPLE PREPARATION

Although a detailed discussion of these techniques appears elsewhere in this text, we will briefly review some of the available methods. The simplest technique is that of the dot- or slot-blot hybridization, wherein purified DNA is applied directly to nitrocellulose paper and then hybridized with the desired probe. This is a very sensitive assay whose lower limits of detection generally fall in the range of 1 to 10 pg of DNA. For a virus the size of varicella-zoster virus (VZV), this corresponds to a range of 7500 to 75,000 copies of the viral genome. The price paid for the simplicity, speed, and sensitivity of this technique comes in the form of specificity. Several factors, such as viscosity of the samples applied to the filter, high G + C contents, and the presence of homologous sequences in contaminating eukaryotic, prokaryotic, and viral genomes, may all result in false positive signals. Thus, careful attention to negative controls is critical for an accurate interpretation of dot-blot results.

Further simplifying the dot-blot procedure, certain samples can be applied directly to the nitrocellulose, thus bypassing the DNA extraction steps. The addition of chaotropic salts, such as NaI, to a sample allows for rapid cell lysis and precipitation of DNA directly onto nitrocellulose filters. Again, the major concern has to do with specificity of the result as the additional protein, mucus, and nucleic acid components of unprocessed samples can only increase the nonspecific binding.

Specificity problems can be substantially reduced by performing a Southern blot hybridization.[1] This entails DNA extraction from a sample, restriction enzyme digestion, agarose gel electrophoresis, transfer to a nitrocellulose membrane, and finally, hybridization with the desired probe. The resulting series of bands which represent the different restriction fragments of the virus (see, for example, Figure 6), is highly specific and is ample justification, in many instances, for employing this more time-consuming technique. Because the signal is divided among several bands, the sensitivity of a Southern blot hybridization is not quite that of a dot blot. Generally, 5 to 20 pg of viral DNA are needed to detect a signal by this technique. Assuming that one can electrophorese 20 μg of total DNA per lane, then a viral sequence could be detected if one copy were present in 5 to 10% of all cells in the sample.

In situ hybridization, though far more complex than the other techniques, offers several distinct advantages: (1) very low copy numbers of a virus can be detected; (2) the signal is not diminished or lost, regardless of the number of uninfected surrounding cells; (3) the specific cell type bearing the infection can be determined by histologic or simultaneous immunocytochemical methods; and (4) an adequate signal can be obtained even if very few cells are available for study. A detailed review of this procedure has recently been published,[2] and many variations in the methodology have appeared in the literature.[3-6] In brief, cells or tissue sections (frozen or embedded) are placed onto adhesive-coated slides, then treated with proteinase so as to expose nuclear and cytoplasmic contents, and finally hybridized with either DNA or RNA probes. Probe size is important in this protocol, as large probes cannot weave their way through cross-linked

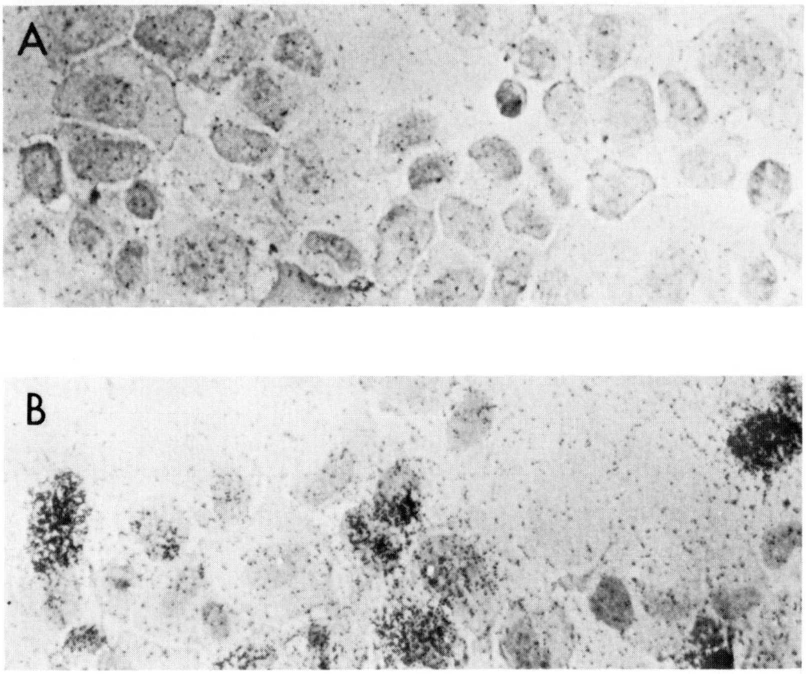

FIGURE 1. *In situ* hybridization of uninfected (A) and VZV-infected (B) human embryonic fibroblasts using a ^{35}S-RNA probe transcribed from a cloned *Sma*I-E insert in a pGEM vector (Promega) using the T7 polymerase. Infected cells were harvested at 20 to 25% CPE. Exposure was for 3 days at 4°C.

cellular components as readily as smaller ones. The optimal size has been shown to be in the range of 50 to 150 nucleotides.[2,3] Following the wash procedure, slides are dipped in radiographic emulsion and exposed for varying lengths of time prior to development and staining. Simultaneous immunochemical staining of samples for a particular antigen can be performed using an avidin-biotin-coupled peroxidase complex in the presence of diaminobenzidine.[7] Figure 1 shows an example of VZV-infected and uninfected human embryonic fibroblasts using an RNA probe from the U_S region of the VZV genome (see also Figure 2).

Quantitation of an *in situ* hybridization signal can be estimated,[4,8] but the accuracy does not compare with that derived from densitometric measurements of Southern blots and slot blots. Nevertheless, to illustrate the power of *in situ* hybridization, which can detect between 1 and 100 copies of a gene or its message (RNA) within a single cell,[8] consider the following: the presence of one VZV-infected cell on a slide with 100 copies of the genome (equal to 1 fg) will give a positive signal by *in situ* hybridization. As discussed above, the limits of detection by blot hybridizations are in the neighborhood of 10 pg, or 10^4-fold more virus than is necessary for *in situ* hybridization. Thus, for the detection of latent viruses such as herpes simplex virus (HSV) or VZV, *in situ* hybridization is the best available approach.

III. PROBE SELECTION

The accumulation of large quantities of data on the structure and sequence of viral genomes allows for the selection of very specific regions to be used as probes. Several considerations go into probe selection, and different probes may be needed to answer different questions. For example, if one is looking for a replicating virus, in which case virtually all of the genome is present in abundance, just about any segment (or segments) of the genome that lacks homology to the cellular DNA can be chosen. If, on the other hand, one is trying to detect a latent virus,

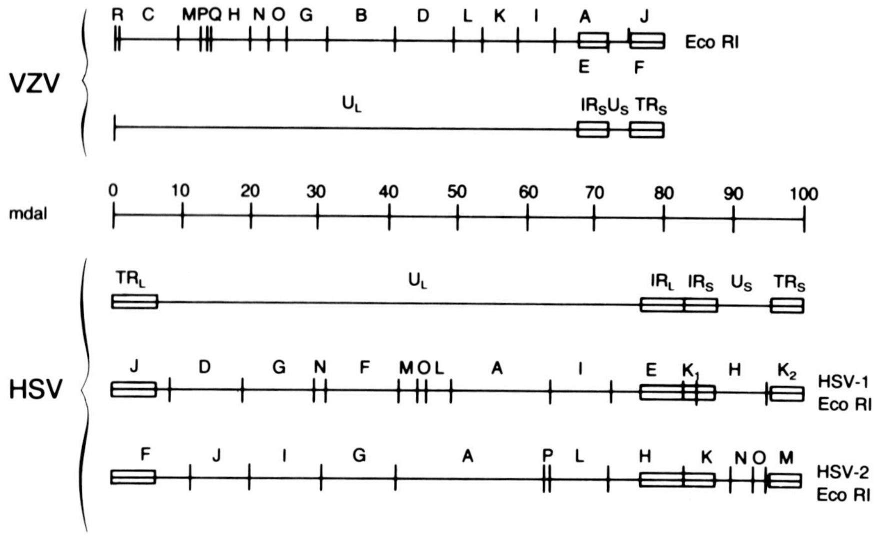

FIGURE 2. Oragnization and *Eco*RI restriction maps of the VZV and HSV genomes.

then an RNA probe complementary to a known or anticipated latent transcript is desirable. This takes advantage of a potential transcriptional amplification of the DNA sequence as well as the more avid binding of RNA:RNA hybrids as compared to DNA:DNA hybrids. If one is trying to distinguish between two different viruses or virus types (i.e., HSV-1 vs. HSV-2), then obviously, specific segments must be chosen. This also becomes a critical issue for the enteric adenoviruses (EAds) as will be discussed in detail.

Before getting to the choice of probes relevant for each individual virus, we will briefly discuss some considerations about the probe label. Radiolabeled nucleotides are the most commonly used markers and are incorporated into DNA probes by a variety of techniques, including nick translation, M13-primed DNA synthesis, and reverse-transcriptase generated cDNA. RNA probes (riboprobes) can be generated from cloned DNA fragments containing bacteriophage transcriptional promoters SP6, T3, and T7. High specific activity probes are desirable for enhancing sensitivity and decreasing the duration of exposure needed to detect a signal. While only a small percent of all possible residues becomes labeled with nick translation, virtually all may be labeled when preparing riboprobes. Thus, despite the lower specific activity of, for instance, ^{35}S-labeled nucleotides compared to those of ^{32}P, riboprobes with specific activities in the range of 5 to 10×10^8 cpm/μg can be generated with ^{35}S. The ^{35}S-labeled probes are desirable for *in situ* hybridization, since the high-energy emissions of ^{32}P-labeled probes result in excessive scatter of the signal.

Another labeling alternative, one which eliminates the need for radioactive compounds, involves the use of biotinylation and an enzymatic detection system.[9,10] This labeling technique is based on the high affinity of streptavidin for biotin and the ability to link biotin both to nucleotides and to enzymes such as peroxidase or alkaline phosphatase. Following hybridization and wash, a complex of streptavidin-biotin-enzyme is allowed to conjugate with the biotinylated probe bound to sequences on the blot or slide. Detection is then accomplished by adding a substrate for the enzyme. The stability and reduced biohazard of these probes makes them very attractive for use in a diagnostic lab.

A. Varicella-Zoster Virus (VZV)

Our knowledge of the molecular biology of VZV lags significantly behind that of its closest human herpesvirus relative, HSV, due largely to two properties of this virus. VZV grows very slowly in tissue culture and is highly cell associated, making the isolation of pure VZV DNA or cell-free virus very difficult. Consequently, cloning of the VZV genome became a prerequisite to detailed biochemical studies and has only recently been accomplished.[11,12]

VZV contains a linear genome that is approximately 125,000 (bp) in length, or 80×10^6 daltons. As shown in Figure 2, the genome consists of long (U_L) and short (U_S) unique sequences which are flanked by inverted repeats. The genome primarily exists in two isomeric forms, dependent upon inversion of the U_S.[12-14] Inversion of the U_L relative to the U_S has also been recently demonstrated to occur in very low frequency.[15] The overall G + C content is 47%, though substantial portions of the inverted repeats as well as mini-repeat elements within *Eco*RI-P may run as high as 65%.[16] Numerous restriction maps of the genome have been generated as, for example, the *Eco*RI map shown in Figure 2. A preliminary map of VZV transcripts reveals the genomic locations of 58 mRNAs,[17] and follow-up studies using undirectional RNA probes identified a number of additional transcripts and assigned their directions of transcription.[18] These results provide a rational basis for the selection of probes to characterize genome transcription in productive and latent infections.

Another approach to probe selection utilizes segments of the viral genome which encode known proteins or glycoproteins. Only a few VZV gene products have been mapped directly including glycoproteins[19] and thymidine kinase.[20] The sequence of the entire VZV genome has recently been determined by Davison and Scott.[21] This will facilitate comparisons of nucleic acid and amino acid sequences of VZV open reading frames to known proteins and glycoproteins of HSV, thus enabling their tentative localization on the VZV genome.

In order for a probe to be useful for the detection of a praticular gene or region within the genome, it must be shown to lack reactivity with other sources of DNA or RNA. Homology of VZV sequences to the human genome has not been observed in numerous DNA-DNA blot hybridizations.[13] During the studies of mRNA directionality, however, Northern blots (RNA-RNA) showed significant hybridization between several VZV segments and 18S and 28S rRNA at T_m – 24°C.[18] In addition, homology of VZV to HSV and other herpesviruses has been demonstrated under conditions of low stringency (T_m – 26°C to T_m – 36°C).[22] Nonspecific binding due to the high G + C content of portions of the herpesvirus genomes may have been responsible for some of the hybridization observed. This is of particular relevance in the evaluation of clinical samples for VZV where it is often important to exclude HSV.

B. Herpes Simplex Virus (HSV)

That HSV-1 and HSV-2 are two distinct viruses is evidenced by clinical, immunologic, and restriction pattern differences. Structurally, the two genomes have been shown to be quite similar. As indicated in Figure 2, the genomes have the same overall organization and have been shown to be colinear with respect to location and orientation of major genes.[22] Each are about 160,000 bp, or 98×10^6 daltons, contain a pair of inverted repeats flanking each unique sequence, and have G + C contents of 67% (HSV-1) and 69% (HSV-2). As in VZV, the repeats have a greater G + C content, approaching 80%. Inversion of the unique regions relative to each other generates four isomeric forms which are present in equimolar amounts.

Cross-hybridization studies demonstrate extensive homology between HSV-1 and HSV-2.[21,23] Nevertheless, it is possible to select probes of sufficiently low relative homology which, when used under appropriately stringent conditions, permit differentiation between HSV-1 and HSV-2.[24,25] In addition to intertypic homology, HSV-1 and HSV-2 hybridize to other herpesviruses, and, to a lesser degree, to mammalian DNA sequences.[26-29] The regions of the genome bearing this homology fall within the inverted repeats as well as in the mid portion of the U_L. One of these regions, that from the inverted repeats flanking the U_L, binds to the mammalian 28S ribosomal gene.[29]

The ability to chose probes based on the locations of specific gene products has obvious benefit when attempting to detect RNA transcripts. As with VZV, a detailed transcription map

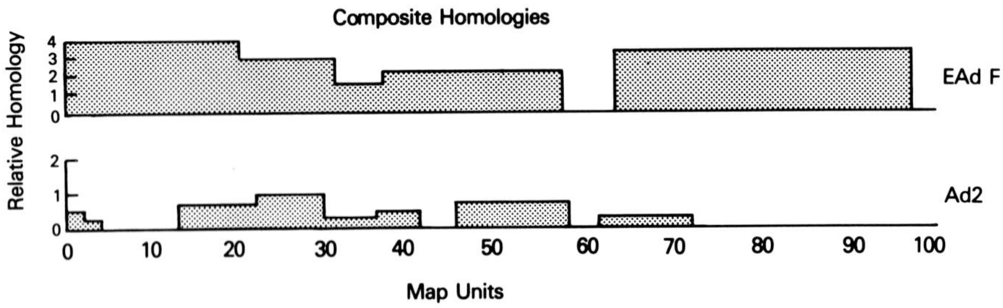

FIGURE 3. Composite maps of homologies between EAd type 41 and EAd F (type 40) or Ad2; obtained by cross-hybridizations of Southern blots done at $T_m - 22°C$. (From Takiff, H. E., Reinhold, W., Garon, C. F., and Straus, S. E., *J. Virol.*, 51, 131, 1984. With permission.)

for HSV is available.[30] Unlike VZV, the map locations for about 20 specific HSV proteins and glycoproteins are known. Furthermore, the time of appearance of many transcripts in the life cycle of the HSV infection (α, β, σ; immediate early, early, and late, respectively) is well defined. Thus, with appropriate selection of probes, questions about the expression of specific genes can be addressed.

C. Enteric Adenoviruses (EAds)

Adenoviruses are currently classified into six groups (A to F) based on DNA homology, and within these groups, 40 serotypes have been identified. The most recent additions to this family of viruses, the enteric adenoviruses, or EAd types 40 and 41, are extremely fastidious in tissue culture, growing poorly in cell lines typically used to support adenovirus infection (i.e., HEK, HeLa, HEp-2). For reasons that are poorly understood, they grow reliably in the adenovirus type 5-transformed 293 cells.[31] These cells are transformed with the left hand 12% of the Ad type 5 genome and express the E1A and E1B genes contained therein. Growth in 293 cells, as well as other diagnostic strategies, has helped implicate the EAds in 3 to 52% of cases in childhood enteritis.[32-36]

The genomes of EAd types 40 and 41 are linear DNA molecules consisting of approximately 34,000 bp and weighing 23×10^6 daltons. They each have a G + C content of 52% and, between them, have 62 to 69% homology based on reassociation kinetics studies.[37] Takiff et al. cloned these adenoviruses into bacterial plasmids and, by cross-hybridizing with Ad2 DNA, demonstrated regions of reduced and non-detectable homology between the EAds and Ad2 as shown in Figure 3.[38] Having recognized the two EAd-specific regions, they selected a probe from the right end of the EAd41 genome, mapping at about 75 to 85 map units (see Figure 4), for use in a dot-blot assay.[39] This probe was shown to hybridize to EAds 5000-fold more avidly than to other Ads. In a similar study, Kidd et al. selected type-specific probes by cross-hybridization studies from the right end of the genome of type 40 (map units 68 to 73), but from the left end of type 41 (map units 11 to 30).[40] Results from both of these studies will be discussed below.

IV. APPLICATIONS

Rather than attempt to catalog all of the applications of nucleic acid hybridization relevant to the study of these three groups of viruses, we will provide a few examples which show the potential of these techniques and which have the greatest bearing on clinical issues.

A. Varicella-Zoster Virus (VZV)

VZV is a highly prevalent pathogen of man that causes a characteristic cutaneous lesion both with primary infection (varicella) and with reactivation (zoster). While this is a relatively benign infection in the normal host, it causes substantial morbidity and mortality among immunocom-

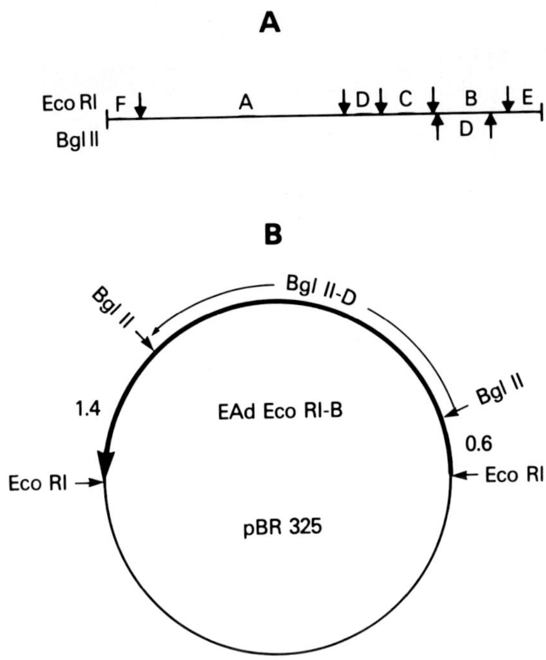

FIGURE 4. (A) *Eco*RI and partial *Bgl*II restriction maps of the EAd type 41 genome. (B) *Eco*RI-B fragment of EAd type 41 inserted into pBR325 in reverse orientation. (From Takiff, H. E., Seidlin, M., Krause, P., Rooney, J., Brandt, C., Rodriguez, W., Yolken, R., and Straus, S. E., *J. Med. Virol.*, 16, 107, 1985. With permission.)

promised patients. Such patients are, therefore, candidates for antiviral therapy with acyclovir or vidarabine.[41,42] A rapid diagnostic test would obviously be useful in this setting. Currently, laboratory confirmation of VZV infection relies upon the development of characteristic cytopathic changes in cell culture, a process which usually requires 4 to 7 days, but may take as long as 14 days. Even in the best of hands, the sensitivity of tissue culture is only 60 to 85%, due largely to the lability of the virus.

As shown by Seidlin et al., a diagnosis based on the detection of viral DNA can be made in 1 to 2 days and does not require viable virions.[43] In the dot-blot protocol used in that study, vesicle aspirates were treated with 1% SDS, while other samples were freeze-thawed and treated with 1% SDS and proteinase K, followed by phenol extraction and ethanol precipitation, before being applied to nitrocellulose filters. Their probe was a nick-translated mixture of cloned *Eco*RI fragments representing almost the entire U_L plus *Eco*RI-E. Hybridization was carried out at T_m – 21°C. Figure 5 shows a typical blot containing 29 retrospectively analyzed samples, of which 9 were considered positive. Overall, they observed a sensitivity of 76% and a specificity of 94%, which compares with their tissue culture results of 58% and 100%, respectively.

Although we can readily recognize and treat some VZV infections, our understanding of their pathogenesis is far from complete. Among the many unresolved questions is that of the role of circulating white blood cells during the course of the infection. The well-recognized viremia of varicella has been shown to involve infection of circulating mononuclear cells.[44-46] The fact that virus can be isolated from the blood of patients during the incubation period as many as 5 days prior to onset of the disease, suggests the potential use of a dot-blot assay for the early detection of infection in susceptible hosts. This may be of particular value in the compromised patient or in the screening of exposed hospital personnel. The existence of circulating virus in patients with zoster has not been clearly defined. Gilden et al. have presented dot-blot hybridizations

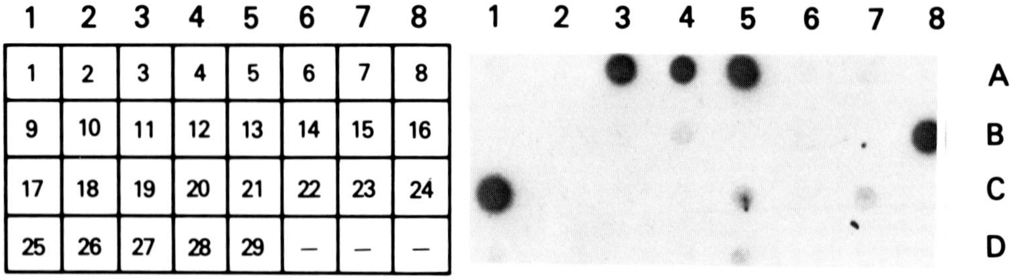

FIGURE 5. Dot-blot hybridization of 29 clinical specimens probed for VZV (see text). Samples 3, 12, 16, 17, and 23, were culture positive while samples 4, 5, and 21 were culture negative, but were from patients with a strong clinical impression of varicella or zoster. Sample 29 was a throat wash from a patient with hypogammaglobulinemia but no signs of VZV infection. (From Seidlin, M., Takiff, H. E., Smith, H. A., Hay, J., and Straus, S. E., *J. Med. Virol.*, 13, 53, 1984. With permission.)

suggesting leukocyte-associated VZV in 4 of 35 immunocompromised patients.[47] We have also looked for VZV in circulating mononuclear and polymorphonculear neutrophilic leukocytes from patients with zoster, as well as immunocompromised patients without zoster, but failed to detect it by dot-blot hybridization in the 25 samples examined.[48] A definitive answer to this question awaits further study.

The pathogenesis of the postzoster encephalitis syndrome is another area where there is a gap in our understanding. Because the literature contains very few indications for a direct involvement of VZV, the conclusion has been that this is an immune-mediated process. A recent report of an AIDS patient with a progressive encephalopathy following his second episode of cutaneous zoster includes the Southern blot hybridization shown in Figure 6, which documents a productive VZV infection.[49] While the contribution of an immunopathogenic process in this setting is not excluded, it does suggest that antiviral therapy may be of value in some patients with this disorder.

The latent state of VZV has proven far more difficult to study than that of HSV. While one can readily recover HSV via cocultivation of sensory ganglia, this has never been accomplished for VZV. To date, two reports have appeared in which evidence of latent VZV has been demonstrated in neural tissue. Hyman et al. performed *in situ* hybridizations of nine ganglia from five autopsies using a nick-translated probe spanning the entire VZV genome. They reported a positive signal in three of the nine specimens.[50] Gilden et al. extracted DNA from dorsal root and trigeminal ganglia and performed Southern blot hybridizations with various cloned VZV fragments. Out of 13, 5 trigeminal ganlgia were reported to contain VZV sequences at an estimated 0.28 to 1.0 copy per cell. They also detected signals from a sacral ganglion with seven different *Bam*HI fragments, thus suggesting the presence of the entire VZV genome.[51]

B. Herpes Simplex Virus (HSV)

HSV types 1 and 2 are extremely comon pathogens of humans and cause substantial morbidity and mortality in a wide range of clinical settings. As with VZV and other herpesviruses, a latent state is established following initial exposure. In the normal host, recurrent infections arising from latent foci result in mild mucocutaneous lesions, although rarely, encephalitis can occur. Blindness from recurrent bouts of herpes keratitis, and the psychologic consequences of recurrent genital herpes add to the impact of this virus. Reactivation of latent virus poses a much greater threat, however, to immunocompromised hosts and to the newborn. Without treatment in these settings, the mortality is considerable.

The need for a rapid diagnostic test for HSV is obvious, particularly in light of the availability and efficacy of acyclovir therapy for many of its presentations. While it is unlikely that nucleic

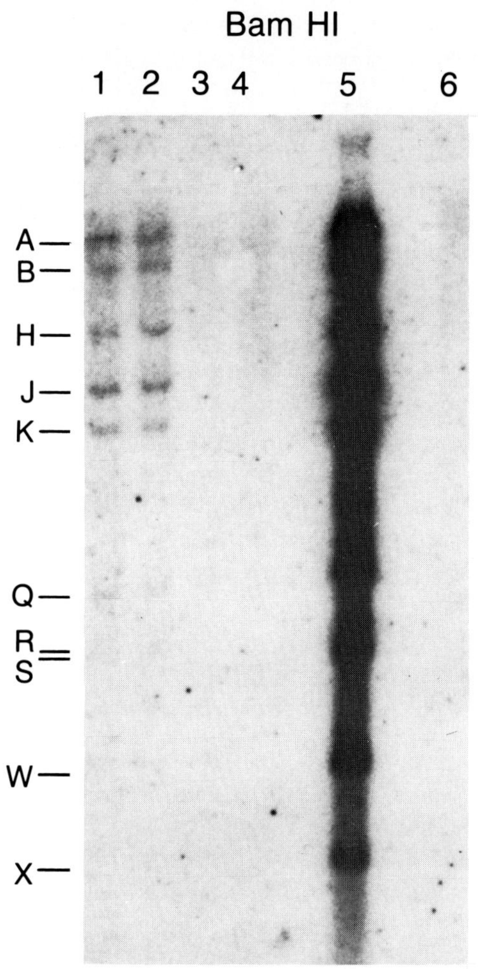

FIGURE 6. Southern hybridization of brain tissue from an AIDS patient who died with progressive encephalopathy following herpes zoster. Lane 1: 22.5 μg of normal human liver DNA plus 50 pg of VZV DNA; lane 2: 19.5 μg of normal human brain DNA plus 50 pg of VZV DNA; lane 3: 15 μg of normal human liver DNA; lane 4: 19.5 μg of normal human brain DNA; lane 5: 20 μg of temporal lobe DNA from the patient with AIDS; lane 5: 20 μg of cerebellum DNA from the patient with AIDS. All samples were digested with *Bam*HI and hybridized with nick-translated VZV fragments *Eco*RI-A and -B at $T_m - 21°C$. The expected homologous *Bam*HI fragments are indicated at the left. Specific hybridization is noted in control lanes 1 and 2 as well as lane 5. Densitometric analysis suggests the presence of 50 to 100 copies of the VZV genome per cell in lane 5. (From Ryder, J. W., Croen, K., Kleinschmidt-DeMasters, B. K., Ostrove, J. M., Straus, S. E., and Cohn, D. L., *Ann. Neurol.,* 19, 182, 1986. With permission.)

acid hybridization studies will surpass culture and immunocytochemical methods for simplicity and speed in diagnosis, there have been some useful applications of this technology in the diagnostic lab.

As mentioned earlier, type-specific regions of the HSV genomes can be used to distinguish between types 1 and 2. Redfield et al.[24] and Richman et al.[25] have taken advantage of this by selecting probes for each virus from the area of the junction between the short and long segments. Using nick-translated probes from these regions, they readily distinguished type 1 from 2 in slot-blot analyses of clinical samples. The ability to distinguish between HSV-1 and HSV-2 is of benefit both epidemiologically and in predicting rates of recurrence.[52]

In a more general application of this technology to the diagnosis of HSV infection, several groups have reported the use of cloned probes in dot-blot assays of clinical samples.[53-55] With careful attention to controls, this approach can compare favorably to other diagnostic tools in terms of sensitivity and specificity. However, because of the greater complexity and labor intensiveness of this approach as compared to other diagnostic tests, it is likely to have limited use. One situation where this may be of particular value, however, is in the diagnosis of HSV when cultures are negative, i.e., in the patient on acyclovir at the time a culture is taken. This may be an increasingly common occurrence as more patients with mucocutaneous disease are placed on chronic suppressive therapy. In addition, dot-blot hybridization can be useful in confirming the growth of HSV in culture.

An ideal clinical application of *in situ* hybridization is in the diagnosis of HSV encephalitis. Since very few infected cells are needed to obtain a signal by this technique, a touch preparation of a diagnostic brain biopsy should be all that is required to detect HSV. Forghani et al.[56] used a biotinylated probe consisting of two HSV-1 fragments and a single HSV-2 fragment (specific clones not specified) for *in situ* hybridization of touch preparations of frozen encephalitis brain tissue. They reported a sensitivity of 94% with a 100% specificity. The cerebrospinal fluid sediment in patients with HSV encephalitis has not yet been examined by *in situ* hybridization, but seems to be ideally suited as a less invasive means for diagnosis.

The first documentation of transmission of genital HSV from an asymptomatic individual to a sexual partner has recently been published.[57] Figure 7 shows the result of a dot-blot hybridization of several patient specimens and controls. Bcause of the handling, storage, and site of specimen collection, many of the samples were contaminated or no longer contained viable virus. Nevertheless, probing the extracted DNA with an HSV-specific probe, a positive signal could still be demonstrated from several samples. This helped to confirm that asymtomatic shedding of HSV had occurred on the day of transmission to a sexual partner. Thus, sample contamination and loss of virus viability, both common obstacles for tissue culture, need not pose insoluble diagnostic problems.

Defining the pathophysiology of HSV infections has been a major research goal, particularly with regard to the latent phase of this infection.[58] Recent studies have identified novel transcripts from a single region of the viral genome during latency while all other genes are inactive.[59,60] These transcripts overlap with a regulatory gene (ICPO) of the virus but in the opposite orientation. This juxtaposition raises the possibility of regulation at the level of an RNA:RNA interaction. Efforts are now underway to determine the role played by these latency transcripts in the pathogenesis of this infection.

C. Enteric Adenoviruses (EAds)

Having only recently been identified, the basic clinical and epidemiologic data on these agents are just now being accrued. In three studies of 1 year duration or greater, in which causes of diarrhea in infants were evaluated, EAds were felt to be responsible for 3.9%, 8.1%, and 15% of cases.[32,33,35] In another study, which covered a 12-week period from late fall to early winter, 52% of infants hospitalized with diarrhea had EAds in their stool.[34] Aspects of the clinical illness were reviewed in two of the studies.[34,35] Both reported a more prolonged course of gastroenteritis

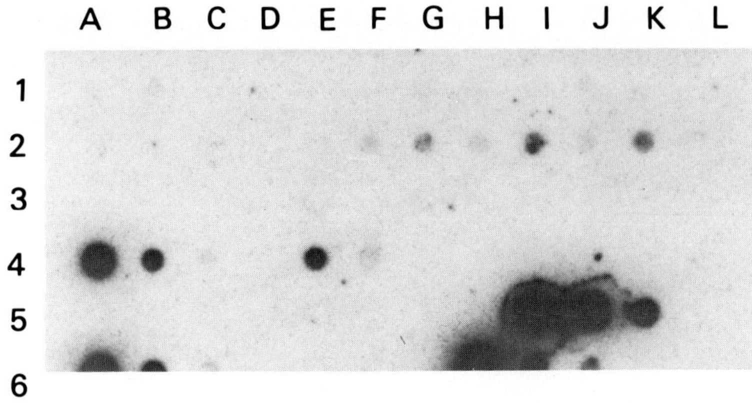

	A	B	C	D	E	F	G	H	I	J	K	L
1	PT. 2 SWABS 9/30	10/1	—	—	PATIENT 2 SWABS 10/4	10/6	10/9	10/10	10/12	10/13	—	—
2	C1	C2	C5	—	C4	C7	PATIENT 2 SWABS 7/11	7/12	7/13	7/14	7/15	7/17
3	—	—	—	—	—	—	—	—	—	—	—	—
4	HSV 1-FAA 1ng	100pg	10pg	—	HSV 2-GL 1ng	100pg	10pg	—	—	—	—	—
5	—	—	—	—	—	—	—	—	PATIENT 2 VIRUS 7/11	7/15	10/2	—
6	HSV-106 DNA 100pg	10pg	1pg	—	VERO DNA 1μg	100ng	—	VERO + HSV-106 100pg	10pg	1pg	—	t-RNA
	A	B	C	D	E	F	G	H	I	J	K	L

FIGURE 7. Dot-blot hybridization of self-culture specimens from a patient and suitable controls. Probed with a 3.4-kb fragment of HSV-1 with 70 to 90% homology to HSV-2. Samples 2H-J were contaminated in tissue culture. sample 2I documents viral shedding on the day of transmission to a sexual partner. (From Rooney, J. F., Felser, J. M., Ostrove, J. M., and Straus, S. E., *N. Engl. J. Med.,* 314, 1561, 1986. With permission)

in the EAd-infected patients, averaging 8 to 12 days as opposed to 4 days among the other cases. Upper respiratory symptoms were noted in 93% of patients in one study, but only 21% in the other. No attempts were made to identify the EAds in respiratory secretions.

Although EAds are cultivable in 293 cells, it may take 14 days or longer for cytopathogenic effects (CPE) to be observed. Thus, a more rapid diagnostic test that can readily distinguish EAds from the other 5 groups of adenoviruses is of great value. Cross-reactivity between antigens from the different adenovirus groups renders immunocytochemical techniques less specific. Takiff et al. used the *Bg*II-D fragment from the right end of the EAd41 genome in a dot-blot assay.[39] Figure 8 confirms the specificity of this fragment, and demonstrates it to possess significantly less cross-reactivity with other adenoviruses than the 600-bp segment immediately to its left on the genome.

In an effort to enhance the sensitivity limit of 20 pg (about 10^5 virions) per gram of stool, samples were inoculated into 293 cells before extracting DNA and blotting. Figure 9 shows the development of a clear signal 25 h after inoculation where none was appreciated earlier, and further shows that this virus cultivation step had no effect on the specificity of the probe. When DNA extracted directly from clinical samples was evaluated by dot-blot hybridization with the *Bg*II-D probe, they detected EAds with 91% sensitivity and excluded other diarrhea-causing

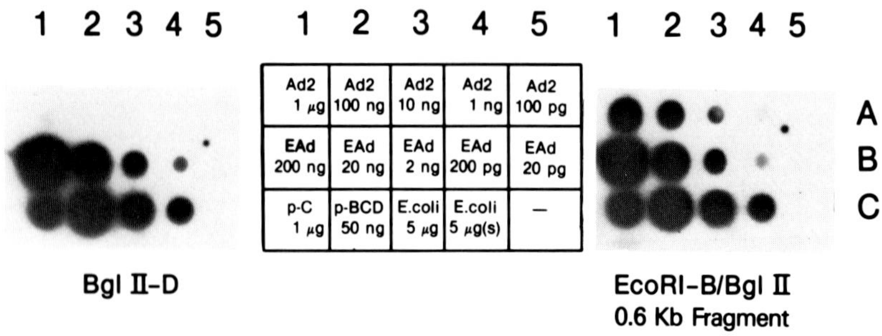

FIGURE 8. Specificity comparison: gel purified *Bgl*II-D vs. adjacent 600-bp fragment (see Figure 4). The 600-bp fragment binds to Ad2 DNA 1000-fold more avidly than *Bgl*II-D. Both probes hybridize to *Eco*RI-C (pC), *Eco*RI-B,C,D (pBCD), and large amounts of *E. coli* DNA, thought to be due to contaminating plasmid and perhaps *E. coli* DNA in the gel purified fragments. (From Takiff, H. E., Seidlin, M., Krause, P., Rooney, J., Brandt, C., Rodriguez, W., Yolken, R., and Straus, S. E., *J. Med. Virol.,* 16, 107, 1985. With permission.)

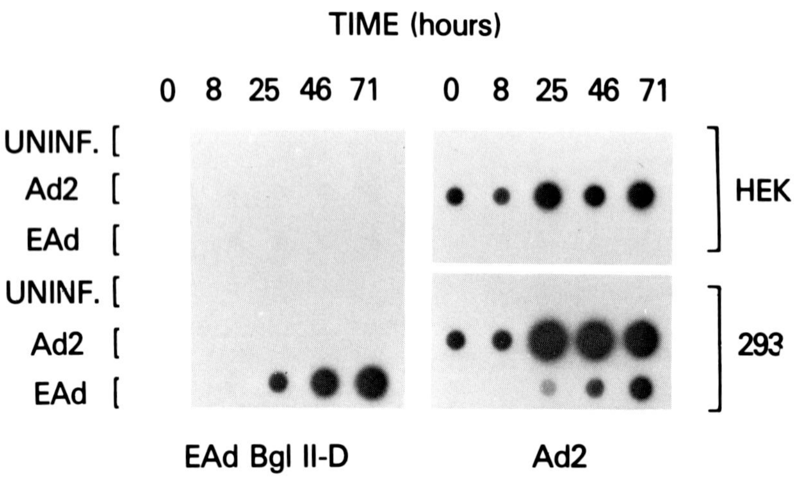

FIGURE 9. Amplification of the EAd type 41 DNA by growth in 293 cells. Roller tubes of HEK cells (top 3 lanes) or 293 cells (bottom 3 lanes) were innoculated with Ad2 or EAd type 41 or left uninfected, and harvested at varying times after infection. Duplicate blots were hybridized either with EAd *Bgl*II-D or Ad2 nick-translated probes. (From Takiff, H. E., Seidlin, M., Krause, P., Rooney, J., Brandt, C., Rodriguez, W., Yolken, R., and Straus, S. E., *J. Med. Virol.,* 16, 107, 1985. With permission.)

agents, including other adenoviruses, with 71% specificity. These numbers compared favorably with tissue culture and virus neutralization assays. Using the identical gel purified *Bg*II-D probe, Hammond et al. have reported 81% sensitivity and 88% specificity with stool samples spotted directly (without prior DNA extraction) on nitrocellulose filters.[61] In a comparable study, Kidd et al. used the type-specific probes noted earlier and reported them to be 100% type specific. When stool samples were directly applied to the nitrocellulose filters, they observed 74% and 88% sensitivity. All false negative samples were subjected to phenol extraction and subsequently gave signals with the expected probe by dot-blot hybridization.[40]

Ground having just been broken on understanding the clinical aspects of this infection, questions of pathophysiology will inevitably follow. Some obvious considerations worthy of investigation include the sites of infection, the role of these agents, if any, in respiratory disease, the ability to establish latency in enteric lymphoid tissue, and the possibility of recurrent disease. At the molecular level, it is of interest that the right hand portion of the adenovirus genome, which in EAds are relatively nonhomologous to the rest of the adenovirus family, includes the region that encodes the fiber protein. This protein is important in cell attachment and thus, may participate in the gastrointestinal tropism of the EAds. The complementation between the left end of Ad5 and the EAds, as evidenced by the ability of 293 cells to support EAd growth, raises several questions about the regulation of viral gene expression and replication. This is just a sample of what are now very approachable problems.

V. CONCLUSION

In this chapter we have touched upon the fundamental issues regarding the use of nucleic acid hybridization for varicella-zoster virus, herpes simplex virus, and the enteric adenoviruses. Although we are at very different levels of sophistication in our understanding of these three DNA viruses, hybridization technologies have a great deal to offer each. An appreciaton of the pathogenesis of acute infection, latency, reativation, and the potential viral role in transformation are among the many goals of such investigations. Ultimately, a fuller understanding of the pathogenesis of these infections will aid in the design of rational strategies for their prevention and treatment.

REFERENCES

1. **Southern, E. M.,** Detection of specific sequences among DNA fragments separated by gel electrophoresis, *J. Mol. Biol.,* 98, 503, 1975.
2. **Haase, A., Brahic, M., Stowing, L., and Blum, H.,** Detection of viral nucleic acids by in-situ hybridization, in *Methods in Virology,* Vol. 7, Maramorosch, K. and Koprowski, H., Eds., Academic Press, New York, 1984, 7.
3. **Cox, K. H., DeLeon, D. V., Angerer, L. M., and Angerer, R. C.,** Detection of mRNAs in sea urchin embryos by *in situ* hybridization using asymmetric RNA probes, *Dev. Biol.,* 101, 485, 1984.
4. **Gendelman, H. E., Narayan, O., Molineaux, S., Clements, J. E., and Ghotbi, Z.,** Slow, persistent replication of lentiviruses: role of tissue macrophages and macrophage precursors in bone marrow, *Proc. Natl. Acad. Sci. U.S.A.,* 8, 7086, 1985.
5. **Harper, M. E., Marselle, L. M., Gallo, R. C., and Wong-Staal, F.,** Detection of lymphocytes expressing human T-lymphotropic virus type III in lymph nodes and peripheral blood from infected individuals by *in situ* hybridization, *Proc. Natl. Acad. Sci. U.S.A.,* 83, 772, 1986.
6. **Singer, R. H. and Ward, D. C.,** Actin gene expression visualized in chicken muscle tissue culture by using *in situ* hybridization with a biotinated nucleotide analog, *Proc. Natl. Acad. Sci. U.S.A.,* 79, 7331, 1982.
7. **Brahic, M., Haase, A. T., and Cash, E.,** Simultaneous *in situ* detection of viral RNA and antigens, *Proc. Natl. Acad. Sci. U.S.A.,* 81, 5445, 1984.
8. **Brahic, M. and Haase, A. T.,** Detection of viral sequences of low reiteration frequency by *in situ* hybridization, *Proc. Natl. Acad. Sci. U.S.A.,* 75, 6125, 1978.
9. **Langer, P. R., Waldrop, A. A., and Ward, D. C.,** Enzymatic synthesis of biotin-labeled polynucleotides: novel nucleic acid affinity probes, *Proc. Natl. Acad. Sci. U.S.A.,* 78, 6633, 1981.
10. **Leary, J. J., Brigati, D. J., and Ward, D. C.,** Rapid and sensitive colorimetric method for visualizing biotin-labeled DNA probes hybridized to DNA and RNA immobilized on nitrocellulose: bio-blots, *Proc. Natl. Acad. Sci. U.S.A.,* 80, 4045, 1983.
11. **Straus, S. E., Owens, J., Ruyechan, W. T., Takiff, H. E., Casey, T. A., VandeWoude, G. F., and Hay, J.,** Molecular cloning and physical mapping of varicella-zoster virus DNA, *Proc. Natl. Acad. Sci. U.S.A.,* 79, 993, 1982.
12. **Ecker, J. R. and Hyman, R. W.,** Varicella zoster virus DNA exists as two isomers, *Proc. Natl. Acad. Sci. U.S.A.,* 79, 156, 1982.
13. **Straus, S. E., Aulakh, H. S., Ruyechan, W. T., Hay, J., Casey, T. A., VandeWoude, G. F., Owens, J., and Smith, H. A.,** Structure of varicella-zoster virus DNA, *J. Virol.,* 40, 516, 1981.

14. **Dumas, A. M., Geelen, J. L. M. C., Weststrate, M. W., Wertheim, P., and Noordaa, J.,** Xba I, Pst I, and Bgl II restriction enzyme maps of the two orientations of the varicella-zoster virus genome, *J. Virol.*, 31, 390, 1981.
15. **Kinchington, P. R., Reinhold, W. C., Casey, T. A., Straus, S. E., Hay, J., and Ruyechan, W. T.,** Inversion and circularization of the varicella-zoster virus genome, *J. Virol.*, 56, 194, 1985.
16. **Kinchington, P. R., Remenick, J., Ostrove, J. M., Straus, S. E., Ruyechan, W. T., and Hay, J.,** Putative glycoprotein gene of varicella-zoster virus with variable copy number of a 42-base-pair sequence has homology to herpes simplex virus gycoprotein C, *J. Virol.*, 59, 660, 1986.
17. **Ostrove, J. M., Reinhold, W., Fan, C.-M., Zorn, S., Hay, J., and Straus, S. E.,** Transcription mapping of the varicella-zoster virus genome, *J. Virol.*, 56, 600, 1985.
18. **Reinhold, W. C., Straus, S. E., and Ostrove, J. M.,** Directionality and further mapping of varicella-zoster virus, *Virus Research*, 9, 249, 1988.
19. **Ellis, R. W., Keller, P. M., Lowe, R. S., and Zivin, R. A.,** Use of a bacterial expression vector to map the varicella-zoster virus major glycoprotein gene, gC, *J. Virol.*, 53, 81, 1985.
20. **Sawyer, M. H., Ostrove, J. M., Felser, J. M., and Straus, S. E.,** Mapping of the varicella zoster virus deoxypyrimidine kinase gene and preliminary identification of its transcript, *Virology*, 149, 1, 1986.
21. **Davison, A. J. and Scott, J. E.,** A complete DNA sequence of varicella-zoster virus, *J. Gen. Virol.*, 67, 1759, 1986.
22. **Davison, A. J. and Wilkie, N. M.,** Location and orientation of homologous sequences in the genomes of five herpesviruses, *J. Gen. Virol.*, 64, 1927, 1983.
23. **Kieff, E., Hoyer, B., Bachenheimer, S., and Roizman, B.,** Genetic relatedness of type 1 and type 2 herpes simplex viruses, *J. Virol.*, 9, 738, 1972.
24. **Redfield, D. C., Richman, D. D., Cleveland, P. H., Albanil, S., and Oxman, M. N.,** Detection and typing of herpes simplex virus by DNA hybridization, presented at 23rd Interscience Conf. Antimicrobial Agents and Chemotherapy, Las Vegas, October 24 to 26, 1985, 741.
25. **Richman, D. D., Cleveland, P. H., Redfield, D. C., Oxman, M. N., and Wahl, G. M.,** Rapid viral diagnosis, *J. Inf. Dis.*, 149, 298, 1984.
26. **Maitland, N. J., Kinross, J. H., Busuttil, A., Ludgate, S. M., Smart, G. E., and Jones, K. W.,** The detection of DNA tumour virus-specific RNA sequences in abnormal human cervical biopsies by *in situ* hybridization, *Gen. Virol.*, 55, 123, 1981
27. **Peden, K., Mounts, P., and Hayward, G. S.,** Homology between mammalian cell DNA sequences and human herpesvirus genomes detected by a hybridization procedure with high-complexity probe, *Cell*, 31, 81, 1982.
28. **Puga, A., Cantin, E. M., and Notkins, A. L.,** Homology between murine and human cellular DNA sequences and the terminal repetition of the S component of herpes simplex virus type 1 DNA, *Cell*, 31, 8, 1982.
29. **Jones, T. R., Parks, C. L., Spector, D. J., and Hyman, R. W.,** Hybridization of herpes simplex virus DNA and human ribosomal DNA and RNA, *Virology*, 144, 384, 1985.
30. **Wagner, E. K.,** Individual HSV transcripts: characterization of specific generes, in *The Herpesviruses*, Vol. 3, Roizman, B., Ed., Plenum Press, New York, 1985, chap. 2.
31. **Takiff, H. E., Straus, S. E., and Garon, C. F.,** Propagation and in vitro studies of previously non-cultivable enteral adenoviruses in 293 cells, *Lancet*, 2, 832, 1981.
32. **Brandt, C. D., Kim, H. W., Yolken, R. H., Kapikian, A. Z., Arrobio, J. O., Rodriguez, W. J., Wyatt, R. G., Chanock, R. M., and Parrott, R. H.,** Comparative epidemiology of two rotavirus serotypes and other viral agents associated with pediatric gastroenteritis, *Am. J. Epidemiol.*, 110, 243, 1979.
33. **Retter, M., Middleton, P. J., Tam, J. S., and Petric, M.,** Enteric adenoviruses: detection, replication, and significance, *J. Clin. Microbiol.*, 10, 574, 1979.
34. **Yolken, R. H., Lawrence, F., Leister, F., Takiff, H. E., and Straus, S. E.,** Gastroenteritis associated with enteric type adenovirus in hospitalized infants, *J. Pediatr.*, 101, 21, 1982.
35. **Uhnoo, I., Wadell, G., Svensson, L., and Johansson, M. E.,** Importance of enteric adenoviruses 40 and 41 in acute gastroenteritis in infants and young children, *J. Clin. Microbiol.*, 20, 365, 1984.
36. **Brown, M., Petric, M., and Middleton, P. J.,** Diagnosis of fastidious enteric adenoviruses 40 and 41 in stool specimens, *J. Clin. Microbiol.*, 20, 334, 1984.
37. **van Loon, A. E., Rozijn, Th. H., de Jong, J. C., and Sussenbach, J. S.,** Physicochemical properties of the DNAs of the fastidious adenovirus species 40 and 41, *Virology*, 140, 197, 1985.
38. **Takiff, H. E., Reinhold, W., Garon, C. F., and Straus, S. E.,** Cloning and physical mapping of enteric adenoviruses (cadidate types 40 and 41), *J. Virol.*, 51, 131, 1984.
39. **Takiff, H. E., Seidlin, M., Krause, P., Rooney, J., Brandt, C., Rodriguez, W., Yolken, R., and Straus, S. E.,** Detection of enteric adenoviruses by dot-blot hybridization using a molecularly cloned viral DNA probe, *J. Med. Virol.*, 16, 107, 1985.
40. **Kidd, A. H., Harley, E. H., and Erasmus, M. J.,** Direct detection of fastidious enteric adenovirus DNA in stools and specific typing as Ad40 or Ad41 by dot-blot hybridization, *J. Clin. Microbiol.*, 22, 934, 1985.

41. **Schepp, D. H., Dandliker, P. S., and Meyers, J. D.,** Treatment of varicella-zoster virus infection in severely immunocompromised patients, a randomized comparison of acyclovir and vidarabine, *N. Engl. J. Med.,* 314, 208, 1986.
42. **Whitley, R., Hilty, M., Haynes, R., Bryson, Y., Connor, J. D., Soong, S.-J., Alford, C., and NIAID Collaborative Antiviral Study Group,** Vidarabine therapy of varicella in immunosuppressed patients, *J. Pediatr.,* 101, 125, 1982.
43. **Seidlin, M., Takiff, H. E., Smith, H. A., Hay, J., and Straus, S. E.,** Detection of varicella-zoster virus by dot-blot hybridization using a molecularly cloned viral DNA probed, *J. Med. Virol.,* 13, 53, 1984.
44. **Feldman, S. and Epp, E.,** Detection of viremia during incubation of varicella, *J. Pediatr.,* 94, 748, 1979.
45. **Asano, Y., Itakura, N., Hiroishi, Y., Hirose, S., Nagai, T., Ozaki, T., Yazaki, T., Yamanishi, K., and Takashi, M.,** Viremia is present in incubation period in nonimmunocompromised children with varicella, *J. Pediatr.,* 106, 69, 1985.
46. **Ozaki, T., Ichikawa, T., Matsui, Y., Kondo, H., Nagai, T., Asano, Y., Yamanishi, K., and Takahashi, M.,** Lymphocyte-associated viremia in varicella, *J. Med. Virol.,* 19, 249, 1986.
47. **Gilden, D. H., Hayward, A. R., Crupp, J., Hunter-Laszlo, M., Huff, J. C., and Vafai, A.,** Varicella-zoster virus infection of human mononuclear cells, *Virus Res.,* 7, 117, 1987.
48. **Croen, K. D.,** unpublished observation, 1986.
49. **Ryder, J. W., Croen, K., Kleinschmidt-DeMasters, B. K., Ostrove, J. M., Straus, S. E., and Cohn, D. L.,** Progressive encephalitis three months after resolution of cutaneous zoster in a patient with AIDS, *Ann. Neurol.,* 19, 182, 1986.
50. **Hyman, R. W., Ecker, J. R., and Tenser, R. B.,** Varicella-zoster virus RNA in human trigeminal ganglia, *Lancet,* 2, 814, 1983.
51. **Gilden, D. H., Vafai, A., Shtram, Y., Becker, Y., Devlin, M., and Wellish, M.,** Varicella-zoster virus DNA in human sensory ganglia, *Nature (London),* 306, 478, 1983.
52. **Corey, L. and Spear, P.,** Infections with herpes simplex viruses, *N. Engl. J. Med.,* 314, 686, 1986.
53. **Stalhandske, P., Hyypia, T., Gadler, H., Halonen, P., and Pettersson, U.,** The use of molecular hybridization for demonstration of adenoviruses in human stools, *Curr. Top. Microbiol. Immunol.,* 104, 299, 1983.
54. **Redfield, D. C., Richman, D. D., Albanil, S., Oxman, M. N., and Wahl, G. M.,** Detection of herpes simplex virus in clinical specimens by DNA hybridization, *Diagn. Microbiol. Infect. Dis.,* 1, 117, 1983.
55. **Schuster, V., Matz, B., Wiegand, H., Polack, A., Corsten, B, and Neumann-Haefelin, D.,** Nucleic acid hybridization for detection of herpes viruses in clinical specimens, *J. Med. Virol.,* 19, 277, 1986.
56. **Forghani, B., Dupuis, K. E., and Schmidt, N. J.,** Rapid detection of herpes simplex virus DNA in human brain tissue by in situ hybridization, *J. Clin. Microbiol.,* 22, 656, 1985.
57. **Rooney, J. F., Felser, J. M., Ostrove, J. M., and Straus, S. E.,** Acquisition of genital herpes from an asymptomatic sexual partner, *N. Engl. J. Med.,* 314, 1561, 1986.
58. **Hill, T. J.,** Herpes simplex virus latency, in *The Herpesviruses,* Vol. 3, Roizman, B., Ed., Plenum Press, New York, 1985, 4.
59. **Stevens, J. G., Wagner, E. K., Devi-Rao, G. B., Cook, M. L., and Feldman, L. T.,** RNA complementary to a herpesvirus α gene mRNA if prominent in latently infected neurons, *Science,* 235, 1056, 1987.
60. **Croen, K. D., Ostrove, J. M., Dragovic, L. J., Smialek, J. E., and Straus, S. E.,** Latent herpes simplex virus in human trigeminal ganglia: detection of an immediate-early gene "anti-sense" transcript by in situ hybridization, *N. Engl. J. Med.,* 317, 1427, 1987.
61. **Hammond, G., Mauthe, G., Lukes, H., Thompson, L., Fischer, K., and Hannan, C.,** DNA hybridization for the detection of enteric adenoviruses 40 and 41, presented Int. Comparative Virology Meeting, Lake Louise, Alberta, May 1986.

Chapter 15

PROGRESS TOWARD THE DEVELOPMENT OF A PAN-ENTEROVIRAL NUCLEIC ACID PROBE

Harley A. Rotbart and Myron J. Levin

TABLE OF CONTENTS

I. INTRODUCTION

The enteroviruses (EV) comprise one genus in the family Picornaviridae. They are unbiquitous human pathogens, responsible for diseases ranging in severity from trivial to fatal. Infections due to the prototypic EV, the polioviruses (PV), have been largely controlled in the developed world with the introduction of vaccines in the late 1950s. These agents continue, however, to cause substantial morbidity and mortality in underdeveloped countries where 4 of every 1000 school-age children contract paralytic poliomyelitis.[1] In the U.S., PV still pose a finite risk, particularly among the immunocompromised and nonimmunized populations. The nonpolio EV have received far less study than the PV — and far less progress has been made in understanding and controlling them. There are more than 60 serotypes of nonpolio EV with widely divergent clinical and laboratory characteristics. These agents are estimated to cause 1 to 3×10^7 infections annually in the U.S.* and are among the most common causes of meningitis (7.5×10^4 cases per year), encephalitis (7.5×10^2/year), myocarditis (5.0×10^2/year), nonspecific febrile and exanthematous illnesses (5×10^6/year), stomatitis, pleurodynia, and neonatal sepsis. Recently identified serotypes cause epidemic hemorrhagic conjunctivitis[2] and, perhaps of greatest concern, poliomylitis indistinguishable from poliomyelitis caused by the PV.[3] Furthermore, EV can establish persistent infections lasting for years in immunocompromised patients[4] and have been postulated as etiologic agents in such chronic diseases as diabetes mellitus,[5] dermatomyositis,[6,7] amyotrophic lateral sclerosis,[8] and congenital hydrocephalus.[9]

Recent work in our laboratory and elsewhere has demonstrated significant genetic homology among the EV. This is noteworthy because, although the viruses have similar physicochemical and biological properties (resulting in their categorization as a single genus), they are antigenically (and hence serologically) extremely heterogeneous. The demonstration of genomic homology among the EV makes feasible the use of cloned subgenomic fragments of one or a few EV as nucleic acid probes in the study of many or all EV serotypes. Our recent work in preparing and testing such "pan-enteroviral" probes is the subject of this chapter.

II. TRADITIONAL TAXONOMIC AND DIAGNOSTIC TECHNIQUES

EV are currently classified as such on the basis of their similar physicochemical properties.[10] Subgroups within the EV genus were originally determined on the basis of host susceptibilities; thus, classically, echoviruses produce tissue culture cytopathology but no murine disease, whereas coxsackieviruses cause illness in newborn mice. The patterns of murine pathology distinguish coxsackie A from coxsackie B serotypes. These somewhat arbitrary subclassifications have become blurred, however, and new isolates are simply numbered as enterovirus 68, 69, etc. Hepatitis A virus has recently been classified as enterovirus 72 because of its physicochemical similarity to the others.[11] The mainstay of laboratory diagnosis is tissue culture, a technique which is only 65% sensitive and can take days to weeks before results are known.[12] Serotyping of EV tissue culture isolates is achieved by the use of cross-neutralizing pools of antisera.[10,11] Suckling mice inoculation, the only method for recovering certain EV serotypes, is cumbersome and rarely performed. Attempts at developing immunoassays have been hampered by the wide antigenic heterogeneity among the EV.[13-15]

III. GENOMIC HOMOLOGY AMONG EV SEROTYPES

Although antigenic variations among the EV are great, there were reasons to suspect that a significant degree of genetic conservation existed across serotype (species) lines. First, the

* All annual incidence figures by personal communication with Larry Anderson, M. D., Centers for Disease Control, Atlanta, Ga.

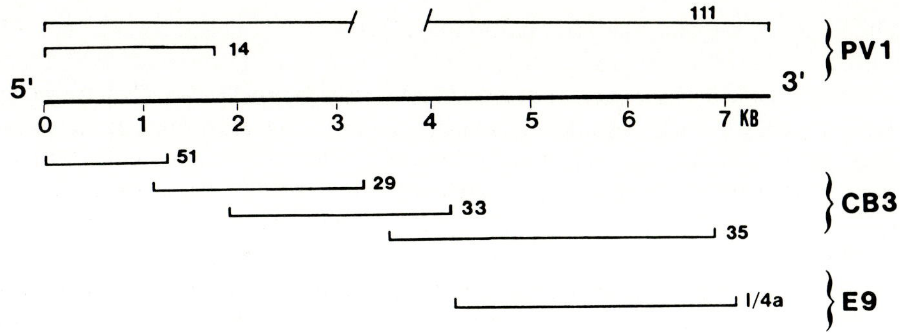

FIGURE 1. Diagram of cloned enteroviral fragments used in preparing cDNA and RNA probes with their locations superimposed on the 7500 nucleotide enteroviral genome. Abbreviations used: PV1 = poliovirus type 1 (clone designation pDS), CB3 = coxsackievirus B3 (clone designation pCBIII), and E9 = echovirus 9 (clone designation pE C9). *Bam*220, not shown on this figure, was derived by a *Bam*HI digestion of pDS14 at nucleotide 220 (from the 5′ end) and religation.

physicochemical similarities which resulted in different isolates being classified as EV are undeniable and unlikely to be coincidental. Additionally, the biological characteristics of these agents overlap impressively — most observed clinical disease due to one EV serotype has also been reported associated with many different serotypes. Even classical poliomyelitis has been sporadically caused by several nonpolio EV. Finally, early hybridization and sequencing data detected small amounts of homology — now recognized to have been an underestimation of true genetic conservation.

The first convincing molecular evidence of genomic homology was presented in 1973.[16] Young demonstrated 3 to 5% homology among the most distantly related serotypes tested by RNA-RNA annealing experiments. Subsequently, oligonucleotide mapping techniques confrmed genetic similarities among the EV.[17] In the past few years, specific sequence information has demonstrated that the three PV share more than 70% of their nucleotide sequences.[18] Particularly strong homology is present at the 5′ and 3′ ends of the genomes — untranslated regions thought to serve regulatory functions.[19] Much less is known about the genetic structure of the nonpolio EV. Coxsackie B (CB)1 and PV1 share 17 of their first 20 5′ nucleotides,[20] and CB3 and PV1 are approximately 70% homologous in their polymerase gene regions and in their 5′ noncoding regions.[21,22]

Within the past 2 years, several studies using cDNA hybridization probes for the EV have greatly expanded the spectrum of EV serotypes now known to share specific nucleotide sequences. We used three well-characterized, cloned subgenomic fragments of PV1 as hybridization probes to detect tissue culture infections with representative members of each of the major subgroups of EV.[23] Clone pDS111 (containing nearly the entire PV1 genome except for the nucleotide segment 3235 to 3925; Figure 1) and clone pDS14 (containing the first 1809 5′ nucleotides of PV1) are contained in the pBR325 plasmid vector and were kindly provided to us by Dr. B. Semler.[24] We prepared clone *Bam*220 by restriction digest of the pDS14 clone with *Bam*HI and religation. The probes were evaluated in LLC-MK2 monkey kidney cells infected with PV and nonpolio EV. When 1 to 2+ cytopathic effect was noted, cell lysates were prepared, applied to membrane filters, and tested in dot hybridization assays. Nick-translated ^{32}P-labeled pDS111 and pDS14 whole plasmids each detected all serotypes tested (PV1, PV2, PV3, CB1, coxsackie A9, echovirus 11) at high stringency, whereas *Bam*220 detected only the three PV under the same conditions. We concluded that the region between nucleotides 220 and 1809 on the PV1 genome was significantly homologous with other EV tested. Support for this hypothesis is found in the more recent genomic mapping work by Nomoto et al. in which a highly conserved region among the PV was found between nucleotides 400 and 650.[18] Stanway et al. have shown absolute sequence homology between bases 509 and 573 of PV1 and 3.[25] Semler et al. have shown that a region of 5′ nucleotides, from bases 220 to 625 can be substituted from CB3 into PV1, resulting in infectious progeny virus,[26] and further testifying to the homology among these agents. The reading frame for PV1 begins at nucleotide 743.[19] We suspect that the homologous sequences detected by our experiments with pDS14 largely lie between nucleotides 220 and the beginning of the reading frame (which begins the coding for capsid proteins, known to be highly divergent among the EV and resulting in the antigenic diversity noted earlier).

In similar experiments, Hyypiä et al. used a probe derived from the 3′ end of CB3 to detect tissue culture infections with a variety of coxsackie and two non-coxsackie EV.[27] Subsequently, that probe was characterized by Stalhandske et al. and was found to contain (approximately) the 4500 3′ nucleotides of CB3, the terminal 1514 of which were sequenced.[21] The gene thought to code for the polymerase was within this sequenced segment and was found to be 68% homologous to the known sequence of the PV1 polymerase gene. Less homology to PV1 was found in the 3′ terminal noncoding sequences.

Tracy has recently prepared several cloned subgenomic fragments of the CB3 genome for homology studies among the CB virus group.[28,29] While all of the six CB viruses share some homologous sequences (as demonstrated by dot hybridization assays), CB1, 2, and 3 appear more related to each other than to CB4 and 6; CB5 appears to share approximately equal homology with all of the other CB serotypes. Finally, using the PV1-derived clones described above (pDS111 and pDS14) and four subgenomic cloned fragments of CB3 (pCBIII51, pCBIII29, pCBIII33, and pCBIII35, kindly provided by Dr. S. Tracy) (Figure 1), we prepared additional probes and extended our EV homology studies.[30] Hybridization asssays were performed using nick-translated ^{32}P-labeled probes singly and in combinations. Examples of taxonomical relationships which were derived from these preliminary experiments include: echoviruses 2, 4, 6, and 11 are more closely related to CB3 than to PV1; coxsackievirus A16 was detected by both PV1-derived probes, but only by the 5′ CB3 probe, whereas coxsackie A9 hybridized well with both 5′ and 3′ CB3 probes. The segment of CB3 represented by one of the clones, pCBIII29, appeared to be very specific, hybridizing only weakly with CB6 and echovirus 4. This confirmed work by Tracy showing pBCIII29 to be relatively specific for CB3.[28] Probe combinations including a 5′ end and a 3′ end segment gave stronger hybridization signals with a broader spectrum than those obtained with the probes used singly. Of note, none of the cDNA probes detected echovirus 22 under the stringent hybridization conditions employed, consistent with other evidence that this virus is unusual in several respects and may be less like other EV than previously thought.[31] Similarly, we have not found detectable hybridization of our cDNA probes with hepatitis A virus under stringent conditions,[31a] a finding validated recently by sequence information showing no significant hepatitis A genome homology with poliovirus.[32] Several rhinoviruses do hybridize with the PV1 and CB3 fragments to a greater extent than would be expected by their classification as a separate genus of picornaviruses. We used probes derived from PV1 and CB3 to detect $<10^4$ $TCID_{50}$ rhinoviruses 2, 14, and 50, but were not able to detect rhinovirus B1.[31a] Recent sequencing data confirm the presence of significant homology between the EV and rhinoviruses 2 and 14.[33-35]

It is clear that future sequencing and cross-hybridization studies will further clarify the evolution of EV and their appropriate taxonomical relationships. However, significant evidence for broad homology among the EV already exists, as summarized above, and warrants the further pursuit of rapid EV diagnosis with nucleic acid hybridization technology.

IV. ESTABLISHING OPTIMAL HYBRIDIZATION CONDITIONS FOR CLINICAL SPECIMENS

The most vexing EV infections with which the physican must grapple are those of the central nervous system (CNS). Meningitis and encephalitis are among the most common manifestations of EV infection and clinically are often difficult to distinguish from infections due to bacteria and herpes simplex — particularly in young children. Whereas CNS EV infections generally have an excellent prognosis, fear of a more serious etiology often results in unnecessary hospitalization and antibiotic therapy.[12] Furthermore, antiviral drugs effective in vitro against the EV are available and might be extremely useful in the more serious infections due to these agents.[36] Unfortunately, overhospitalization and overtreatment are unavoidable, and clinical trials of antiviral drugs are not yet possible because of the difficulty in making a laboratory diagnosis of EV infection. For these reasons, as well as the difficulty encountered by Hyypiä et al. in detecting EV by hybridization of stool specimens,[27] we focused our efforts on the cerebrospinal fluid (CSF).

We performed CSF reconstruction experiments to determine (1) optimal handling, storage, and processing conditions for infected CSF specimens; (2) the effect of the elements of the host inflammatory response on detection of EV RNA in the CSF; and (3) the sensitivity and specificity of nick-translated ^{32}P-labeled hybridization probes derived from PV1 and CB3, used singly and in combination, in detecting EV RNA in CSF.[30] We found that EV in CSF were stable for at least 4 days at a variety of storage temperatures (–70°, –20°, 4°, 23°C). While cellular elements of the inflammatory response (polymorphonuclear and mononuclear white blood cells) did not interfere with the hybridization assay, gamma globulin added to CSF in concentrations present during infection dramatically attenuated the signals. This problem was overcome by adding proteinase K, which restored the hybridization signals to their original strength. The mechanism of gamma globulin interference with hybridization appears to be physical rather than immunologic.[37] Neither red blood cells nor albumin adversely affected the assay. Each of six cloned fragments of PV1 or CB3 produced a unique pattern of hybridizations with the test viruses, however, combining a 5′ end-containing with a 3′ end-containing probe produced the broadest spectrum and greatest sensitivity of EV RNA detection. Sensitivity of the dot hybridization assay ranged from 10^2 to 10^5 $TCID_{50}$ for the serotypes tested* (except echovirus 22). Little is known regarding the actual titers of EV in the CSF of patients with meningitis and/or encephalitis. Sketchy reports suggest, however, as little as 10^1 to 10^3 $TCID_{50}$ may be present.[38]

V. TRIAL OF cDNA PROBES ON CLINICAL CSF SPECIMENS

Having established what we hoped would be ideal handling and hybridization conditions for infected CSF, we tested aliquots of CSF from 35 children for EV by routine tissue culture and by dot-blot hybridization.[38a] Of 12 children with aseptic meningitis or encephalitis and positive cultures for EV, only 4 were positive by hybridization (33% sensitivity). Ten children with aseptic meningitis or encephalitis during the summer months had negative cultures for EV — two of those CSF specimens were positive by hybridization. Ten children had spinal tap results not suggestive of any infection and negative by culture — one of those was hybridization positive (specificity = 90%). Finally, CSF specimens from children with bacterial meningitis were tested for EV by hybridization. One specimen, from a child with *Hemophilus influenzae* meningitis, was positive by dot-blot hybridization with EV probes. While the specimen was not cultured for viruses, we presume this to be a false positive hybridization.

The results of this clinical trial of cDNA probes confirmed our suspicion that their utility is limited by the predictable insensitivity. Although two probable tissue culture false negatives were detected by hybridization, two-thirds of tissue culture positives were missed. Furthermore, false positivity with CSF from a patient with bacterial meningitis is unacceptable.

* EV serotypes tested include: PV types 1, 2, 3, coxsackievirus A9, A16, B1, B6, and echovirus 2, 4, 6, 11, 22.

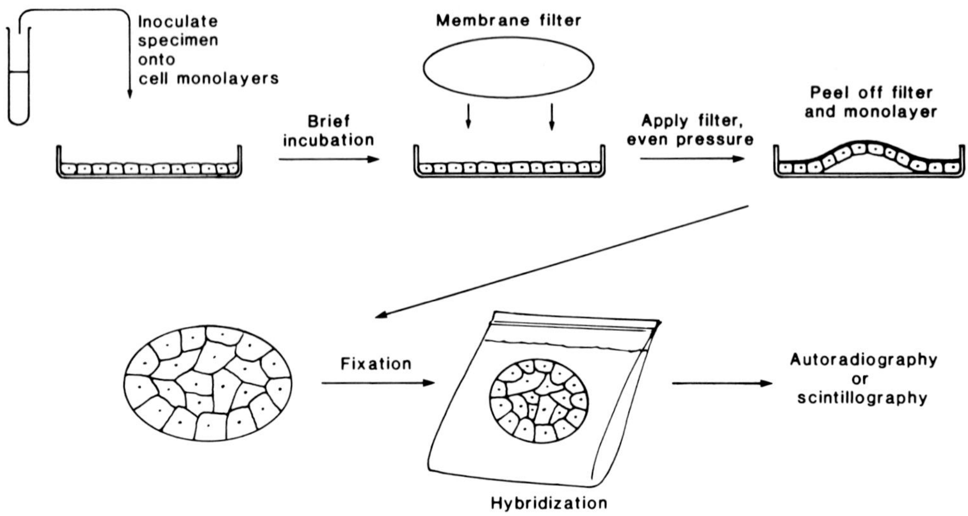

FIGURE 2. Schematic representation of monolayer blot technique. A membrane filter is directly applied to an infected monolayer of cells. The cell sheet is peeled off onto the filter, which is then fixed and hybridized.

VI. STRATEGIES TO OVERCOME SHORTCOMINGS IN SENSITIVITY AND SPECIFICITY

A. Target Amplification

We hypothesized that detectable levels of RNA replication (by hybridization) may precede visual detection of cytopathic effect (CPE) in tissue culture cells by hours to days. Particularly for "slow-growing" serotypes, RNA may "accumulate" in infected cells in just a few hours, permitting a slighly less rapid, but still useful, diagnosis by hybridization of cell cultures which are briefly incubated with CSF. To test this approach, we used a monolayer blot technique, as schematized in Figure 2. LLC-MK2 cells were infected with $<10^2$ $TCID_{50}$ of various EV (an innoculum below the limit of sensitivity for direct dot-blot hybridizations, as noted earlier) for varying lengths of time. Medium was decanted and membrane filters were applied directly to the infected monolayer. The entire cell sheet was peeled off the culture dish onto the membrane, which was then fixed with glutaraldehyde, paraformaldehyde, and salt, and hybridized with the cDNA probes by our standard solid phase hybridization techniques. Autoradiography and scintillography were then used to quantitate the relative amounts of target RNA detected. Results of a typical experiment are summarized in Table 1 and Figure 3. CPE due to $<10^2$ coxsackie A9 virions was first visualized at 60 h of tissue culture incubation. In contrast, the presence of RNA was detectable by monolayer blot hybridization at 36 h (by scintillography) or 42 h (by autoradiography). Further enhancement of this technique might permit even briefer incubations prior to hybridization with serotypes which grow well in tissue culture. It is also possible, but as of yet untested, that EV serotypes which do not "grow" in tissue culture by the criterion of producing CPE, might replicate and produce RNA detectable by hybridization.

B. Preparation of Single-Stranded RNA Probes

The results of our trial of nick-translated cDNA probes on clinical CSF specimens emphasized the importance of optimal probe reagents — inadequately sensitive and specific probes, despite optimized hybridization conditions, produced suboptimal results. Therefore, we considered the use of single-stranded RNA probes which have several theoretical advantages compared with cDNA probes: RNA:RNA hybrids have greater affinity than do DNA:RNA hybrids; with negative-sense, single-stranded RNA probes, there is no self-annealing to compete

Table 1
MONOLAYER BLOT AMPLIFICATION OF $<10^2$ TCID$_{50}$ COXSACKIE A9 VIRIONS FOLLOWING INCUBATION IN TISSUE CULTURE FOR VARYING LENGTHS OF TIME

Method	Hours											
	12	**18**	**24**	**30**	**36**	**42**	**48**	**54**	**60**	**66**	**72**	**C[a]**
CPE[b]	–	–	–	–	–	–	–	–	+	++	+++	–
Autoradiography[c]	–	–	–	–	–	+	+	+	+	+	+	–
Scintillography[d]	159	175	168	160	257*	272	279	425	880	2147	1815	146

[a] Control (uninfected specimen).
[b] Cytopathic effect seen visually.
[c] Results shown in Figure 3.
[d] Counts (* first positive blot).

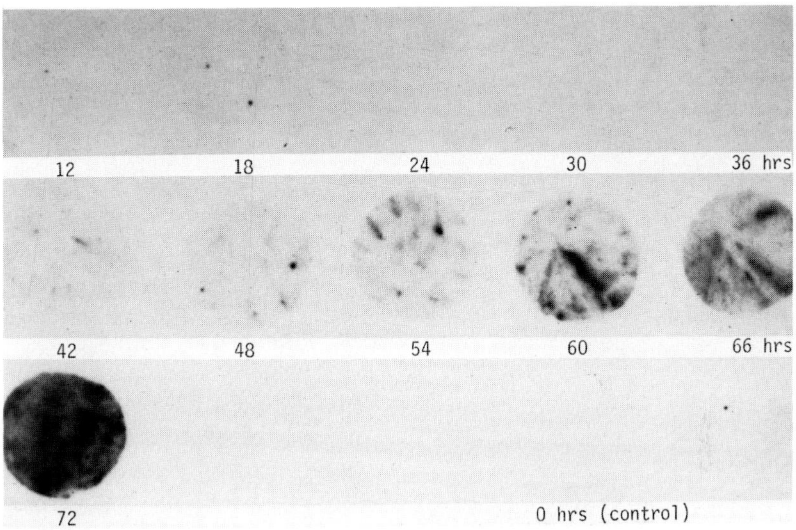

FIGURE 3. Results of monolayer blot technique in detecting coxsackie A9 infection. Blots are obtained every 6 h from identical culture wells. An inoculum of $<10^2$ TCID$_{50}$ is first detected by autoradiography following 42 h of incubation in culture, in contrast to the first visible cytopathic effect seen at 60 h (not shown).

with target for probe strands; finally, there are no vector sequences present to cause nonspecific background hybridization or to hybridize with plasmid-containing bacteria in the clinical specimens.[39] The commercial availability of transcription promoter-containing plasmids makes the synthesis of single-stranded RNA probes convenient and efficient.[40]

We excised cloned cDNA fragments of 3 EV-polio 1 (pDS14; provided by B. Semler), coxsackie B3 (pCBIII35; provided by S. Tracy), and echo 9 (pEC9 I/4a; provided by B. Rosenwirth[41]) (Figure 1) from their original vectors by restriction endonuclease digestion. We then separated the insert fragments by agarose gel electrophoresis, purified them, and recloned them into vectors containing the bacteriophage promoters SP6 and T7. Using promoter-specific polymerases, negative-sense RNA transcripts were generated and radioactively labeled.[42]

Log dilutions of PV1 (10^1 to 10^7 TCID$_{50}$ virions) and coxsackie A9 (10^1 to 10^6 TCID$_{50}$ virions) were applied to nylon membranes and hybridized with either nick-translated ^{32}P-labeled pDS14

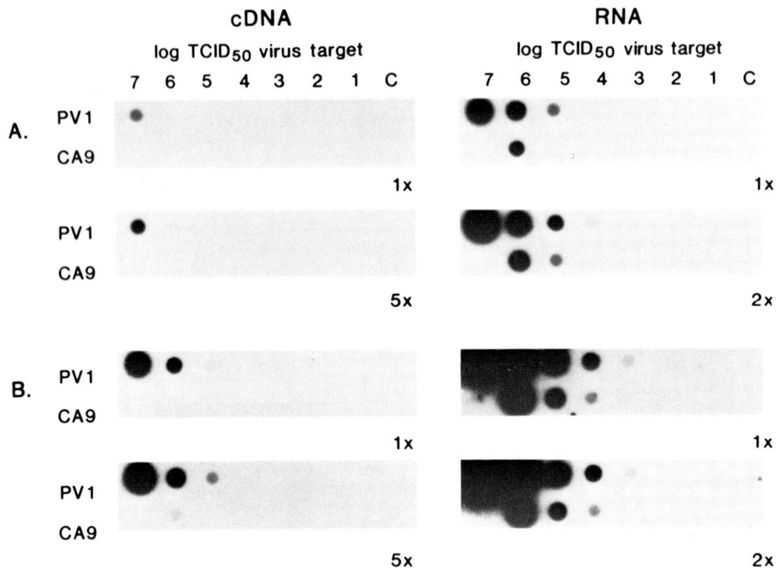

FIGURE 4. Comparison of PV1-derived cDNA and RNA probes in detecting log dilutions of PV1 (10^1 to 10^7 $TCID_{50}$ virions) and coxsackie A9 (10^1 to 10^6 $TCID_{50}$ virions). Probes were used in standard (1×) or higher (2×, 5×) concentrations. Autoradiography exposures were 1 h (A) and 12 h (B).

or the ^{32}P-labeled RNA transcript of the same insert. The results are shown in Figure 4. The RNA probe was found to be 10- to 100-fold more sensitive than its cDNA counterpart, with results obvious by autoradiography after only 1 h of exposure. A fivefold excess of cDNA probe resulted in hybridization signals still far inferior to those obtained with the RNA probe.

In the interest of ultimate adaptation of a diagnostic laboratory setting, we sought a more "user-friendly" labeling system. ^3H-labeled pDS14-derived RNA probe was prepared with ^3H-UTP and tested in slot-blot hybridization experiments with serial dilutions of PV1, coxsackie A9, and CB6. Hybridization conditions for this preliminary experiment were those we had used previously for cDNA probes. The slots were cut out of the membrane filter following standard washes and counted by scintillography. The results, shown in Figure 5, suggest the potential utility of these probes even with the relatively low energy ^3H label.

Experiments to optimize RNA-RNA hybridization conditions were similarly performed using ^3H-labeled RNA probes and are summarized in Figures 6 to 10. Hybridizidation time of 24 h was superior to 12 h and 4 h in detecting serial dilutions of PV1 with the pDS14-generated RNA probe (Figure 6). Formamide concentrations made surprisingly little difference in hybridization efficiency, with 50% concentration showing only slight advantage over 0% and 25% (Figure 7). Hybridization temperatures, on the other hand, significantly affected the results of PV1 detection with pDS14-derived RNA, with 50°C superior to both higher and lower temperatures (Figure 8). Phenol-chloroform extraction and ethanol precipitation of CSF "spiked" with PV1 did not improve upon results obtained with the crude CSF preparation (Figure 8). Probe preparation techniques were similarly evaluated. The crude transcription mix of pDS14-derived RNA probe performed at least as well as "cleaned-up" preparations (which had be treated with DNase alone or with DNase followed by phenol-chloroform extraction and ethanol precipitation) in detecting serial dilutions of coxsackie A9 (Figure 9). Probe volume in the hybridization mix was found to be an important variable. ^3H-labeled pCBIII35-derived RNA probe in high concentration was superior to lower quantities of the same probe in hybridizing to serial dilutions of CB6, without increasing background noise. As few as 10 $TCID_{50}$ virions were clearly detected with this protocol (Figure 10).

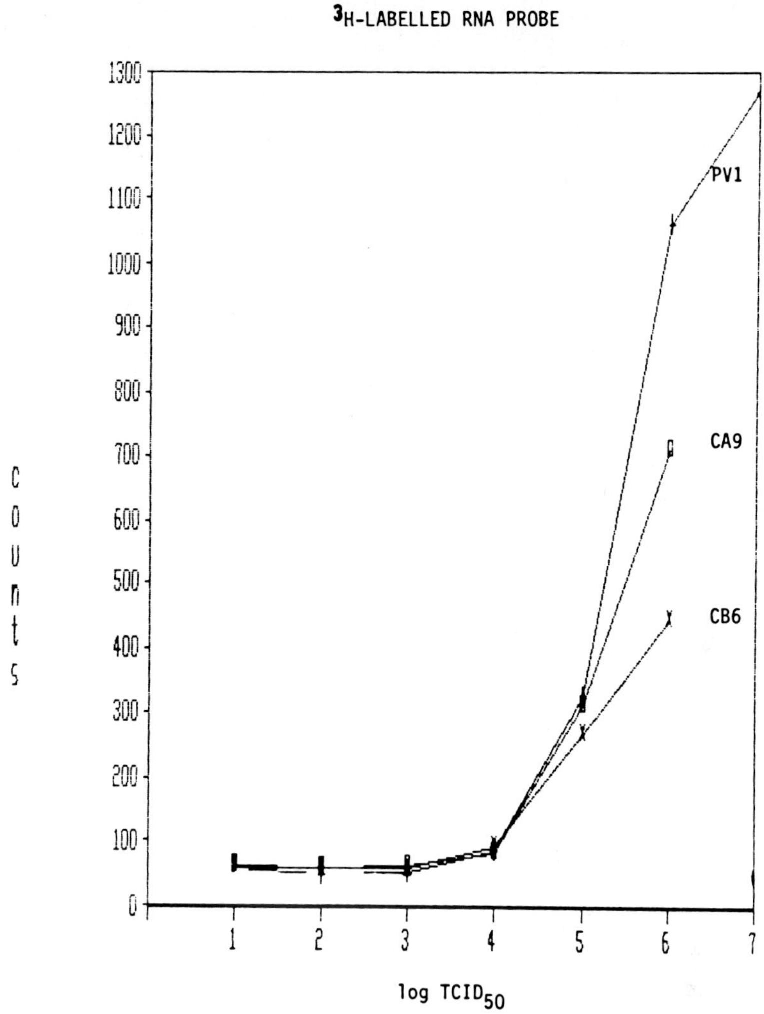

FIGURE 5. ^{3}H-labeled RNA probe (PV1-derived) used to detect log dilutions of 3 EV serotypes: poliovirus 1 (PV1), coxsackie A9 (CA9), and coxsackie B6 (CB6). Scintillography-determined counts are shown.

Finally, a mixture of ^{3}H-labeled probes derived from pDS14, pCBIII35, and pEC91/4a was used to determine the potential spectrum of EV serotype detection (Table 2). A positive hybridization was defined as counts (by scintillography) at least three times those obtained with uninfected CSF controls. All 16 EV serotypes tested were detected using the optimal hybridization conditions determined above, including 2 serotypes that do not grow in tissue culture at all (coxsackie A4 and A17), several that grow sluggishly, and 3 that we could not detect with nick-translated cDNA probes (echovirus 22, coxsackie A4 and A17). Thus, conditions have been established under which nucleic acid hybridization may be a useful adjunct to traditional EV diagnostic techniques.

VII. CONCLUSIONS

EV are important human pathogens, causing substantial morbidity and mortality worldwide. Moreover, their clinical mimicry of non-EV diseases results in confusion and misjudgements

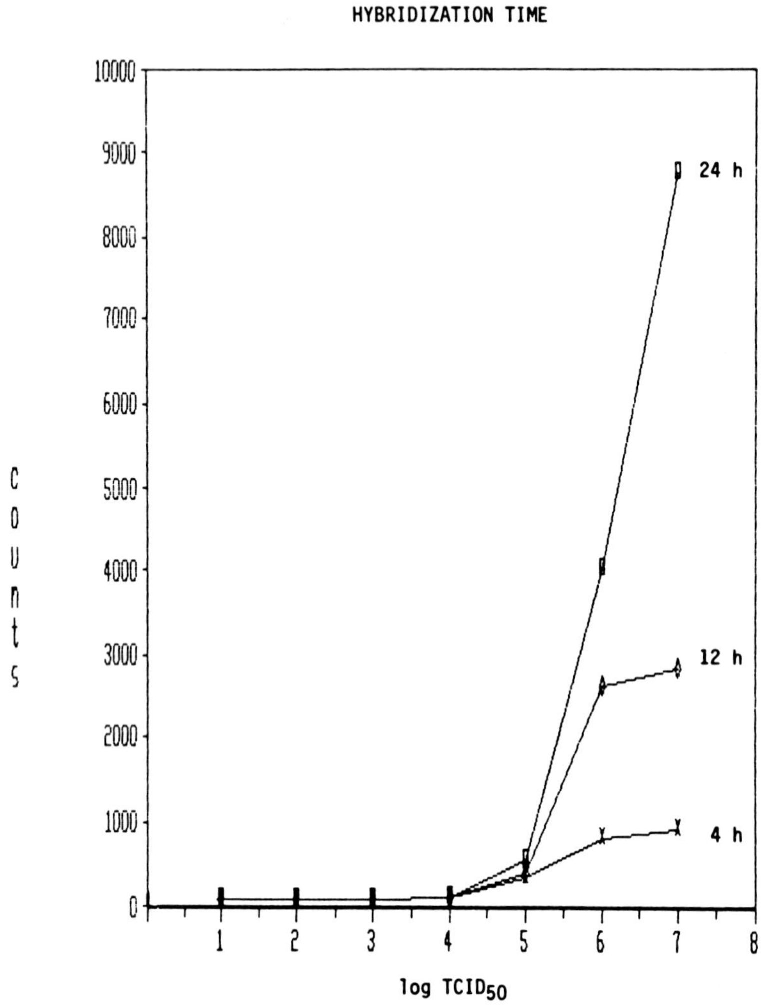

FIGURE 6. PV1-derived RNA probe hybridized to log dilutions of PV1 virions for various hybridization times. Counts determined by scintillography.

in patient care. Potentially effective therapies for serious EV infections sit on the shelf, untried in humans despite their efficacy in animals.[43,44] The antigenic heterogeneity among the many EV serotypes has impeded the adaptation of rapid immunoassays, so successful for other viral diagnosis, to the EV. Finally, the incriminating evidence to associate the EV with a number of serious chronic disorders awaits an accurate technique to trace EV "footprints" in affected tissues.

The data presented in this chapter summarize the progress made to date in using nucleic acid probes to meet the challenges of EV diagnosis. The widely demonstrable genomic homologies and the successes of reconstruction experiments using CSF are encouraging. The mediocre results of the first clinical trials and the seemingly inherent limit of sensitivity characteristic of all hybridization systems necessarily temper our enthusiasm. Three important variables remain to be determined before the ultimate success of probes for EV diagnosis can be predicted: the amount of target viral RNA present in infected fluids and tissues, the true limit of sensitivity of our best probe constructions, and our ability to amplify or enhance either of those factors if necessary. Our personal optimism for this technology lies largely in the promising results

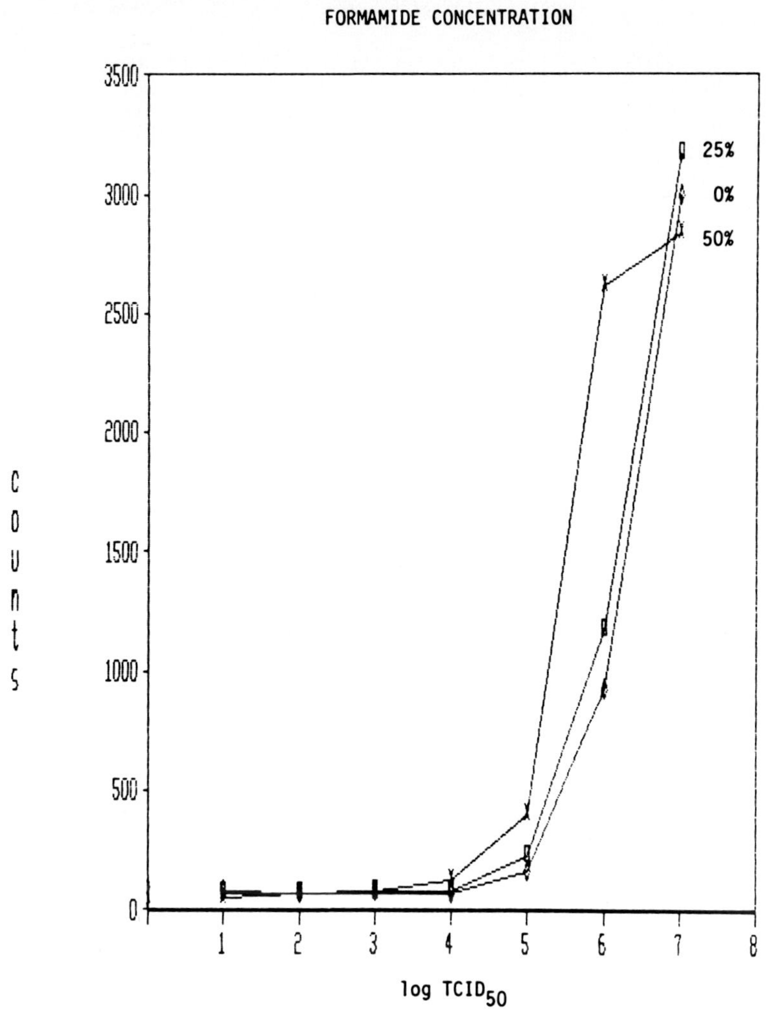

FIGURE 7. Same as Figure 6, but with varying formamide concentrations, each for 12 h hybridization time.

obtained with the "first generation" techniques described above and in our appreciation of how rapidly new and innovative technical advances have changed this field over the past few years. Future years surely promise the same.

ACKNOWLEDGMENTS

The authors gratefully acknowledge Dr. Lewis Pizer for advice and encouragement, Neva L. Murphy for technical assistance, and Rose Bradley for preparing the manuscript.

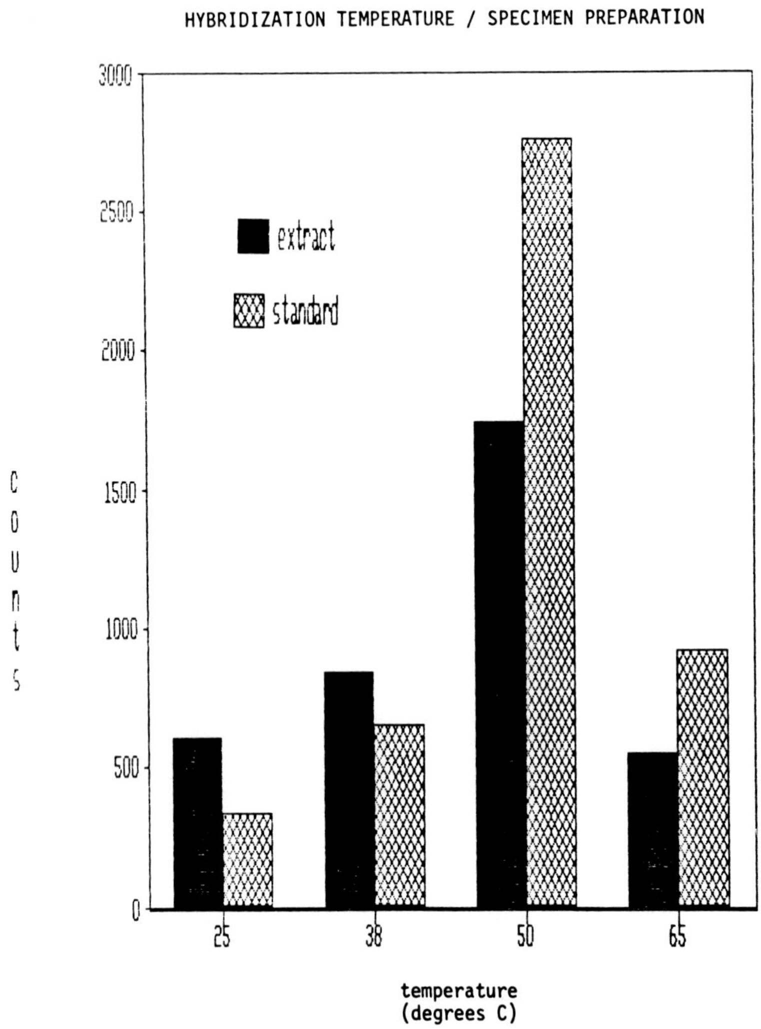

FIGURE 8. Same as Figure 6, but with varying hybridization temperatures, each for 12 h hybridization time and 50% formamide concentration. Target virion titers are 10^5 $TCID_{50}$ for all determinations. Shaded bars indicate extracted/precipitated specimens; cross-hatched bars indicate crude specimen (our standard technique).

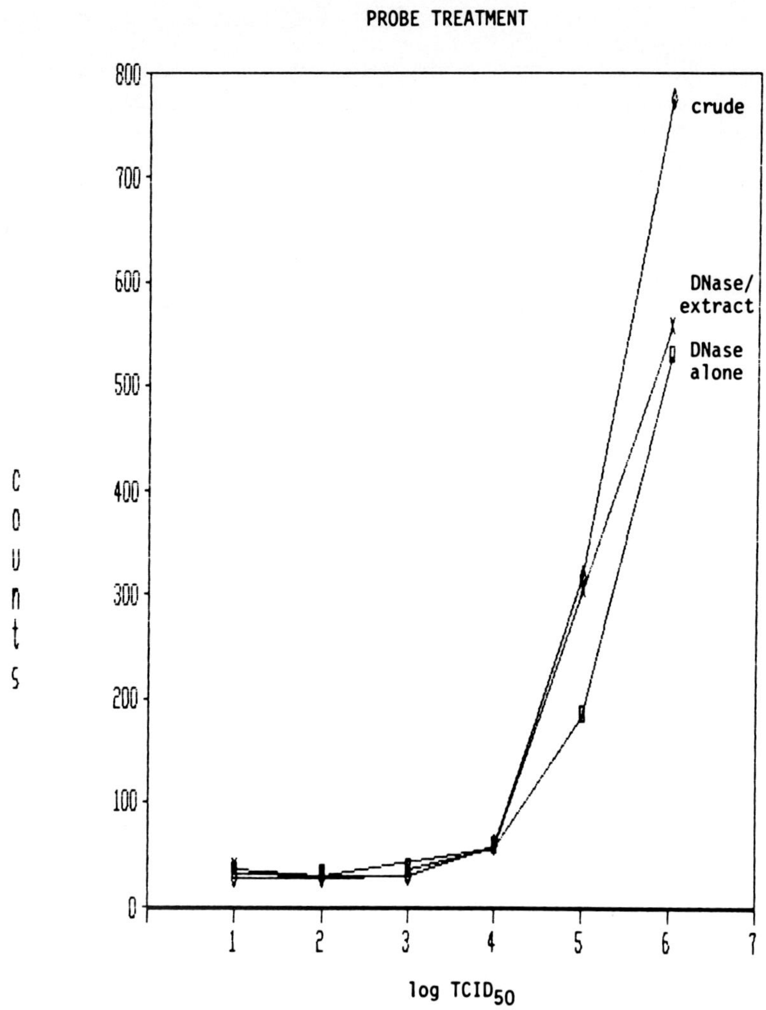

FIGURE 9. PV1-derived RNA probe hybridized to log dilutions of coxsackie A9 virions with various probe preparation techniques. Counts determined by scintillography.

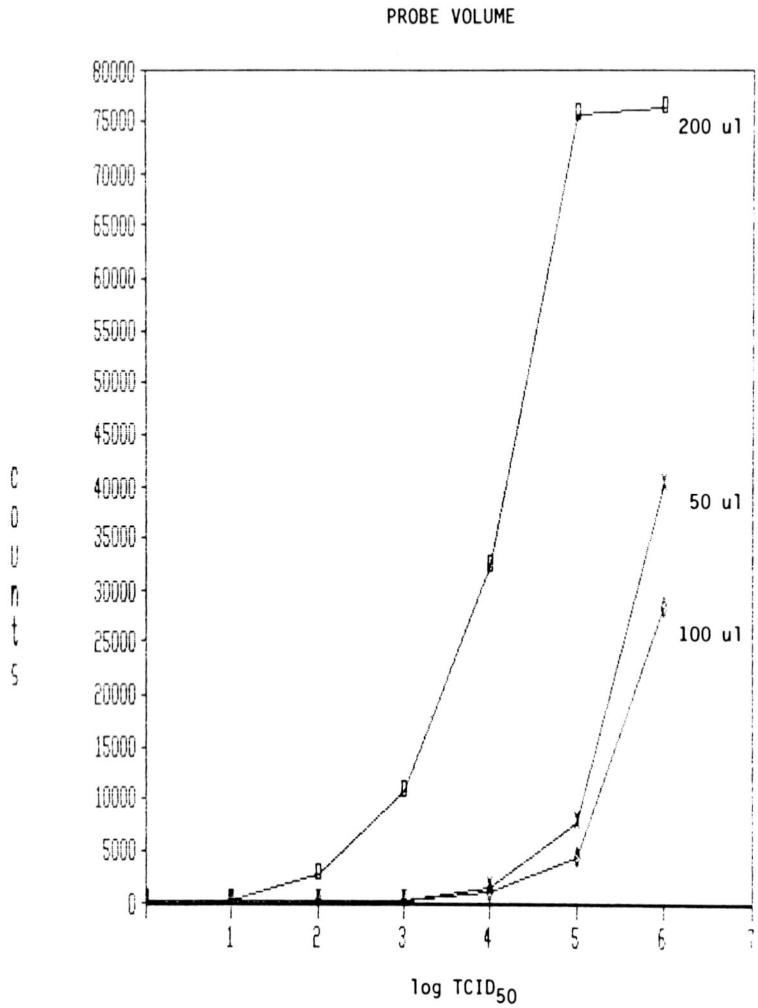

FIGURE 10. Coxsackie B3-derived RNA probe, in various concentrations, hybridized to log dilutions of coxsackie B6 virions. Counts determined by scintillography.

Table 2
HYBRIDIZATION OF VARIETY OF ENTEROVIRUS SEROTYPES WITH A MIXTURE OF RNA PROBES DERIVED FROM POLIO 1, COXSACKIE B3, AND ECHO 9

Serotype[a]	^{3}H counts[b]
PV1	4,820
PV2	2,762
PV3	4,829
CB1	924
CB6	1,914
CA4[c]	5,845
CA9	3,988
CA14[d]	1,671
CA16	3,590
CA17[c]	1,639

Table 2 (continued)
HYBRIDIZATION OF VARIETY OF ENTEROVIRUS SEROTYPES WITH A MIXTURE OF RNA PROBES DERIVED FROM POLIO 1, COXSACKIE B3, AND ECHO 9

Serotype[a]	^{3}H counts[b]
CA24[d]	710
E2	1,835
E4	30,607
E6	1,543
E11	471
E22	303
Control	89

Note: Counts are measured by scintillography.

[a] Abbreviations used: PV = poliovirus; CA = coxsackievirus A; CB = cosackievirus B; E = echovirus.
[b] Target virion titers are 10^5 $TCID_{50}$ unless noted otherwise.
[c] $TCID_{50}$ is unknown for these serotypes.
[d] $TCID_{50}$ for these serotypes is 10^4.

REFERENCES

1. **Assaad, F. and Ljungars-Esteves, K.,** World overview of poliomyelitis: regional patterns and trends, *Rev. Infect. Dis.,* 6, S302, 1984.
2. **Mirkovic, R. R., Kono, R., Yin-Murphy, M., Schier, R., Schmidt, N. J., and Melnick, J. L.,** Enterovirus type 70: the etiologic agent of pandemic hemorrhagic conjunctivitis, *Bull. WHO.,* 49, 341, 1973.
3. **Melnick, J. L.,** Enterovirus type 71 infections: a varied clinical pattern sometimes mimicking paralytic poliomyelitis, *Rev. Infect. Dis.,* 6, S387, 1984.
4. **Wilfert, C. M., Buckley, R. H., Mohanakumar, T., Griffith, J. F., Katz, S. L., Whisnant, J. K., Eggleston, P. A., Moore, M., Treadwell, E., Oxman, M. N., and Rosen, F. S.,** Persistent and fatal central nervous system echovirus infections in patients with agammaglobulinemia, *N. Engl. J. Med.,* 296, 1485, 1977.
5. **Yoon, J., Austin, M., Onadera, T., and Notkins, A. L.,** Virus-induced diabetes mellitus: isolation of a virus from the pancreas of a child with diabetic ketoacidosis, *N. Engl. J. Med.,* 300, 1173, 1979.
6. **Travers, R. L., Hughes, G. R. V., Cambridge, G., and Sewell, J. R.,** Coxsackie B neutralisation titers in polymyositis/dermatomyositis, *Lancet,* 1, 268, 1977.
7. **Gyorkey, F., Cabral, G. A., Gyorkey, P. K., Uribe-Botero, G., Dreesman, G. R., and Melnick, J. L.,** Coxsackievirus aggregates in muscle cells of a polymyositis patient, *Intervirology,* 10, 69, 1978.
8. **Fallis, R. J. and Weiner, L. P.,** Further studies in search of a virus in amyotrophic lateral sclerosis, in *Human Motor Neuron Diseases,* Rowland, L. P., Ed., Raven Press, New York, 1982, 355.
9. **Gauntt, C. J., Gundvangen, J. R., Brans, Y. W., and Marlin, A. E.,** Coxsackievirus group B antibodies in the ventricular fluid of infants with severe anatomic defects in the central nervous system, *Pediatrics,* 76, 68, 1985.
10. **Rueckert, R. R.,** On the structure and morphogenesis of picornaviruses, in *Comprehensive Virology,* Vol. 6, Fraenkel-Conrat, H. and Wagner, R. R., Eds., Plenum Publishing, New York, 1982, 131.
11. **Melnick, J. L.,** Enteroviruses, in *Viral Infections of Human-Epidemology and Control.,* 2nd ed., Evans, A. S., Ed., Plenum Medical Book Company, New York, 1982, 251.
12. **Chonmaitree, T., Menegus, M. A., and Powell, K. R.,** The clinical relevance of "CSF viral culture": a two-year experience with aseptic meningitis in Rochester, N.Y., *JAMA,* 247, 1843, 1982.
13. **Yolken, R. H. and Torsch, V. M.,** Enzyme-linked immunosorbent assay for detection and indentification of coxsackie B antigen in tissue culture and clinical specimens, *J. Med. Virol.,* 6, 45, 1980.
14. **Yolken, R. H. and Torsch, V. M.,** Enzyme-linked immunosorbent assay for detection and identification of coxsackieviruses A., *Infect. Immun.,* 31, 742, 1981.

15. **Herrmann, J. E., Hendry, R. M., and Collins, M. F.,** Factors involved in enzyme-linked immunoassay of viruses and evaluation of the method for identification of enteroviruses, *J. Clin. Microbiol.*, 10, 210, 1979.
16. **Young, N. A.,** Polioviruses, coxsackieviruses, and echoviruses: comparison of the genomes by RNA hybrdization, *J. Virol.*, 11, 832, 1973.
17. **Frisby, D. P., Newton, C., Carey, N. H., Fellner, P., Newman, J. F. E., Harris, T. J. R., and Brown, F.,** Oligonucleotide mapping of picornavirus RNAs by two-dimensional electrophoresis, *Virology*, 71, 379, 1976.
18. **Nomoto, A. H., Toyoda, Y., Kataoka, M., Kohara, T., Suganuma, T., Omata, T., and Imura, N.,** Approaches to the development of new poliovirus vaccines based on molecular genetics, *Rev. Infect. Dis.*, 6, 494, 1984.
19. **Dorner, A. J., Dorner, L. F., Larsen, G. R., Wimmer, E., and Anderson, C. W.,** Identification of the initiation site of poliovirus polyprotein synthesis, *J. Virol.*, 42, 1017, 1982.
20. **Hewlett, M. J. and Florkiewicz, R. Z.,** Sequence of piconarvirus RNAs containing a radioiodinated 5′-linked peptide reveals a conserved 5′ sequence, *Proc. Natl. Acad. Sci. U.S.A.*, 77, 303, 1980.
21. **Stalhandske, O. K., Lindberg, M., and Pettersson, U.,** Replicase gene of cocksackievirus B3, *J. Virol.*, 51, 742, 1984.
22. **Tracy, S., Liu, H., and Chapman, N. M.,** Coxsackievirus B3: primary structure of the 5′ non-coding and capsid protein-coding regions of the genome, *Virus Res.*, 3, 263, 1985.
23. **Rotbart, H. A., Levin, M. J., and Villarreal, L. P.,** Use of subgenomic poliovirus DNA hybridization probes to detect the major subgroups of enteroviruses, *J. Clin. Microbiol.*, 20, 1105, 1984.
24. **Semler, B. L., Dorner, A. J., and Wimmer, E.,** Production of infectious poliovirus from cloned cDNA is dramatically increased by SV40 transcription and replication signals, *Nucleic Acids Res.*, 12, 5123, 1984.
25. **Stanway, G., Cann, A. J., Hauptmann, R., Hughes, P., Clarke, L. D., Mountford, R. C., Minor, P. D., Schild, G. C., and Almond, J. W.,** The nuclotide sequence of polivirus type 3 leon $12a_1b$: comparison with poliovirus type 1, *Nucleic Acids Res.*, 11, 5629, 1983.
26. **Semler, B. L., Johnson, V. H., and Tracy, S.,** A chimeric plasmid from cDNA clones of poliovirus and coxsackievirus produces a recombinant virus that is temperature-sensitive, *Proc. Natl. Acad. Sci. U.S.A.*, 83, 1777, 1986.
27. **Hyypiä, T. P., Stalhandske, R., Vainionpaa, R., and Pettersson, U.,** Detection of enteroviruses by spot hybridization, *J. Clin. Microbiol.*, 19, 436, 1984.
28. **Tracy, S.,** A comparison of genomic homologies among the coxsackievirus B group: use of fragments of the cloned coxsackievirus B3 genome as probes, *J. Gen. Virol.*, 65, 2167, 1984.
29. **Tracy, S.,** Comparison of genomic homologies in the coxsackievirus B group by use of cDNA dot-blot hybridization, *J. Clin. Microbiol.*, 19, 371, 1985.
30. **Rotbart, H. A., Levin, M. J., Villarreal, L. P., Tracy, S. M., Semler, B. L., and Wimmer, E.,** Factors affecting the detection of enteroviruses in cerbrospinal fluid with coxsackievirus B3 and poliovirus type 1 cDNA probes, *J. Clin. Microbiol.*, 22, 220, 1985.
31. **Seal, L. A. and Jamison, R. M.,** Evidence of secondary structure within the virion RNA of echovirus 22, *J. Virol.*, 50, 641, 1984.

31a. **Rotbart, H. A. and Levin, M. J.,** unpublished data.

32. **Linemeyer, D. L., Menke, J. G., Martin-Gallardo, A., Hughes, J. V., Young, A., and Mitra, S. W.,** Molecular cloning and partial sequencing of hepatitis A viral cDNA, *J. Virol.*, 54, 247, 1985.
33. **Skern, T., Sommergruber, W., Blaas, D., Gruendler, P., Fraundorfer, F., Pieler, C., Fogy, I., and Kuechler, E.,** Human rhinovirus 2: complete nucleotide sequence and proteolytic processing signals in the capsid protein region, *Nucleic Acids Res.*, 13, 2111, 1985.
34. **Stanway, G., Hughes, P. J., Mountford, R. C., Minor, P. D., and Almond, J. W.,** The complete nucleotide sequence of a common cold virus: human rhinovirus 14, *Nucleic Acids Res.*, 12, 7859, 1984.
35. **Skern, T., Sommergruber, W., Blaas, D., Pieler, C. H., and Kuechler, E.,** Relationship of human rhinovirus strain 2 and poliovirus as indicated by comparison of the polymerase gene regions, *Virology*, 136, 125, 1984.
36. **McSharry, J. J., Caliguiri, L. A., and Eggers, H. J.,** Inhibition of uncoating of poliovirus by Arildone, a new antiviral durg, *Virology*, 97, 307, 1979.
37. **Rotbart, H. A., Levin, M. J., and Villarreal, L. P.,** Nucleic acid probes in the diagnosis of enteroviruses, in *Medical Virology IV — Proc. 1984 Int. Symp. Medical Virology*, de la Maza, L. and Peterson, E. M., Eds., Lawrence Erlbaum Associates, 1985, 109.
38. **Wilfert, C. M., Lehrman, S. N., and Katz, S. L.,** Enteroviruses and meningitis, *Pediatr. Infect. Dis.*, 2, 333, 1983.

38a. **Rotbart, H. A.,** Human enterovirus infections — molecular approaches to diagnosis and pathogenesis, in *Molecular Aspects of Picornaviruses Infection and Detection*, Semler, B. and Ehrenfeld, E., Eds., Am. Soc. Microbiol., Washington, D.C, in press, 1988.

39. **Ambinder, R. F., Charache, P., Staal, S., Wright, P., Forman, M., Hayward, S. D., and Hayward, G. S.,** The vector homology problem in diagnostic nucleic acid hybridization of clinical specimens, *J. Clin. Microbiol.*, 24, 16, 1986.

40. **Melton, D. A., Krieg, P. A., Rebagliati, M. R., Maniatis, T., Zinn, K., and Green, M. R.,** Efficient in vitro synthesis of biologically active RNA and RNA hybridization probes from plasmids containing a bacteriophage SP6 promoter, *Nucleic Acids Res.,* 12, 7035, 1984.
41. **Werner, G., Rosenwirth, B., Bauer, E., Seifert, J., Werner, F., and Besemer, J.,** Molecular cloning and sequence determination of the genomic regions encoding protease and genome-linked protein of three picornaviruses, *J. Virol.,* 57, 1084, 1986.
42. **Rotbart, H. A., Abzug, M. J., and Levin, M. J.,** Development and application of RNA hybridization probes for the study of picornoviruses, *Mol. Cell. Probes,* in press, 1988.
43. **Steinberg, B. A., Visosky, A. A., and McKinlay, M. A.,** Oral administration of WIN 51711 prevents poliovirus induced paralysis and death in mice, Proc. 24th Interscience Conf. Antimicrobial Agents and Chemotherapy, Washington, D.C., 1984, Abstr. 432.
44. **Steinberg, B. A., Visosky, A. A., Frank, J. A., Jr., and McKinlay, M. A.,** WIN 51711-2 prevents ECHO-9 induced paralysis in suckling mice, Proc. 24th Interscience Conf. Antimicrobal Agents and Chemotherapy, Washington, D.C., 1984, Abstr. 433.

Chapter 16

THE APPLICATION OF DNA HYBRIDIZATION TO THE UNDERSTANDING OF HUMAN PARVOVIRUS DISEASE

Jonathan P. Clewley

TABLE OF CONTENTS

I. INTRODUCTION

The members of two genera of the Parvoviridae family infect higher animals: those that require a helper virus (the dependovirus genus), and those capable of autonomous replication (the parvovirus genus).[1] The dependoviruses are otherwise known as the adeno-associated viruses (AAV), and five human serotypes have been described, though these have not been shown to be the cause of any disease. DNA hybridization studies on clinical materials using AAV-specific probes have not been reported to date (mid 1986), though there are at least two reasons for starting such projects. The first is that their genome structure and behavior in tissue culture suggest that the AAVs are able to integrate in the host chromosome and so establish a latent infection.[2] If such integration occurred during infection of a host with an AAV, it might be possible by hybridization to identify an AAV as the primary causal agent of an ensuing disease. The second reason for developing a hybridization assay for these viruses is that AAV-5 was isolated from a genital condylomatous lesion.[3,4] Although this may represent an adevntitious association, the involvement of the AAVs in cervical and other cancers warrants investigation.

Members of the genus of the autonomously replicating parvoviruses have been found to infect rats, mice, cats, mink, dogs, cattle, swine, rabbits, chickens, and geese, and cause a range of diseases (e.g., panleukopenia, hemorrhagic encephalopathy, cerebellar ataxia, myocarditis, hepatitis, enteritis, osteolytic syndrome, and fetal death and congenital malformation).[5] Included in this genus is the human virus B19, and this review will concentrate on the use of probes for hybridization with materials suspected of harboring that virus DNA. The possibilities of investigations using AAV probes has been mentioned above. Other candidate parvoviruses suspected of involvement in human disease are RA-1,[6] which has been claimed as a cause of rheumatoid arthritis, and small round viruses (SRVs) implicated in cases of nonbacterial gastroenteritis.[7,8] More work needs to be done with the SRVs to characterize their nature and disease potential before probes are likely to become available.

II. BACKGROUND

A. The Disease

Apart from the AAVs, there is only one well-characterized parvovirus that infects humans, and that is called B19. It was discovered by Cossart and co-workers[9] in 1975 during the routine counter-immunoelectrophoresis (CIE) screening of blood donations for hepatitis B. Other laboratories in France and Japan discovered the virus independently using a similar CIE test system.[10,11] In the following years, the involvement of B19 in human disease became apparent, and the virus has changed its name several times. Cossart and colleagures[9] did not give the virus a specific name, referring to it as parvovirus-like agent (PVLA), serum parvovirus-like virus (SPLV), human serum parvovirus (HSPV), and human parvovirus (HPV). The name B19, which refers to the identification number of an original serum specimen, was proposed by the appropriate committee[1] and is now generally accepted.

The pathogenic potential of B19 has become clear over the last 10 years, and it is now apparent that most people become infected with the virus in childhood during outbreaks of erythema infectiosum or fifth disease (slapped cheek syndrome).[12,13] If adults are infected, the illness may be asymptomatic or feverish (headache, pyrexia, malaise, fatigue, myalgia).[7,14] Additionally, there may be purpura,[15,17] vasculitis,[18] and arthritis-like symptoms, particularly when adult women are infected.[19-21] Though the reason for these different manifestations of B19 infection is unclear, the genetic constitution (e.g., HLA type) of the infected person may be a contributing factor. This idea is supported by observations on B19 infections of people with one or other types of genetically determined hemolytic anemia.[22-29] A severe aplastic crisis can develop in individuals with sickle cell anemia, pyruvate kinase deficiency, hereditary spherocytosis, or related inherited disorders. This is probably a consequence of the virus replicating in the bone marrow and killing progenitor cells of erythrocytes, as shown by the observation that B19 inhibits erythropoietic colony formation in vitro,[30] and the recent demonstration of virus replication in suspension cultures of human erythroid bone marrow cells.[31]

A further complication of parvovirus infections arises if the host is pregnant. Initial reports have associated B19 infection with hydrops fetalis,[32] stillbirth,[33] and spontaneous abortion.[34-38] Fetal death is not an inevitable result of virus infection during pregnancy, however, and there are reports of normal outcomes following infection.[39] As will be detailed later, the investigation of B19 replication in fetal tissues has benefited from the availability of a DNA hybridization test.[40]

B. The Virus

The first reports of B19 characterized it as parvovirus-like on the basis of its buoyant density, size, and morphology as determined by CsCl density gradient centrifugation and electron microscopy.[7,9] Only later were studies undertaken on the nucleic acid and proteins of the virus particle.[41-43] After purification of virus from the serum of viremic individuals, nucleic acid was extracted and analyzed by gel electrophoresis. The virion particles were found to contain single-stranded (ss) DNA of either polarity, which on extraction annealed to form double-stranded (ds) molecules. This is a property of parvoviruses of all genera, though the majority of the autonomously replicating viruses package only the minus strand in the virion particle. The significance of this strand packaging is unclear, but the ease of preparing ds molecules of B19 DNA has facilitated its cloning, which has been reported by several groups.[43-47] Internal fragments generated by restriction enzyme digestion were cloned by Anderson et al.[45] and Morinet et al.[47] Cotmore and Tattersall[43] cloned the genome in two pieces by synthesis of the second strand, *Bam*HI digestion at a single internal site, and addition of *Bam*HI linkers to the ends before insertion into pAT153. Clewley[44] cloned half the genome by tailing with dCTP and inserting into the dG-tailed *Pst*I site of pBR322. Other cloning experiments have generated nearly full-length clones by exonuclease digestion to remove the terminal repeats and subsequent addition of linkers.[48] Molecular cloning has allowed an investigation into the relatedness of B19 to other parvoviruses, and it was found that there was cross-hybridization between B19 and the genomes of some autonomously replicating parvoviruses (LuIII, H-1, KRV, and MVM).[43] This result indicates that it might be possible to search for parvovirus sequences in clinical materials using probes derived from nonhuman parvoviruses, although it is difficult to decide what probes to construct and what specimens to use.

C. The Test

Hybridization assays for the detection of viral nucleic acid (Figure 1) usually take about 2 days, and the limit of detection is about 1 pg of DNA of 10^4 to 10^5 viral particles.[49] False positive results are possible by means of cross-hybridization of vector sequences with contaminating bacteria. Virus-positive specimens may be missed if less that 10^4 virions are present, whereas, by growth in cell culture, it may be possible to detect the presence of much lower numbers of particles. Tests for the presence of virus antigen may be equally as sensitive as DNA hybridization and give a result in a shorter time. These are all drawbacks of the current DNA hybridization tests, but there are, nevertheless, good reasons for the use of such tests, and B19 provides a useful illustration of their value and application.

First, a virus may have no easily available tissue culture system. B19 has not been grown in tissue culture and, therefore, is one of those fastidious viruses for which a hybridization test is useful (e.g., the human papillomaviruses, hepatitis B virus, adenovirus types 40 and 41, some rotaviruses, and SRVs associated with gastroenteritis). In such cases, the only supply of antigen may be material from virus-positive individuals. This can mean that the antigen for raising monoclonal antibody is in short supply, and a probe, once available, can be used to search for

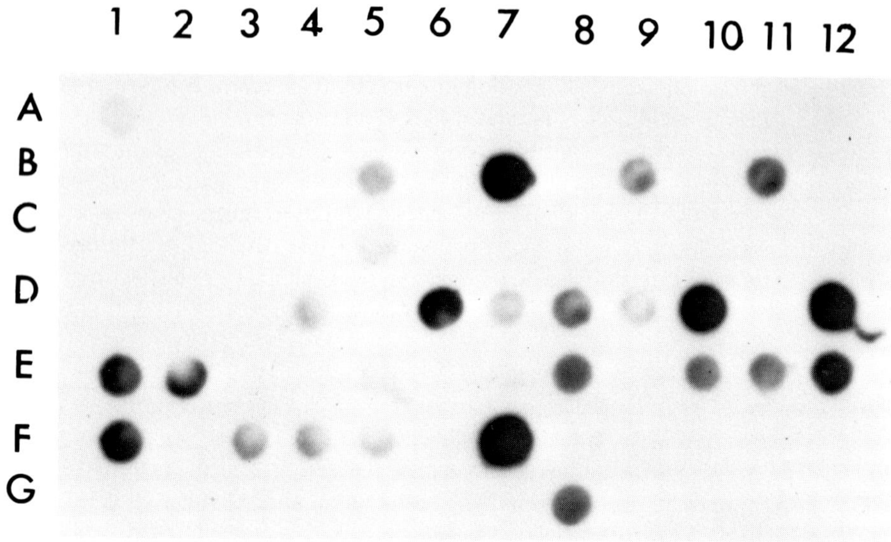

FIGURE 1. A typical dot hybridization result obtained from a range of samples applied to a nylon filter using a 96-well Schleicher and Schuell "minifold" apparatus. The intensity of darkening of the spots on the autoradiogram corresponds with the amount of viral specific DNA present.

more virus. Also, once cloned, ways of expressing viral proteins by recombinant DNA techniques in vitro can be explored.

Second, there may be no available antigens. Upon infection, B19 multiplies rapidly, and there is a viremia at about day 7. A day or two later, anti-B19-specific IgM is detectable, and this can bind to viral particles. Such immune complexes may not be detectable using an antibody test, but are readily found by hybridization. Some viruses are able to escape the surveillance of the immune system by integrating in the host genome or by replicating as an episome (retroviruses, papovaviruses, hepadnaviruses). This may have an immediately deleterious effect on the host, or there may be future excision and virus activation leading to disease. In either case, DNA detection by Southern blotting is a useful method of looking for viral sequences. Although B19 has not been shown to integrate in the human genome, it is worth screening appropriate tissues to investigate this. If, during primary infection, a small number of progenitor cells had B19 DNA integrated into their chromosome, it is possible that virus antigens would be expressed on the cell surface leading to attack by circulating antibodies. Or the virus DNA might be excised, leading to another round of replication and antigenic interaction with the immune system. Both of these processes could take place in the absence of evidence of infection and, conceivably, have undesirable consequences.

Third, the nature of the specimen to be tested may determine which tests are feasible. For example, a fresh-frozen tissue or serum specimen is equaly suitable for antigen detection, electron microscopy, and DNA hybridization. A formalin-fixed or embedded tissue specimen is likely to prove to be resistant to the former two techniques, but accessible to hybridization.[40] DNA in small fragments can persist in old specimens and should be detectable with a specific hybridization probe, given the necessary controls. Both the survival and loss of DNA have been illustrated in these circumstances by the recent cloning of DNA fragments from an Egyptian mummy and from an extinct quagga, and the failure to do so from an Iron Age body preserved in peat (Lindow man).[50]

Fourth, DNA hybridization techniques can be highly convenient and versatile. The preparation of a dot blot from up to 96 specimens, using a commercial "minifold" apparatus, gives a filter that may be probed several times. For example, a blot of serum or synovial specimens from patients with arthritis might be probed initially with B19 DNA and then with DNA from any of several other viruses thought to be involved in the onset of arthritis. Likewise, blots may be prepared in "the field" or in another laboratory such as a blood transfusion laboratory and — provided there is a convenient means of transport — subsequently probed in the research laboratory.

Fifth, *in situ* hybridization is capable of locating nucleic acid sequences in tissues allowing the pathogenesis of viral infections to be studied. Possibly, slides prepared for antigen detection by fluoresence could also be used for DNA/RNA analysis. Fetal tissues from cases of intrauterine death where B19 involvement is suspected, and from relevant biopsy material, could be investigated to determine the sites of replication of the virus.

D. Other Tests for B19 Infection

The use of laboratory tests for the diagnosis of infection with B19 has recently been reviewed by Cohen,[51] and the reader is referred to that comprehensive discussion of the evolution and use of tests for the virus. The diagnosis of B19 infection can be made in two ways: either by directly detecting virus or by detecting specific antibody. Apart from nucleic acid hybridization, B19 has been detected by radioimmunoassay (RIA) for viral antigen using monoclonal antibodies; by immune electron microscopy (IEM) using specific antiserum; and by CIE, also using specific detector serum.[7,9,52] RIA is the most sensitive method for detecting antigen and is comparable with hybridization. It does not suffer from the problem of false positives due to the presence of contaminating bacteria harboring plasmids which cross-hybridize with vector sequences. However, in some cases where particles were aggregated by IgM, RIA was unable to detect viral antigen, whereas DNA was detectable.[44,51] For these reasons, RIA and hybridization are best viewed as complementary tests, and whether either or both are used in the demonstration of parvovirus B19, a follow-up or confirmatory test will probably be desirable.

Any laboratory diagnosing B19 infection will need to perform test for both IgM- and IgG-specific antibody as well as antigen. It is frequently the case that a patient presents with symptoms after virus has been cleared from the blood, and in such situations a diagnosis is made by measuring specific IgM and IgG levels.[51] Conversely, virus can be present in the absence of specific immunoglobulin, and this may be revealed by testing serum from asymptomatic contacts of infected individuals. Like the screening of blood donors, this may provide a sizable quantity of infected blood, which is useful in maintaining and developing laboratory tests.

III. SELECTED RESULTS

A. Dot Hybridization of Serum Samples

Obtaining a molecularly cloned probe[43] is the first step toward setting up a hybridization test, and two papers have specifically described such tests for B19 DNA.[44,45] Another[47] discussed the use of this technique. Anderson and colleagues[45] found that 0.3 to 0.5 pg of viral DNA, equivalent to 10^4 virus particles, could be detected in human serum using a 700-base pair (bp) *Pst*I fragment cloned in the vector pAT153. This was estimated to be 200 to 300 times more sensitive that the solid phase RIA described by Cohen and associates.[52] Four specimens were found to contain parvovirus by dot hybridization but not by RIA, and this was attributed either to the difference in sensitivity of the two assays or to antiparvovirus antibody present in the serum. This test showed also that viremia may be detectable at any time from 11 days prior to the onset of symptoms to 11 days after. The dot hybridization test described by Clewley[44] used a 27,000-bp *Pst*I insert cloned in the vector pBR322, and the test was found to be of comparable sensitivity to RIA. This test was also able to detect about 1 pg of viral DNA. In two specimens in which immune complexes were known to be present, the antigen could not be detected by CIE and was only weakly reactive by RIA, but was readily detected by dot hybridization. Refinements of this test have included cloning the B19 genome in lambda and the "riboprobe" vectors pGFM-1 and pGEM-2.[48] Morinet and co-workers[47] used a 32,000-bp *Hin*dIII-*Bam*HI fragment cloned in pBR322 and were also able to detect as little as 1 pg of B19 DNA in serum. In this study, the specimens that were found to be positive by dot hybridization were also positive by CIE and IEM, and so it was not possible to determine if hybridization was a more sensitive test in these cases.

B. Dot Hybridization of Body Fluids and Tissue Specimens

The dot hybridization test on serum samples usually works well on as little as 1 μl of serum diluted in an alkaline solution to free the DNA from protein and ensure that it remains denatured. The specimen is simply applied to a nylon or nitrocellulose membrane, often using a commercial multiwell apparatus for this purpose. Detection of viral-specific DNA in other specimens is not so straightforward as this for two main reasons. One is that the DNA is less easily accessible and may have to be recovered by phenol extraction; the other is that body fluids and similar materials are more likely to give false positive results. These arise because of contaminating bacterial DNA which cross-reacts with residual vector sequences in the probe, and because of nonspecific binding of ^{32}P-labeled DNA to charged polymers present in some samples (e.g., feces). Nevertheless, throat, nasal wash, stool, buffy-coat blood leukocytes, urine, synovial fluid and membrane, placenta, and various fetal tissue specimens have been examined by hybridization for B19 DNA.[13,32,35,36,38,40,53] These investigations have had two ends: the determination of the routes of transmission of the virus, and the establishment of the role of B19 in diseases in which it was not previously known to be involved. In a volunteer study in which adults were inoculated intranasally with B19,[53] DNA was detected in specimens from the upper respiratory tract (nasal washes and gargle specimens), as well as in serum. DNA was not detected in 10% extracts of stool specimens, and only in one case was it detected in urine. This urine was, however, contaminated with menstrual blood, and this was the probable source of the virus. This study, therefore, suggests that transmission of the virus is primarily via the respiratory tract.

B19 DNA has been detected in fetal tissues in cases of hydrops fetalis,[32,35,36,38,40] where the mother was infected with the virus during pregnancy. Initially, only formalin-fixed tissues were available for analysis, and the only test able to demonstate the presence of virus in these tissues was dot hybridization. Viral antigen does not survive formalin fixation in a form recognizable by RIA or IEM,[40,54] but sufficiently large fragments of DNA remain to detect by dot hybridization. Genome-sized DNA is usually not demonstrable by Southern blotting, however. When unfixed frozen tissue was available from such cases, it was in some instances possible to show genome-sized DNA by Southern blotting and, additionally, virus antigen was detected by RIA, and viral particles were observed by electron microscopy.[36,38,40] These studies point to the liver as the site of virus replication in the fetus, since the majority of hybridizing sequences are found in that organ. This is compatible with the pathological findings, and it is known that the proposed target cells of the virus develop in the fetal liver.

C. Confirmation of True Positives, and the Problem of False Positives

Assuming that further investigation is suggested by a positive dot hyridization result, the two most useful "different technology" tests are an RIA for viral antigen and electron microscopy for viral particles. Recent work demonstrating the presence of B19 DNA in fetal tissues from cases of maternal infection during pregnancy[32,35] has been confirmed in other cases by these two techniques.[38,40]

There is also a need to confirm a dot hybridization test by "same technology" methods and, as instanced above, the definitive means is a Southern blot. This will demonstrate the presence of genome-sized DNA. It is unlikely that any contaminating cross-hybridizing sequences will have DNA of the same size as B19. In some cases, more that one band of B19 DNA is present. These are the 5.4-kilobase pair (kbp) ds DNA form, the 5.4-kb ss form, and sometimes bands of greater size than the ds form resulting from the formation of concatamers due to the repeat

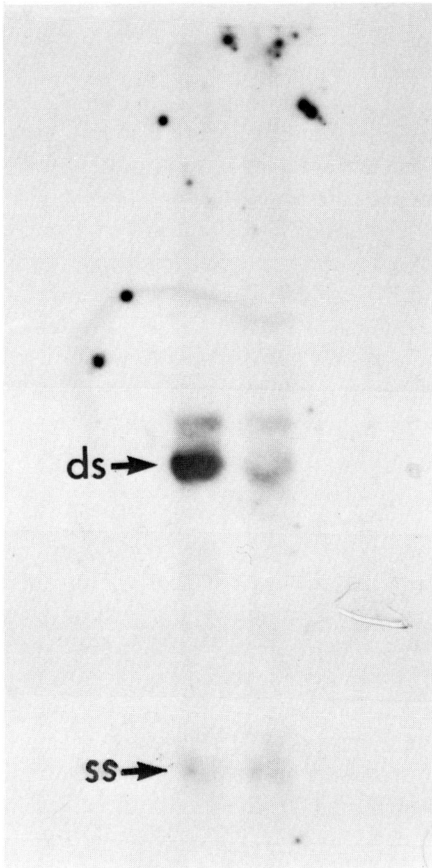

FIGURE 2. A Southern blot confirmatory test showing the electrophoretic profile often observed for B19 DNA. After agarose gel electrophoresis, DNA was transferred to a filter which was then probed with labeled B19 DNA. The single-stranded (ss) genome and the double-stranded (ds) self-annealed DNA are arrowed. The band above the ds DNA is probably due to multimers.

sequences at the end of the genome. These bands give a characteristic electrophoretic profile which can be thought of as "diagnostic" for B19 (Figure 2). Digestion of the DNA sample with restriction enzymes prior to Southern blotting can be used to give recognizable bands, and comparisons of different isolates by this method have been undertaken.[47,55] If searching for B19 DNA integrated in the human chromosome from, for instance, bone marrow cells or cells from arthritic immune complexes, it would be a good idea to digest with one of the enzymes known usually not to cut B19 DNA (*Eco*RI, *Sal*I, *Xho*I), as well as ones expected to give characteristic internal fragments.

Southern blotting cannot always be used to confirm a dot hybridization result. There simply may not be enough specimen to extract DNA from. There is about a tenfold difference in sensitivity between the two methods; about 10 pg of B19 DNA is required for a successful Southern blot. Also, as evidenced by the studies on fixed and unfixed fetal tissues, parvovirus-specific DNA may be present in a specimen in a degraded form which is detectable by dot blotting, but which is of too low molecular weight to give an interpretable band on a Southern blot. In these cases, confirmation that the signal is due to B19 DNA may be done in two ways.

The first is to hybridize the same dot blot with different probes: vector DNA alone to rule out cross-hybridization, and a probe prepared from B19 DNA purified from viremic serum to confirm specificity. The second is to prepare a dot blot of "control" fetal tissues to check that the putative B19 DNA does not cross-hybridize with human DNA.

The problem of contamination of serum by bacteria that cross-hybridize with the probe is well illustrated by two specimens recently analyzed at the Virus Reference Laboratory.[55] One of these gave a positive result in a dot hybridization test using a probe prepared from a viral insert that had been electrophoretically separated from the vector DNA. However, other test results (RIA) were negative. Therefore, DNA was extracted from this serum and analyzed by electrophoresis and ethidium bromide staining of the gel. Three bands were seen, one slow-migrating form typical of bacterial genome DNA, and two others of the size often observed for plasmid DNA. On culture of the specimen, an atypical *Escherichia coli* was isolated, confirming the presence of contaminating bacteria. The second specimen proved to contain both B19 DNA and bacterial DNA. Both cases emphasize the need for care in interpreting the result of a dot hybridization test. False positivies can and will occur.

IV. CONCLUSIONS AND PROSPECTS

DNA hybridization has found a place in the diagnosis of disease caused by parvovirus B19. It complements other methods and can contribute information where antigen testing by RIA, CIE, or IEM is, for one reason or another, unsuccessful. A positive dot hybridization result will often be the starting point for further, more sophisticated experiments. The most commonly used test is, at present, direct hybridization using ^{32}P-labeled probes. Biotinylated probes also work,[55] though like ^{32}P, they have their drawbacks. Refinements of the test are likely to include the development of "sandwich" hybridization methods, the use of universal probes, the use of improved nonradioactive labeling methods, the use of alternative radioactive labeling methods, and the improvement of the sensitivity by amplifying both the substrate and the signal. Finally, a semiautomated test suitable for serum or blood samples in a microtiter format may be developed if there is a need to screen large number of blood donations.

REFERENCES

1. **Siegl, G., Bates, R. C., Berns, K. I., Carter, B. J., Kelly, D. C., Kurstak, E., and Tattersall, P.,** Characteristics and taxonomy of Parvoviridae, *Intervirology,* 23, 61, 1985.
2. **Berns, K. I. and Hauswirth, W. W.,** Adeno-associated virus DNA structure and replication, in *The Parvoviruses,* Berns, K. I., Ed., Plenum Press, New York, 1984, 1.
3. **Bantel-Schaal, U. and Zur Hausen, H.,** Characterization of the DNA of a defective human parvovirus isolated from a genital site, *Virology,* 134, 52, 1984.
4. **Georg-Fries, G., Biederlack, S., Wolf, J., and Zur Hausen, H.,** Analysis of proteins, helper dependence, and seroepidemiology of a new human parvovirus, *Virology,* 134, 67, 1984.
5. **Siegl, G.,** Biology and pathogenicity of autonomous parvoviruses, in *The Parvoviruses,* Berns, K. I., Ed., Plenum Press, New York, 1984, 297.
6. **Simpson, R. W., McGinty, L., Simon, L., Godzeski, C. W., and Boyd, R. J.,** Association of parvoviruses with rheumatoid arthritis of humans, *Science,* 223, 1425, 1984.
7. **Paver, W. K. and Clarke, S. K. R.,** Comparison of human fecal and serum parvo-like viruses, *J. Clin. Microbiol.,* 4, 67, 1976.
8. **Caul, E. O. and Appleton, H.,** The electron microscopical and physical characteristics of small round human fecal viruses: an interim scheme for classification, *J. Med. Virol.,* 9, 257, 1982.
9. **Cossart, Y. E., Field, A. M., Cant, B., and Widdows, D.,** Parvovirus-like particles in human sera, *Lancet,* 1, 72, 1985.
10. **Courouce, A.-M., Ferchal, F., Morinet, F., Muller, A., Drouet, J., Soulier, J. P., and Perol, Y.,** Human parvovirus infections in France, *Lancet,* 1, 160, 1984.

11. **Okochi, K., Mori, R., Miyazaki, M., Cohen, B. J., and Mortimer, P. P.,** Nakatani antigen and human parvovirus (B19), *Lancet,* 1, 160, 1984.
12. **Anderson, M. J., Jones, S. E., Fisher-Hoch, S. P., Lewis, E., Hall, S. M., Bartlett, C. L. R., Cohen, B. J., Mortimer, P. P., and Pereira, M. S.,** The human parvovirus, the cause of erythema infectiosum (fifth disease)?, *Lancet,* 1, 1378, 1983.
13. **Plummer, F. A., Hammond, G. W., Forward, M. D. K., Sekla, L., Thompson, L. M., Jones, S. E., Kidd, I. M., and Anderson, M. J.** An erythema infectiosum-like illness caused by human parvovirus infection, *N. Engl. J. Med.,* 313, 74, 1985.
14. **Shneerson, J. M., Mortimer, P. P., and Vandervelde, E. M.,** Febrile illness due to a parvovirus, *Br. Med. J.,* 280, 1580, 1981.
15. **Lefrere, J.-J., Boutard, P., Courouce, A.-M., Lacaze, T., and Girot, R.,** Familial human parvovirus infections, *Lancet,* 2, 333, 1985.
16. **Lefrere, J.-J., Courouce, A.-M., Muller, J.-Y., Clark, M., and Soulier, J.-P.,** Human parvovirus and purpura, *Lancet,* 2, 730, 1985.
17. **Mortimer, P. P., Cohen, B. J., Rossiter, M. A., Fairhead, S. M., and Rahman, A. F. M. S.,** Human parvovirus and purpura, *Lancet,* 2, 730, 1985.
18. **Li Loong, T. C., Coyle, P. V., Anderson, M. J., Allen, G. E., and Connolly, J. H.,** Human serum parvovirus associated vasculitis, *Postgrad. Med. J.,* 62, 493, 1986.
19. **White, D. G., Woolf, A. D., Mortimer, P. P., Cohen, B. J., Blake, D. R., and Bacon, P. A.,** Human parvovirus arthropathy, *Lancet,* 1, 419, 1985.
20. **Reid, D. M., Reid, T. M. S., Brown, T., Rennie, J. A. N., and Eastmond, C. J. M.,** Human parvovirus-associated arthritis: a clinical and laboratory description, *Lancet,* 1, 422, 1985.
21. **Cohen, B. J., Buckley, M. M., Clewley, J. P., Jones, V. E., Puttick, A. H., and Jacoby, R. K.,** Human parvovirus infection in early rheumatoid and inflammatory arthritis, *Ann. Rheum. Dis.,* 45, 832, 1986.
22. **Pattison, J. R., Jones, S. E., Hodgson, J., Davis, L. R., White, J. M., Stroud, C. F., and Murtaza, L.,** Parvovirus infections and hypoplastic crises in sickle cell anaemia, *Lancet,* 1, 664, 1981.
23. **Serjeant, G. R., Topley, J. M., Mason, K., Serjeant, B. E., Pattison, J. R., Jones, S. E., and Mohamed, R.,** Outbreak of aplastic crisis in sickle cell anaemia associated with parvovirus-like agent, *Lancet,* 2, 595, 1981.
24. **Kelleher, J. F., Luban, N. L. C., Mortimer, P. P., and Kamimura, T.,** The human serum "parvovirus": a specific cause of aplastic crisis in hereditary spherocytosis, *J. Pediatr.,* 102, 720, 1983.
25. **Rao, K. R. P., Patel, A. R., Anderson, M. J., Hodgson, J., Jones, S. E., and Pattison, J. R.,** Infection with parvovirus-like virus and aplastic crisis in chronic hemolytic anemia, *Ann. Intern. Med.,* 98, 930, 1983.
26. **Duncan, J. R., Potter, C. G., Cappellini, M. D., Kurtz, J. B., Anderson, M. J., and Weatherall, D. J.,** Aplastic crisis due to parvovirus infection in pyruvate kinase deficiency, *Lancet,* 2, 14, 1983.
27. **Kelleher, J. F., Luban, L. C., Cohen, B. J., and Mortimer, P. P.,** Human serum parvovirus as the cause of aplastic crisis in sickle cell disease, *Am. J. Dis. Child.,* 138, 401, 1984.
28. **Young, N. S., Mortimer, P. P., Moore, J. G., and Humphries, R. K.,** Characterization of a virus that causes transient aplastic crisis, *J. Clin. Invest.,* 73, 224, 1984.
29. **Bertrand, Y., Lefrere, J.-J., Leverger, G., Courouce, A. M., Feo, C., Clark, M., Schaison G., and Soulier, J. P.,** Autoimmune haemolytic anemia revealed by human parvovirus linked erythroblastopenia, *Lancet,* 2, 382, 1985.
30. **Mortimer, P. P., Humphries, R. K., Moore, J. G., Purcell, R. H., and Young, N. S.,** A human parvovirus-like virus inhibits haematopoietic colony formation in vitro, *Nature (London),* 302, 426, 1983.
31. **Ozawa, K., Kurtzman, G., and Young, N.,** Replication of the B19 parvovirus in human bone marrow cell cultures, *Science,* 233, 883, 1986.
32. **Brown, T., Anand, A., Richie, L. D., Clewley, J. P., and Reid, T. M. S.,** Intrauterine parvovirus infection associated with hydrops fetalis, *Lancet,* 2, 1033, 1984.
33. **Knott, P. D., Welply, G. A. C., and Anderson, M. J.,** Serologically proved intrauterine infection with parvovirus, *Br. Med. J.,* 289, 1660, 1984.
34. **Mortimer, P. P., Cohen, B. J., Buckley, M. M., Cradock-Watson, J. E., Ridehalgh, M. K. S., Burkhardt, F., and Schilt, U.,** Human parvovirus and the fetus, *Lancet,* 2, 1012, 1985.
35. **Gray, E. S., Anand, A., and Brown, T.,** Parvovirus infections in pregnancy, *Lancet,* 1, 208, 1986.
36. **Bond, P. R., Caul, E. O., Usher, J., Cohen, B. J., Clewley, J. P., and Field, A. M.,** Intrauterine infection with human parvovirus, *Lancet,* 1, 448, 1986.
37. **Lefrere, J.-J., Dumez, Y., Courouce, A.-M., and Deschene, G.,** Intrauterine infection with human parvovirus, *Lancet,* 1, 449, 1986.
38. **Anand, A., Gray, E. S., Brown, T., Clewley, J. P., and Cohen, B. J.,** Human parvovirus infection in pregnancy and hydrops fetalis, *N. Engl. J. Med.,* 316, 183, 1986.
39. **Wright, E. P., Dyson, A. J., and Alaily, A.,** Infection with parvovirus during pregnancy, *Br. Med. J.,* 290, 241, 1985.
40. **Clewley, J. P., Cohen, B. J., and Field, A. M.,** Detection of parvovirus B19 DNA, antigen, and particles in the human fetus, *J. Med. Virol.,* 23, 367, 1987.

41. **Summers, J., Jones, S. E., and Anderson, M. J.,** Characterization of the genome of the agent of erythrocyte aplasia permits its classification as a human parvovirus, *J. Gen. Virol.*, 64, 2527, 1983.
42. **Clewley, J. P.,** Biochemical characterization of a human parvovirus, *J. Gen. Virol.*, 65, 241, 1984.
43. **Cotmore, S. F. and Tattersall, P.,** Characterization and molecular cloning of a human parvovirus genome, *Science*, 226, 1161, 1984.
44. **Clewley, J. P.,** Detection of human parvovirus using a molecularly cloned probe, *J. Med. Virol.*, 15, 173, 1985.
45. **Anderson, M. J., Jones, S. E., and Minson, A. C.,** Diagnosis of human parvovirus infection by dot-blot hybridization using cloned viral DNA, *J. Med. Virol.*, 15, 163, 1985.
46. **Shade, R. O., Blundell, M. C., Cotmore, S. F., Tattersall, P., and Astell, C. R.,** Nucleotide sequence and genome organization of human parvovirus B19 isolated from the serum of a child during aplastic crisis, *J. Virol.*, 58, 921, 1986.
47. **Morinet, F., Tratschin, J.-D., Perol, Y., and Siegl, G.,** Comparison of 17 isolates of the human parvovirus B19 by restriction enzyme analysis, *Arch. Virol.*, 90, 165, 1986.
48. **Mori, J., Beattie, P., Melton, D. W., Cohen, B. J., and Clewley, J. P.,** Structure and mapping of the DNA of human parvovirus B19, *J. Gen. Virol.*, 68, 2797, 1987.
49. **Clewley, J. P.,** Diagnostic methods based on viral genome detection, in *Public Health Virology—12 Reports*, Mortimer, P. P., Ed., Public Health Laboratory Service, London, 1986, 14.
50. **Hughes, M. A., Jones, D. S., and Connolly, R. C.,** Body in the bog but no DNA, *Nature (London)*, 323, 208, 1986.
51. **Cohen, B. J.,** Laboratory tests for the diagnosis of infection with B19 virus, in *Parvoviruses and Human Disease*, Pattison, J. R., Ed., CRC Press, Boca Raton, Fla., 1986, 69.
52. **Cohen, B. J., Mortimer, P. P., and Pereira, M. S.,** Diagnostic assays with monoclonal antibodies for the human serum parvovirus-like virus (SPLV), *J. Hyg.*, 91, 113, 1983.
53. **Anderson, M. J., Higgins, P. G., Davis, L. R., Willman, J. S., Jones, S. E., Kidd, I. M., Pattison, J. R., and Tyrell, D. A. J.,** Experimental parvoviral infection in humans, *J. Infect. Dis.*, 152, 257, 1985.
54. **Macartney, M. and Macartney, C. M.,** Canine parvovirus: development of immunofluorescence and immunoperoxidase techniques, *Res. Vet. Sci.*, 40, 201, 1986.
55. **Mori, J. and Cohen, B. J.,** personal communication, 1986.

Chapter 17

NUCLEIC ACID SANDWICH HYBRIDIZATION IN THE DIAGNOSIS OF INFECTIONS CAUSED BY *CHLAMYDIA TRACHOMATIS* AND OTHER MICROBES

Marjut Ranki and Airi Palva

TABLE OF CONTENTS

I. INTRODUCTION

A. Hybridization to Solve Diagnostic Problems

The most characteristic feature of the nucleic acid hybridization methods is their exquisite specificity. Detection of an infectious agent based on this principle is not dependent on the viability of the organism to be identified, the presence of antibodies potentially complexed with the target antigens, or the regular expression of the genetic material. All these properties, together with the possibility of producing probes in large quantities, have made hybridization methods attractive tools for microbiologists.

The hybridization principle has been used in research laboratories for more than 20 years. The challenge of the field has been to convert the research tool into a practical method applicable to diagnostic laboratories handling large numbers of specimens. Such tests should be easy to perform, and the results should be obtained as numerical values with clear distinction between positive and negative.

For practical performance, the decisive question is how to separate the reaction product, the hybrid, from the unreacted probe. This has often been solved by immobilizing one of the reaction components onto a solid carrier which most often is a nitrocellulose filter. In the classical dot blot, the nucleic acid of the specimen is denatured into single-stranded form and attached to the filter. The filter is incubated with the probe in the hybridization solution, and hybrids are separated from the excess probe by removing the filter from the solution and washing it with buffer. This method, however, has some problems: (1) specimen pretreatment is tedious, often requiring isolation and purification of the nucleic acid before immobilization; (2) in spite of purification, background problems, due to specimen impurities such as mucopolysaccharides, occur; and (3) interpretation of results is subjective when assessing intensities of radioactive or color spots.

B. Sandwich Hybridization Technique

In our laboratory, we have developed a sandwich hybridization method,[1] which circumvents the above-mentioned problems. Sandwich hybridization utilizes a solid carrier onto which hybrids are collected, but keeps the specimen in solution making it possible to test the samples in a crude form without elaborate pretreatment.[2-7]

The sandwich hybridization method uses two separate nucleic acid fragments derived from the target. One of them is immobilized onto the carrier, functioning as a capture reagent, the other is the labeled probe. A target, homologous to both reagents, will hybridize both to the capture reagent and the probe, thereby mediating the labeling of the carrier. The hybridization reaches 75% of its completion in 18 h.[1] Therefore, the test requires overnight incubation. By combining multiple probe pairs for the detection of one target, the sensitivity of the test has been improved,[3,8] allowing detection of close to 10^5 target molecules in an overnight reaction. The test is quantitative over a wide range of target concentrations and can be used for quantification purposes.[4,6,9,10]

The probes are usually labeled with radioactive isotopes of high specific activity. Hybrid detection by nonradiometric methods is also possible. Using biotinylated probes and streptavidin-mediated enzymatic detection, the sensitivity of the sandwich hybridization is reduced about tenfold. A similar decrease in sensitivity is a general finding in all hybridization methods. Using time-resolved fluorometry for detection in sandwich hybridization, the sensitivity was, however, comparable to that obtained using radioactively labeled probes.[10]

C. Present Diagnosis of *C. trachomatis*

The incidence of *Chlamydia trachomatis* infections has been rapidly increasing during the last few years; it is now the most common agent of sexually transmitted diseases in Western countries.[11-13] *C. trachomatis* is a taxonomically divergent prokaryotic organism belonging to

Table 1
OUTLINE OF THE SANDWICH HYBRIDIZATION TEST

I. Production of DNA reagents
 Clone two adjacent, nonoverlapping DNA fragments with desired specificity from the microbe of interest into two nonhomologous vectors, e.g., pBR322 and M13[1,9]
II. DNA reagents in the test
 Immobilize the catching DNA probe on a solid support using 4×10^{10} to 2×10^{11} denatured recombinant molecules per one test filter 1 cm in diameter[1]
 Label the probe-DNA by chemical iodination[9] (single-stranded DNA) or by nick translation[46] (double-stranded DNA)
III. Sample pretreatment
 Boil for 5 min in alkaline sodium dodecyl sulfate, releasing the target nucleic acid in single-stranded form[6]
IV. Performance of the test
 Hybridize the denatured sample DNA and the labeled probe and filter DNA in:
  4 × SSC (1 × SSC = 0.15 *M* NaCl, 0.015 *M* sodium citrate)
  50 m*M* phosphate buffer pH 7
  1 × Denhardt's solution
  200 mg/ml denatured carrier DNA
  0.25% SDS
  2×10^5 cpm of labeled probe at 65°C for 18 hr
 Wash the filters in: 0.1 × SSC 0.2% SDS (at 50°C for 1 hr with one change of buffer)
V. Hybrid detection
 Count the filters in an appropriate counter depending on the label used
 Quantify the hybrids formed by comparing the signal with the standard curve
 Use as cutoff value, e.g., the 99% confidence limits, calculated according to a sufficient number of negative controls in the test

the genus *Chlamydia,* which contains only one other species, i.e., *C. psittaci.*[14] Both species are obligate intracellular parasites, but unrelated according to DNA homology, which is only 10%.[15,16] In contrast, the strains of *C. trachomatis,* classified into 15 serotypes, are highly related and share extensive DNA homology. The chlamydial serotypes from D to K are associated with infections of sexually transmitted origin, i.e., cervicitis and urethritis, and the serotypes L1, L2, and L3 give rise to lymphogranuloma venereum.[17] Many complications, e.g., epididymitis, Reiter's syndrome, proctitis, endometritis, salpingitis, and perihepatitis, are seen in adults and are associated with the primary urethritis or mucopurulent cervicitis. Additional syndromes, including neonatal pneumonia and inclusion conjunctivitis, emphasize the severe nature of this pathogen for all age groups.[18,19]

Despite the importance of *C. trachomatis* as a pathogen, its diagnosis is currently difficult and often expensive. The laboratory diagnosis is done mainly by isolation in cell cultures, which requires facilities of specialized laboratories, is cumbersome, and usually takes several days.[20] Alternative immunological methods, not relying on the viability of the microbe, have been developed for the direct detection of *C. trachomatis* from clinical specimens. These include detection of chlamydial antigens in genital and conjunctival smears by immunofluorescence using monoclonal antibodies,[21-26] a solid-phase enzyme immunoassay,[27,28] and an enzyme-amplified immunoassay.[29] Nucleic acid hybridization methods for the identification of chlamydial DNA in clinical specimens have been described.[30-32] Below, we will discuss the application of nucleic acid sandwich hybridization for the diagnosis of *C. trachomatis.*

II. SANDWICH HYBRIDIZATION IN THE DETECTION OF *C. TRACHOMATIS*

A. Outline of the Sandwich Hybridization Test (Table 1)

The two DNA probes, utilized as a reagent pair in each sandwich hybridization assay, are

Table 2
DNA PROBES FOR THE DETECTION OF *C. TRACHOMATIS*

Origin	Size (kb)	DNA-fragment: Cloning vector	Use	Ref.
C. trachomatis, serotype L2	2.9	pAT153	Labeled probe	3
Chromosomal DNA	3.0	pAT153	Labeled probe	3
	2.1	M13 mp8	Filter DNA	3
		M13 mp8	Filter DNA	3
Cryptic plasmid	5.9	pBR327	Labeled probe	35
DNA pLGV 125 of	1.3	M13 mp10	Filter DNA	35
C. trachomatis[34]		M13 mp11	Filter DNA	35

produced using recombinant DNA techniques. One is cloned into the plasmid vector pBR322 or pAT153, and the other segment is cloned into the single-stranded bacteriophage M13. The two DNA fragments must be adjacent in the genome of the microbe of interest. Using this strategy, purification of the specific inserts from the cloning vectors is not required. In the hybridization assay, M13 recombinant DNAs serve as filter reagents, and the nick-translated plasmid clones serve as probes. The ladder formation between the strands of the double-stranded probes most likely increases the hybridization signal.

B. Construction of a Test Specific for *C. trachomatis*

The chromosomal and cryptic plasmid reagents constructed for the detection of *C. trachomatis* are summarized in Table 2. The chromosomal reagents were derived from a shotgun-cloned recombinant plasmid (pKTH1220) containing a chromosomal DNA fragment from serotype L2 subcloned into the pAT153 and M13 vectors. The specificity of the chromosomal probes was confirmed by hybridizing them to DNA from seven antigenically distinct chlamydial serotypes and with 41 unrelated microbes. The probes were highly specific for *C. trachomatis* under stringent conditions, detecting all chlamydial serotypes equally well but none of the unrelated organisms.[3] *C. psittaci* was negative in a direct spot hybridization test using the plasmid pKTH1220 as the probe.[31]

The sensitivity obtained using the chromosomal fragments (Table 2) was 10^6 target molecules in a standard overnight sandwich assay.[3] On the basis of a pilot study, this sensitivity was considered to be unsatisfactory for testing clinical specimens for *Chlamydia*. To increase the detection sensitivity, the cryptic plasmid of *C. trachomatis*,[33] shown to be present in several copies in all serotypes,[34] was used as an additional target sequence. A reagent pair was constructed from the cloned cryptic plasmid, pLGV125[30] (Table 2). The plasmid probes, tested as described for chromosomal probes, were shown to be specific for *C. trachomatis*.[49] By combining the chromosomal and the cryptic plasmid probes, the sensitivity of detection was increased to 10^5 target DNA molecules and to less than 10^5 elementary bodies.[35] The copy number of the plasmid was determined from six female endocervical specimens (Table 3). It was between four and nine, which is in accordance with the results of Palmer and Falkow.[36]

C. Effect of Crude Specimens on Hybridization Results

We have studied the ability of probes to detect *C. trachomatis* from clinical specimens using both spot and sandwich hybridization methods.[31,35] The sample pretreatments for optimal analysis of *Chlamydia* DNA by these two methods are summarized in Table 4. In spite of extensive purification, nucleic acids isolated from crude specimens still caused background problems in the spot test. The spot method using autoradiography was able to detect 10^4 and 5

Table 3
THE COPY NUMBER OF THE CRYPTIC CHLAMYDIAL PLASMID IN FEMALE ENDOCERVICAL SPECIMENS

			Sandwich hybridization[a]				
			Chromosomal		Plasmid		
Experiment	Specimens	Chlamydial culture	cpm[b]	Number of molecules ($\times 10^5$)[c]	Cpm[d]	Number of molecules ($\times 10$)[c]	Copy number[e]
1	1	3+	227	25	1891	200	8
	2	3+	195	19	830	85	4
2	3	3+	262	6	183	38	6
	4	4+	1087	100	2143	670	7
	5	1+	350	15	411	130	9
	6	3+	1030	95	1383	440	5

[a] Hybridization was at 65°C for 17 hr using about 600,000 cpm of nick-translated chromosomal and plasmid probes using as label ^{32}P-dCTP (see Table 2).
[b] Counts per minute on the chromosomal filter (see Table 2); background in exp. 1: 118 cpm, in exp. 2: 147 cpm.
[c] Number of target molecules hybridized as calculated according to chromosomal and plasmid standard curves, respectively.
[d] Counts per minute on the plasmid filter (Table 2); background in exp. 1: 37 cpm, in exp. 24: 28 cpm.
[e] Number of plasmid molecules per number of chlamydial genome equivalents.

Table 4
PROCEDURES FOR THE DETECTION OF CLINICAL CHLAMYDIAL SPECIMENS BY SPOT AND SANDWICH HYBRIDIZATION

Spot hybridization[31]	Sandwich hybridization[35]
Chlamydial swab specimen in 2SP-buffer[a]	Clinical swab specimen in 10 m*M* Tris-HCl, 1 m*M* EDTA, pH 7 containing SDS and carrier DNA
1. Proteinase K digestion	
2. Phenol extraction	
3. Phenol-chloroform extraction	
4. Diethyl-ether extraction	
5. DEAE-chromatography	
6. Denaturation in 0.3 *M* NaOH at 100°C for 5 min	1. Boiling in 150 m*M* NaOH for 5 min
7. Neutralization with ammonium acetate	2. Neutralization with acetic acid
8. Spotting onto the filter	
9. Baking of the filter at 80°C for 2 hr	
10. Prehybridization for 1 hr	
11. Hybridization for 2 to 17 hr[b]	3. Hybridization with pretreated filters overnight[b]
12. Washing of the filter	4. Washing of filters
13. Autoradiography for 1 to 3 days	5. Counting for 5 min

[a] 0.2 *M* sucrose, 0.02 *M* phosphate buffer, 3% fetal calf serum, 10 µg/ml gentamicin, 100 µg/ml vancomycin, and 25 IU/ml nystatin.
[b] The probes were labeled with ^{32}P-dCTP.

Table 5
COMPARISON OF THE SENSITIVITY OF THE SANDWICH AND SPOT HYBRIDIZATION TESTS FOR THE DETECTION OF *C. TRACHOMATIS*

	Sensitivity[a] (number of target molecules detected)	
Sample	**Spot hybridization[31]**	**Sandwich hybridization[35]**
Pure DNA	10^4 after autoradiography for 24 hr 5×10^3 after autoradiography for 72 hr	10^5
Clinical specimens of cervical and urethral origin	10^5 after autoradiography for 16 hr	10^5

[a] Probes were labeled by nick translation with ^{32}P-dCTP (2×10^8 cpm/μg). 10^7 cpm/ml and 6×10^5 cpm/0.4 ml assays were used in the spot test and in the sandwich hybridization, respectively.

$\times 10^3$ molecules of pure standard DNA after 24 and 72 h, respectively[31] (Table 5). However, less than 10^5 target molecules isolated rom the clinical specimens could be detected using autoradiography, since exposures over 16 h also increased the background.

By contrast, the pretreatment of clinical specimens taken from the urethra and the uterine cervix was simple and rapid using sandwich hybridization. Boiling samples in an alkaline sodium dodecyl sulfate solution was sufficient to disrupt both infected cells and chlamydial elementary bodies releasing the target DNA. Specimens obtained from the genitourinary tract did not cause elevated backgrounds in the sandwich hybridization test.[35] Also, the sensitivity of the test was not affected (Table 5).

An additional advantage of the sandwich hybridization test is the possibility of using cloned DNA fragments directly without the isolation of specific inserts from the cloning vectors. In the spot test, the preparation of "vector-free" chlamydial DNA probe was obligatory because pBR322, the most commonly used cloning vector, reacted with nonchlamydial microbes present in the exudate from genitourinary tract.[31] It was recently reported[32] that a ^{35}S-labeled vector hybridized to 1 of 21 *Chlamydia* negative cervical smear specimens tested by the *in situ* hybridization technique. These, as well as our data, suggest that vector-related sequences are commonly present in specimens taken from locations with a high incidence of bacterial infection.

D. Applicability of the Sandwich Hybridization Test to Clinical Specimens

Various rapid methods for detecting chlamydial antigens have proven to be applicable for diagnosing chlamydia from patients with a high prevalence of infections. Screening of asymptomatic women, however, has been less successful, probably because of the difference in the quantity of the infectious agent in the specimens.[12,37,38] We, therefore, chose to initially check the efficacy of our method by studying specimens collected from patients of a sexually transmitted disease (STD) clinic.[35]

1. Patients, Specimens, and Tests

All together, 188 female and 80 male specimens were collected from the endocervix and urethra, respectively, of patients visiting the outpatient clinic of the University Central Hospital, Helsinki. Two separate specimens were taken with cotton-tipped wooden swabs: one for culture and the other for the sandwich hybridization in a randomized order. The specimens for culture were inoculated into cycloheximide-treated McCoy cells within the same day, and the specimens for hybridization were stored at –70°C until tested. Culture was carried out in small vials after centrifugation, and the chlamydial inclusions visualized by iodine staining after 48 h. The sandwich hybridization utilized the combination of specific chromosomal and plasmid probes

Table 6
DETECTION OF *C. TRACHOMATIS* IN ENDOCERVICAL SPECIMENS BY NUCLEIC ACID SANDWICH HYBRIDIZATION USING CHROMOSOMAL AND PLASMID PROBES[35]

	Female patients		Hybridization[a]			
Culture	**Number**	**(%)**	**+**	**(%)**	**–**	**(%)**
+	36	(19)	28	(78)	8	(22)
–	152	(81)	15	(10)	137	(90)
Total	188	(100)	43	(23)	145	(77)

[a] In each test, 10 negative controls were included. The cutoff value, below which 99.5% of the negatives distribute, was calculated according to 99% confidence limits.

Table 7
DETECTION OF *C. TRACHOMATIS* IN URETHRAL SPECIMENS BY NUCLEIC ACID SANDWICH HYBRIDIZATION[35]

	Male patients		Hybridization[a]			
Culture	**Number**	**(%)**	**+**	**(%)**	**–**	**(%)**
+	29	(36)	22	(76)	7	(24)
–	51	(64)	16	(31)	35	(69)
Total	80	(100)	38	(48)	42	(52)

[a] In each test, 10 negative controls were included. The cutoff value, below which 99.5% of the negatives distribute, was calculated according to 99% confidence limits.

(Table 2) allowing a detection sensitivity of less than 10^5 elementary bodies per specimen (Table 5).

2. Results of the Clinical Testing

The sensitivity of the sandwich hybridization test was 77% as compared with the culture. In addition, 15% of the culture-negative specimens became positive by the hybridization test. Of the 268 specimens, 172 were negative by both tests (Table 6 and 7). The distribution of the culture-negative, but DNA-positive specimens, was such that 10% of the female and 31% of the male specimens showed discrepant results, whereas the sensitivity of the test as compared with culture was similar in both groups.

3. Reliability of the Sandwich Hybridization Results

a. Specificity Criteria

The specificity of the test is affected by two main factors, both of which have been taken under consideration in these studies. First, the recognition specificity of the nucleic acid probes chosen for this study was carefully documented. The probes only hybridized to the desired target and to no other microbial or host cell nucleic acid tested. Second, the cutoff value for positivity used in the test was deliberately set high to reduce false positives. The negative control (i.e., background), used for calculation of the cutoff value, was adequate since no difference in the hybridization signal was observed when testing buffer controls and clinical specimens negative for *Chlamydia*, but containing discharge caused by other pathogens.[35] According to our experience, no background problems have been encountered when applying sandwich hybridization to a multitude of crude specimens.[3-9]

b. Value of the Reference Test

The reference test for our studies was chlamydial isolation by cell culture. The culture is estimated to detect approximately 75% of chlamydial infections,[12,38] even if performed optimally, i.e., by using culture vials, one blind passage, and staining of the inclusions by fluorescein-labeled monoclonal antibodies.[12] Our procedure should have detected 60 to 90% of the culture-positive infections depending on variance in sampling and in the infection prevalence,[12,39,40] thus allowing 30 to 50% of true positives to remain undetected. The relatively large number of culture-negative but hybridization-positive specimens in our study is thus in accordance with the expected number of chlamydial infections in the study population.

III. FUTURE OF HYBRIDIZATION METHODS IN THE DIAGNOSIS OF *C. TRACHOMATIS*

Because of the inadequacy of the culture method, alternative techniques suitable for rapid detection, are urgently needed. Immunochemical and hybridization methods, detecting chlamydial antigen and nucleic acid directly from the specimen, respectively, can be used. How do these techniques compare? More experience is available with the immunochemical methods than with hybridization because many of the tests are commercially available. They allow detection of 60 to 99% of the culture-positive infections,[23,28,40-44] and they perform better if the specimens are collected from patients with high prevalence of sexually transmitted diseases than from asymptomatic persons. Thus, more chlamydiae must be present in infections with symptoms than without symptoms. The immunofluorescence methods are recommended for studying patients from STD clinics if culture facilities are not available, but they are not recommended for screening low-prevalence populations.[4,12,13,40] The reliability of an enzyme immunoassay (Chlamydiazyme, Abbott) detecting the lipopolysaccharide of the bacterium has been questioned, because similar cross-reacting structures are present on other bacteria as well.[45]

The nucleic acid detection methods offer, in principle, the best specificity. Their efficacy in practice is difficult to evaluate due to the limited experience of their use to date. A recent report by Horn et al.[32] describes results of using *in situ* hybridization for the diagnosis of *C. trachomatis*. The efficacy of the method as compared with culture was 90%. In addition, 20% of the culture-negative specimens (5/25) were shown to contain chlamydial DNA by the hybridization method. The results obtained by the sandwich hybridization technique also imply that hybridization is capable of detecting infections that remain negative by culture.

The relative efficacies of the two methods, detecting the chlamydial antigen, on the one hand, or the nucleic acid on the other, cannot be compared, since no studies have been published where both methods, in addition to culture, would have been used to study parallel specimens.

Do we have means for estimating the sensitivity requirement of the rapid diagnostic methods? Sandwich hybridization is a quantitative method. Thus, the number of elementary bodies in the clinical specimens studied here could be calculated.[35] Chlamydiae were contained in 96 of the specimens, using as criteria positivity by culture or by hybridization or both. Of them, 15 (15%) contained less than 10^5 elementary bodies because they remained below the detection limit of the hybridization assay. Fourteen specimens contained more than 10^6 elementary bodies and the majority (67 specimens; 70%) between 10^5 and 10^6 elementary bodies. Asymptomatic patients would probably contain even lower amounts of *Chlamydia*. It is, therefore, clear that the sensitivity of the rapid tests should be improved by at least an order of magnitude to be generally applicable and to replace culture.

Recent development of hybridization technology implies that the sensitivity problems can be circumvented. The assumption is based on the unique property of nucleic acids as the only self-duplicable molecules known. Thus, the number of target molecules in the specimen can be amplified before detection. This is done in vitro with oligonucleotide primers and DNA polymerase. As the cycle is repeated, the number of targets is exponentially increased.[47,48]

IV. CONCLUSIONS

The major advantage of the sandwich hybridization method is the possibility to study a variety of specimens in a crude form without the need for elaborate pretreatment. This is particularly valuable with regard to specimens containing mucus, such as nasopharyngeal and genital specimens. The method has been successfully applied to detection of adenovirus from nasopharyngeal mucus,[4,7] cytomegalovirus from urine,[5] and human papillomavirus from cervical smears.[8] We have also demonstrated that the method is suitable for detecting *C. trachomatis* from urethral and cervical specimens collected from patients with a high prevalence of sexually transmitted diseases. The unique properties of the hybridization principle, its specificity, and the possibility of target amplification will render it a universal and valuable tool in the diagnosis of problematic microbes.

REFERENCES

1. **Ranki, M., Palva, A., Virtanen, M., Laaksonen, M., and Söderlund, H.** Sandwich hybridization as a convenient method for the detection of nucleic acids in crude samples, *Gene,* 21, 77, 1983.
2. **Ranki, M., Virtanen, M., Palva, A., Laaksonen, M., Petterson, R., Kääriäinen, L., Halonen, P., and Söderlund, H.,** Nucleic acid sandwich hybridization in adenovirus diagnosis, *Curr. Top. Microbiol. Immunol.,* 104, 307, 1983.
3. **Palva, A., Jousimies-Somer, H., Saikku, P., Väänänen, P., Söderlund, H., and Ranki, M.,** Detection of *Chlamydia trachomatis* by nucleic acid sandwich hybridization, *FEMS Microbiol. Lett.,* 23, 83, 1984.
4. **Virtanen, M., Palva, A., Laaksonen, M., Halonen, P., Söderlund, H., and Ranki, M.,** Novel test for rapid viral diagnosis: detection of adenovirus in nasopharyngeal mucus aspirates by means of nucleic acid sandwich hybridization, *Lancet,* 1, 381, 1983.
5. **Virtanen, M., Syvänen, A.-C., Oram, J., Söderlund, H., and Ranki, M.,** Cytomegalovirus in urine: detection of viral DNA by sandwich hybridization, *J. Clin. Microbiol.,* 20, 1083, 1984.
6. **Palva, A. and Ranki, M.,** Microbial diagnosis by nucleic acid sandwich hybridization, *Clin. Lab. Med.,* 5, 475, 1985.
7. **Lehtomäki, K., Julkunen, I., Sandelin, K., Salonen, J., Virtanen, M., Ranki, M., and Hovi, T.,** Rapid diagnosis of respiratory adenovirus infections in young adult men, *J. Clin. Microbiol.,* 24, 108, 1986.
8. **Parkkinen, S., Mäntyjärvi, R., Syrjänen, K., and Ranki, M.,** Detection of human papillomavirus DNA by the nucleic acid sandwich hybridization method from cervical scraping, *J. Med. Virol.,* 20, 279, 1986.
9. **Palva, A.,** *omp* A gene in the detection of *Escherichia coli* and other *Enterobacteriaceae* by nucleic acid sandwich hybridization, *J. Clin. Microbiol.,* 18, 92, 1983.
10. **Syvänen, A.-C., Tchen, P., Ranki, M., and Söderlund, H.,** Time-resolved fluorometry: a sensitive method to quantify DNA-hybrids, *Nucleic Acids Res.,* 14, 1017, 1986.
11. **Holmes, K. K.,** The *Chlamydia* epidemic, *J. Am. Med. Assoc.,* 245, 1718, 1981.
12. **Schachter, J.,** Immunodiagnosis of sexually transmitted disease, *Yale J. Biol. Med.,* 58, 443, 1985.
13. **Bell, T. A. and Grayston, J. T.,** Centers for disease control guidelines for prevention and control of *Chlamydia trachomatis* infections, *Ann. Intern. Med.,* 104, 524, 1986.
14. **Becker, Y.,** The *Chlamydia*: molecular biology of procaryotic obligate parasites of eucaryocytes, *Microbiol. Rev.,* 42, 274, 1978.
15. **Kingsbury, D. T. and Weiss, E.,** Lack of deoxyribonucleic acid homology between species of the genus *Chlamydia, J. Bacteriol.,* 96, 1421, 1968.
16. **Weiss, E., Schramek, S., Wilson, N. N., and Newman, L. W.,** Deoxyribonucleic acid heterogeneity between human and murine strains of *Chlamydia trachomatis, Infect. Immun.,* 2, 24, 1970.
17. **Schachter, J.,** Chlamydial infections, *N. Engl. J. Med.,* 298, 540, 1978.
18. **Mårdh, P.-A., Holmes, K. K., Oriel, J. D., Piot, P., and Schachter, J., Eds.,** *Chlamydial Infections,* Elsevier, Amsterdam, 1982.
19. **Darougar, J.,** Chlamydial disease, *Br. Med. Bull.,* 39, 107, 1983.
20. **Evans, R. T. and Woodland, R. M.,** Detection of Chlamydiae by isolation and direct examination, *Br. Med. Bull.,* 39, 181, 1983.

21. **Stamm, W. E., Tam, M., Koester, M., and Cles, L.,** Detection of *Chlamydia trachomatis* inclusions in McCoy cell cultures with fluorescein-conjugated monoclonal antibodies, *J. Clin. Microbiol.,* 17, 666, 1983.
22. **Bell, T. A., Kuo, C.-C., Stamm, W. E., Tam, M. R., Stephens, R. S., Holmes, K. K., and Grayston, J. T.,** Direct fluorescent monoclonal antibody stain for rapid detection of infant *Chlamydia trachomatis* infections, *Pediatrics,* 74, 224, 1984.
23. **Stamm, W. E., Harrison, H. R., Alexander, E. R., Cles, L. D., Spence, M. R., and Quinn, T. C.,** Diagnosis of *Chlamydia trachomatis* infections by direct immunofluorescence staining of genital secretions, *Ann. Intern. Med.,* 101, 638, 1984.
24. **Tam, M. R., Stamm, W. E., Handsfield, H. H., Stephens, R., Kuo, C.-C., Holmes, K. K., Ditzenberger, K., Krieger, M., and Nowinski, R. C.,** Culture-independent diagnosis of *Chlamydia trachomatis* using monoclonal antibodies, *N. Engl. J. Med.,* 310, 1146, 1984.
25. **Taylor, H. R., Agarwala, N., and Johnson, S. L.,** Detection of experimental *Chlamydia trachomatis* eye infection in conjunctival smears and in tissue culture by use of fluorescein-conjugated monoclonal anibody, *J. Clin. Micrbiol.,* 20, 391, 1984.
26. **Uyeda, C. T., Welborn, P., Ellison-Birang, N., Shunk, K., and Tsaouse, B.,** Rapid diagnosis of chlamydial infections with the MicroTrak Direct Test, *J. Clin. Microbiol.,* 20, 948, 1984.
27. **Jones, M. F., Smith, T. F., Houglum, A. J., and Herrmann, J. E.,** Detection of *Chlamydia trachomatis* in genital specimens by the chlamydiazyme test, *J. Clin. Microbiol.,* 20, 465, 1984.
28. **Chernesky, M. A., Mahony, J. B., Castriciano, S., Mores, M., Stewart, I. O., Landis, S. J., Seidelman, W., Sargeant, E. J., and Leman, C.,** Detection of *Chlamydia trachomatis* antigens by enzyme immunoassay and immunofluorescence in genital specimens from symptomatic and asymptomatic men and women, *J. Infect. Dis.,* 154, 141, 1986.
29. **Pugh, S. F., Slack, R. C. B., Caul, E. O., Paul, I. D., Appleton, P. N., and Gatley, S.,** Enzyme amplified immunoassay: a novel technique applied to direct detection of *Chlamydia trachomatis* in clinical specimens, *J. Clin. Pathol.,* 38, 1139, 1985.
30. **Hyypiä, T., Jalava, A., Larsen, S. H., Terho, P., and Hukkanen, V.,** Detection of *Chlamydia trachomatis* in clinical specimens by nucleic acid spot hybridization, *J. Gen. Microbiol.,* 131, 975, 1985.
31. **Palva, A.,** Nucleic acid spot hybridization for detection of *Chlamydia trachomatis, FEMS Microbiol. Lett.,* 28, 85, 1985.
32. **Horn, J. E., Hammer, M. L., Falkow, S., and Quinn, T. C.,** Detection of *Chlamydia trachomatis* in tissue culture and cervical scrapings by in situ DNA hybridization, *J. Infect. Dis.,* 153, 1155, 1986.
33. **Lovett, M., Kuo, C. C., Holmes, K., and Falkow, S.,** Plasmids of the genus *Chlamydia,* in *Current Chemotherapy and Infectious Disease,* Vol. 2, Nelson, J. D. and Crassi, C., Eds., American Society for Microbiology, Washington, D.C., 1980, 1250.
34. **Hyypiä, T., Larsen, S. H., Ståhlberg, T., and Terho, P.,** Analysis and detection of chlamydial DNA, *J. Gen. Microbiol.,* 130, 3159, 1984.
35. **Palva, A., Korpela, K., Lassus, A., and Ranki, M.,** Detection of *Chlamydia trachomatis* from genitourinary specimens by improved nucleic acid sandwich hybridization, *FEMS Microbiol. Lett.,* 40, 211, 1987.
36. **Palmer, L. and Falkow, S.,** A common plasmid of *Chlamydia trachomatis, Plasmid,* 16, 52, 1986.
37. **Forbes, B. A., Bartholoma, N., McMillan, J., Roefaro, M., Weiner, L., and Welych, L.,** Evaluation of a monoclonal antibody test to detect *Chlamydia* in cervical and urethral specimens, *J. Clin. Microbiol.,* 23, 1136, 1986.
38. **Schachter, J.,** Rapid diagnosis of sexually transmitted diseases — speed has a price, *Diagn. Microbiol. Infect. Dis.,* 4, 185, 1986.
39. **Quinn, T. C., Warfield, P., Kappus, E., Barbacci, M., and Spence, M.,** Screening for *Chlamydia trachomatis* infection in an inner-city population: a comparison of diagnostic methods, *J. Infect. Dis.,* 152, 419, 1985.
40. **Shafer, M.-A., Vaughan, E., Lipkin, E. S., Moscicki, B. A., and Schachter, J.,** Evaluation of fluorescein-conjugated monoclonal antibody test to detect *Chlamydia trachomatis* endocervical infections in adolescent girls, *J. Pediatr.,* 108, 780, 1986.
41. **Gallo, D.,** Isolation and identification of *Chlamydia trachomatis*: a comparative study, *Lab. Med.,* 15, 481, 1984.
42. **Mumtaz, G., Mellars, B. J., Ridgway, G. L., and Oriel, J. D.,** Enzyme immunoassay for the detection of *Chlamydia trachomatis* antigen in urethral and endocervial swabs, *J. Clin. Pathol.,* 38, 740, 1985.
43. **Amortegui, A. J. and Meyer, M. P.,** A nonculture test for identification of *Chlamydia trachomatis, J. Reprod. Med.,* 30, 279, 1985.
44. **Lipkin, E. S., Moncada, J. V., Shafer, M.-A., Wilson, T. E., and Schachter, J.,** Comparison of monoclonal antibody staining and culture in diagnosing cervical chlamydial infection, *J. Clin. Microbiol.,* 23, 114, 1986.
45. **Saikku, P., Puolakkainen, M., Leinonen, M., Nurminen, M., and Nissinen, A.,** Cross-reactivity between chlamydiazyme and *Acinetobacter* strains, *N. Engl. J. Med.,* 314, 922, 1986.

46. **Rigby, P. W. J., Dieckmann, M., Rhodes, C., and Berg, P.,** Labeling deoxyribonucleic acid to high specific activity *in vitro* by nick translation with DNA polymerase I, *J. Mol. Biol.,* 113, 237, 1977.
47. **Saiki, R. K., Scharf, S., Faloona, F., Mullis, K. B., Horn, G. T., Erlich, H. A., and Arnheim, N.,** Enzymatic amplification of β-globin genomic sequences and restriction site analysis for diagnosis of sickle cell anemia, *Science,* 230, 1350, 1985.
48. **Saiki, R. K., Bugawan, T. L., Horn, G. T., Mullis, K. B., and Erlich, H. A.,** Analysis of enzymatically amplified β-globin and HLA-DQ DNA with allele-specific oligonucleotide probes, *Nature (London),* 324, 163, 1986.
49. **Ranki, M.,** unpublished results, 1986.

Part VI: Nucleic Acid Probes for Identification of Parasitic Agents

Chapter 18

DETECTION OF HEMOTROPHIC PARASITES BY DNA HYBRIDIZATION

Garret M. Ihler and Allison C. Rice-Ficht

TABLE OF CONTENTS

I. INTRODUCTION

Infectious organisms that gain entry into red blood cells include the malarial parasites of the genus *Plasmodium (P. falciparum, P. vivax, P. malariae,* and *P. ovale*), the malaria-like protozoan parasites of the related genus, *Babesia,* and certain strains of bacteria, especially *Bartonella bacilliformis.*[1] Malaria is a major human disease of worldwide significance. Babesiosis is primarily a disease of nonprimate mammals including domestic livestock such as cattle, sheep, goats, and horses, but human cases especially among splenectomized individuals are known.[2-4] Bartonellosis is an obscure human infection confined to certain valleys in the Andes Mountains, characterized by malaria-like symptoms and a mortality rate in excess of 40% in the preantibiotic era.[5]

The diagnosis of these diseases at the present time is heavily dependent upon identification of the parasites within red cells in stained blood smears. This method is often time consuming and requires a skilled technician for detection and proper species identification. The principal problems with visual detection of the parasites are (1) the low incidence of infected cells (low parasitemia), such as that seen in chronic babesiosis, latent malaria, or individuals undergoing drug therapy, and (2) identification of the species involved. Occasionally *Babesia* is mistaken for *Plasmodium* in peripheral blood smears of humans. Additional problems encountered in rural areas of less-developed countries include the absence of microscopes, or even electricity to power the lamp, and the lack of trained health-care workers. There is also concern that transmission of *Plasmodium* or *Babesia* by blood transfusions may become a more serious problem. At present, due to the technical difficulty of detecting low numbers of parasites, there is no effective way of screening blood donors who may be asymptomatic but infectious.[6]

In cases of low parasitemia, a technician is required to examine hundreds or even thousands of cells in order to locate the parasites. A simple fluorescent assay, based on binding of fluorescent intercalating agents such as ethidium bromide,[7] could simplify identification since red cells do not contain DNA. RNA in reticulocytes can complicate the assay somewhat, and for this approach to readily detect low parasitemias, red cells should be separated from nucleated cells which also are labeled by ethidium bromide. More exotic detection and separation systems involving fluorescent DNA binding compounds and automated flow cytometry have been developed as well.[8-10]

A number of alternative methods for clinical diagnosis of blood parasites are available. These include (1) monitoring antibody titers through ELISA assays,[11] (2) monitoring the presence of parasite antigen on erythrocytes using monoclonal antibody,[12,13] and (3) species identification by isozyme analysis[14] and two-dimensional electrophoresis of labeled parasite proteins.[15,16] The first two methods are rapid, sensitive, and lend themselves well to the diagnostic laboratory. However, the presence of parasite-specific antibody cannot distinguish between previous and active infections, and the presence of host antibody previously bound to infected erythrocytes may interfere with monoclonal antibody detection of parasite antigen.[17] The remaining technique is time consuming and requires a great deal of expertise to implement.

A. DNA Hybridization Probes for Diagnosis

Detection of the parasites by use of specific DNA hybridization probes can offer an alternative diagnostic procedure free of these problems. Adequate amounts of DNA are present in infected erythrocytes. At normal red cell concentrations of 5×10^6 cells/μl, a 0.1% parasitemia with only one malarial parasite per erythrocyte corresponds to about 100 pg of parasite DNA per μl of blood; spotting an aliquot of infected blood on a nitrocellulose filter permits a "dot-blot" DNA hybridization[18] procedure to be used to detect the presence of parasites. Since DNA-based detection systems are effective at the picogram level, a few microliters of infected blood properly treated[17] would provide sufficient DNA for detection purposes. In any case, the assay could be readily scaled up by using more blood, provided that the amounts of hemoglobin or leukocyte DNA present do not compete with DNA immobilization on the solid support.[19]

Species identification using DNA probes is readily accomplished. Species identification is an important aspect of any diagnostic procedure since the clinical course of malaria produced by the four species which infect humans is substantially different, falciparum malaria being the most life threatening.

Within species, different strains or geographical isolates can be identified with a little more effort, and such identification could be useful in epidemiological studies. It has been demonstrated that specific geographical isolates or species of parasite may be differentiated by Southern blot analysis using common sequences.[20-22] This analysis is based on alterations in restriction patterns of genomic DNA due to rapid sequence evolution or to gross DNA polymorphisms. Southern blot analysis is more time consuming than DNA dot-blot analysis, but may prove to have routine clinical application.

B. Other Potential Clinical Uses for DNA Probes

DNA probes may also be useful in detecting genes encoding drug resistance, particularly if drug resistance is mediated by gene amplification,[23] or transmissible plasmids or transposons, or genes associated with virulence, thus providing a rapid assay which would not require parasite culture methods. Drug sensitivity tests are now recommended for any patient with falciparum malaria before therapy is initiated.[24]

DNA probes may be also useful in detecting the presence of genes correlated with virulence. The gene coding for a histidine-rich protein, which accumulates in cytoplasmic granules, provides an example of the potential usefulness of such a probe. This protein is associated with the production of knob-like protrusions on membranes of erythrocytes infected with *P. falciparum* which are believed to cause binding of the erythrocytes to endothelium, preventing removal of the cells during circulation through the spleen and causing pathological effects including the symptoms of cerebral malaria. Messenger RNA corresponding to this gene is not found in knobless clones of *P. falciparum*,[25] although the gene is present. Hybridization assays using genomic DNA and nucleic acid preparations which include the mRNA could distinguish between strains of malaria which do or do not contain the gene and also strains which do or do not express the gene. Use of a probe for a particular gene would depend ultimately on the use of the information in diagnosis or therapeutics, but the presence or expression of certain genes might be sufficiently correlated with progonosis to make a DNA-based test worthwhile.

It should be possible to detect parasites in tissue sections, as well as blood, using labeled probe and *in situ* hybridization for the purposes of postmortem examination or for basic research. For example, the technique could be used to examine the role of exoerythrocytic forms of *Plasmodium* in the liver, which has been reviewed by Hollingdale.[26] Among the areas needing further investigation are the small dormant forms (hypnozoites) of *P. vivax* persisting in hepatocytes which are responsible for relapses and which were not discovered until 1980. DNA hybridization should be useful for investigation of dormant forms of parasites as well as in active infections. Additional applications include cytochemical hybridization of rhodamine-labeled rRNA with *P. falciparum*-infected erythrocytes,[27] indicating that it may be possible to conduct cytochemical hybridization of infected cells using DNA probe sequences to detect specific RNA species and to correlate synthesis of the RNA with differentiation. Preliminary studies involving synthesis of RNA and DNA indicate substantial differences at different stages of development.[28]

II. PLASMODIUM DETECTION USING DNA HYBRIDIZATION

A. Total Genomic DNA Probes

A probe for some organisms can readily be prepared by nick translating the total genomic DNA, although such a probe may not have the specificity required to permit closely related organisms to be easily distinguished. This procedure was used by Pollack et al.,[29] who labeled *P. falciparum* DNA to a specific activity of about 2 to 3 × 10^8 cpm/μg for use as a probe. Blood samples (10 μl) were applied to nitrocellulose filters and hybridized with 7 ng/ml of ^{32}P-labeled

P. falciparum probe DNA. The assay specifically detected *Plasmodium* DNA and successfully identified 20 of 21 patients infected with *P. falciparum*. A patient with gametocytes, however, was missed. As few as 50 parasites (10^{-4}% parasitemia) could be detected in a 10-μl aliquot.

Pollack et al.[29] reported that they could detect parasite DNA in infected blood samples containing one parasite per 10^6 red cells. As few as 50 cpm over background of hybridized radioactive DNA can be detected by autoradiography in a 24-h exposure using X-ray film and as few as 10 cpm can be detected if intensifying screens are used; however, film should be preflashed if quantitation with ^{32}P at low numbers of cpm is desired.[30] In the experiments reported, it seems likely that more parasite DNA was spotted onto the filters than was added as probe. In addition, some of the added probe presumably self-hybridized and was not available for hybridization to the DNA on the filter. Thus, the assay sensitivity could probably be improved even above the very sensitive level reported. The spot intensity, as measured by the area under the curve scanned by a densitometer, was proportional to the quantity of *P. falciparum* DNA added in a reconstruction experiment over a tenfold range of DNA concentrations. At higher DNA concentrations (above 5 ng in the experiment reported), the spot intensity was not proportional to the amount of DNA, although the intensity continued to increase with increasing amounts of DNA. Among the potential causes of nonlinearity are competition of protein with DNA for nitrocellulose binding sites,[19] saturation of the filter with DNA,[18] or a nonlinear response of the autoradiography film.[30]

Greater sensitivity for total genomic DNA probes than that reported by Pollack et al.[29] can be obtained by increasing the specific activity of the probe, for example, by a recently developed technique which employs the priming of DNA synthesis with a random set of oligonucleotides generated by DNAse I digestion of calf thymus DNA. This synthesis does not depend on the nick-translating activity of *Escherichia coli* DNA polymerase I and has been observed to yield significantly higher specific activity with genomic DNA substrates.[31] Sensitive, nonradioactive detection procedures would be the most desirable for clinical diagnostic laboratories and are under development.

Much of the DNA of closely related species is nearly identical in sequence or is sufficiently closely related to permit cross-hybridization. In general, DNA from different species cannot be reliably differentiated by DNA hybridization using probes composed of total genomic DNA. For this reason, shorter, more specific DNA probes have been sought which hybridize to only one species.

A total genomic probe would be expected to be more sensitive than a probe composed of only part of the genome or of a particular short sequence, since a total genomic probe can hybridize with all sequences present in the target DNA. Conversely, probes composed of unique DNA sequences, present only once per haploid genome, are probably too insensitive to be routinely used for diagnostic purposes. However, a probe composed of a repeated DNA sequence present many times in the haploid genome may be both sensitive and specific.

B. Repeated DNA within the *Plasmodium* Genome

1. Chromosomes and DNA of Plasmodium

The chromosomes of *P. falciparum*, like some other lower eukaryotes, fail to condense sufficiently during mitosis to permit visualization by cytological techniques,[32,33] and so the number and length of individual chromosomes is not known. Utilizing the technique of pulsed-field gradient gel electrophoresis,[32] seven discrete size groups have been resolved in *P. falciparum*.[34] Hybridization of cloned genes to the separated chromosomes has permitted their assignment to specific chromosomes.[35] Chromosome-size polymorphisms are readily observed in culture lines and in natural infections, even within a single endemic area,[36,37] although molecular karyotypes remain constant through the sexual phases of the life cycle.[34] In another protozoan parasite, *Leishmania*, chromosomal polymorphisms detected using this same technique have been exploited for diagnostic purposes.[38]

The haploid *P. falciparum* genome has been reported to be 1 to 2.2×10^7 bp[39,40] with one much higher value reported.[41] Similar, but slightly higher, values of 2 to 3×10^7 bp have been reported for *P. berghei* and *P. knowlesi*.[42,43] The plasmodial genome size is, therefore, about the same as yeast and about five times that of *E. coli*. Buoyant densities for several *Plasmodium* species have been determined.[40,44] The data indicate an A + T content of 80 to 82%, which is an unusually high value; methylated bases were not detected.[40]

2. Repetitive DNA Sequences

Genomic DNA sequences in eukaryotes are conveniently grouped into three classes based on sequence complexity. These categories are highly repetitive DNA, moderately repetitive DNA, and unique sequences. An additional genomic fraction consists of DNA which reanneals with zero-order kinetics and consists of palindromic DNA;[45,46] these inverted repeat sequences can be found within each of the three repetitive categories listed above. When selecting a sequence for a diagnostic probe, a highly repetitive or moderately repetitive sequence would yield a higher level of hybridization than a unique sequence. To permit species identification, it is desirable to utilize a sequence which is not only repeated but has undergone evolutionary divergence.

Highly repetitive sequences are, in general, short sequences ranging from a few base pairs to a few hundred. These sequences are repeated as much as 500,000 times in some eukaryotic genomes, may be associated exclusively with centromeric or telomeric regions of the chromosome or dispersed, and are rarely transcribed. An extensive examination of mouse satellite DNA,[47,48] which is centromeric, has revealed tandem duplication of sets of short sequences combined with base substitutions. These repeats appear to have been generated either by unequal crossover during recombination or through saltatory replication accompanied by point mutation. The smallest units are, therefore, similar but nonidentical in both length and base composition. A second group of highly repeated sequences which have undergone divergence along with duplication includes the human "Alu family",[49] the primate "*Kpn*I family",[50] and the mouse "R family",[51] which are highly repeated, interspersed with unique sequences, and occasionally transcribed. The Alu and R families consist of repeating units of approximately 300 and 475 bp, respectively; the *Kpn*I family has a unit minimum size of 1200 to 1900 bp.

Repetitive DNA within the *Plasmodium* genome has been estimated at 18% for *P. berghei*[42] and 10% for *P. falciparum*[41] as determined by C_0t analysis. One analysis for the *P. falciparum* genome indicated the genome size to be 3.8×10^8 bp and the repetitive portion to be 3.6×10^7 bp, with a measured kinetic complexity of 3.8×10^5 bp; this suggests an average repetition frequency in the haploid genome for this group of sequences to be 95.

A repetitive sequence which is Alu-like in its distribution has been partially characterized in *P. falciparum* by Guntaka et al.[52] The data suggest that a highly repeated element is present and dispersed in clusters throughout the genome. The same sequence is absent in both human and mouse DNA based on hybridization under high stringency conditions, although no data is available on its presence in other *Plasmodium* genomes.

Another group of simple, highly repetitive sequences is found at chromosomal ends or in telomeric DNA. Sequences from several lower eukaryotes have been examined by Blackburn and others[53] and reveal the presence of tandemly repeated five to nine nucleotide sequences. The sequences are highly conserved, presumably for telomere function, and contain single-strand discontinuities. The highly conserved nature of repeated sequences among a variety of lower eucaryotes does not necessarily preclude their use as a species-specific diagnostic probe, since even a single base difference can have a major effect on hybridization of short probe sequences, but clearly close attention to hybridization and washing conditions would be required.

Analysis of cloned telomeric sequences from *P. berghei*[54] indicates the presence of a heptanucleotide repeated at least 70 times per telomere. It is the only telomeric repeat known to contain all four nucleotides, but otherwise conforms in size, location, and flanking sequence to telomeric regions characterized in other organisms.[55,56]

Also among the candidates for a hybridization probe are the sequences repeated a moderate number of times within the genome; this category includes DNA encoding ribosomal genes and mitochondrial DNA, either of which may be amplified during various stages of development.[39,57-59] Muliple copies of genes corresponding to rRNA are present in *Plasmodium* as in most organisms; however, in all cases studied, the number is small and the gene organization somewhat different from other eukaryotes. In *P. falciparum,*[60] *P. berghei,*[61,62] and *P. lophurae*[63] there are 8 copies of rRNA genes, which is small since there are 7 copies even in *E. coli* and about 140 copies in *Saccharomyces cerevisiae.*[57] The data is consistent with the typical rDNA gene order $_{5'}$18S-5.8S-28S$_{3'}$ for all three species. Two of the three, *P. lophurae* and *P. falciparum,* appear to contain introns in 18S and 28S rDNA and in 28S rDNA, respectively. In *P. falciparum,* the A + U content of the RNA is 60%, consistent with the known conservation of rRNA sequences, but the noncoding sequences are much higher in A + T.[60] In *P. lophurae,* the large rDNA contains an intron of 240 bp and probably another of 110 bp, and the small rDNA contains an intron of 230 bp.[63] Although rRNA sequences are typically conserved and may not be good candidates for a hybridization probe, the spacer regions associated with these genes may be successfully used. For example, the large nontranscribed spacer of the trematode *Schistosoma haematobium* has been employed as a species-specific probe via Southern blot analysis.[26] Most ribosomal DNAs examined to date are characterized by divergence in spacer regions and demonstrate species specificity of the sequence in these areas.[64]

Unique sequences within the eukaryotic genome are less useful as hybridization probes, since any oligomer chosen from a unique sequence would hybridize only once to the genome. This consideration may be of less importance in *Plasmodium* than most other eukaryotes due to the small size of the genome. Nevertheless, in any organism, the maximum sensitivity which can be obtained is directly related to the number of copies of the target DNA sequence in the genome. In *Plasmodium,* some unique genes may contain multiple copies of related DNA sequences, since several major proteins contain repeated epitopes. Two such genes found on the surface of *P. falciparum*[65] and *P. knowlesi*[66] sporozoites are species specific. The *P. falciparum* protein contains 41 tandem repeats of a four-amino acid sequence, although the corresponding DNA sequence repeat contains base pair changes; these point changes may limit the use of the sequence as a hybridization probe. The *P. knowlesi* gene contains 12 repeats of a 12-amino acid sequence which also could be considered for use as a hybridization probe.[66] The S-antigens, or soluble antigens described by Coppel et al. from *P. falciparum,* also vary from species to species and even within a species;[67] one S-antigen examined contains 23 tandem repeats of 11 amino acids and was not detectable by hybridization in two *P. falciparum* isolates.[65] Low copy number sequences in the genome can be employed as a means of parasite detection and species identification, although the sensitivity may not be sufficiently high for routine diagnosis. Coppel et al. have used the genes encoding S-antigens to differentiate between *Plasmodium* isolates in Southern blot analysis.[27]

C. Diagnostic Probes for *Plasmodium*

By using an empirical approach, several good diagnostic probes have been isolated for various *Plasmodium* species. In each case, a highly repeated genomic fragment has been selected by hybridizing nick-translated genomic DNA to libraries of *Plasmodium* DNA. Those giving the most intense signal were selected and further characterized. Franzen et al.[68] reported construction of a *P. falciparum* library by cloning 10- to 15-kb *Sau*3A partial-restriction fragments into bacteriophage lambda. Plaques were screened by hybridization with radioactive *P. falciparum* DNA; DNA derived from a plaque displaying strong hybridization, presumably due to a high level of repetitive DNA, was further cut, and the two restriction fragments which displayed the strongest hybridization (a 3.5-kb *Hpa*II and a 1.7-kb *Hin*dIII fragment) were subcloned. The *Hin*dIII fragment was directly shown to contain repetitive DNA since it hybridized strongly to many bands of *Hin*dIII-digested *P. falciparum* total DNA. Sequencing revealed the presence of 21 bp tandem repeats, two of which were $_{5'}$AGGTCTTAACTTGACT$^{A}_{T}$AC$^{A}_{T}$T$_{3'}$. Although *Bgl*II has a six-base recognition site, $_{5'}$AGATCT$_{3'}$, and so would be expected to cut relatively infrequently, there were many *Bgl*II sites in the original cloned fragment; this is consistent with the presence of a repetitive, slightly variable DNA sequence as suggested by the two sequences reported.

The probe isolated by Franzen et al.[68] efficiently detected *Plasmodium* DNA in a diagnostic assay for *Plasmodium* and did not cross-hybridize with a 10,000-fold excess of human DNA. Under their conditions, the probe could detect about 25 pg of parasite DNA, a somewhat high value which, however, could be improved on easily, since the specific activity of the probe DNA used was only 3×10^7 cpm/μg DNA. The greater sensitivity reported for other assays, for example, by Pollack et al.,[29] seems, in part, to result from using higher specific-activity probes. The sensitivity of the assay can easily be increased by increasing the specific activity of the probe. Synthetic DNA, corresponding to one of the two sequences reported by Franzen et al., was shown to be usable as a hybridization probe and to detect as little as 10 pg of DNA by dot-blot analysis.[68]

Additional sequence data for this repeated DNA is now available.[70,71] Aslund et al.[71] have extended the sequence analysis of this probe and found many tandemly arranged 21 bp imperfect repeats having the consensus sequence $_{5'}$ACTAA..TAGGTCTTA...T.$_{3'}$. The two sequences reported originally by Franzen et al.[68] are members of this collection of related sequences. It should be noted that the probe containing cloned DNA used by Franzen et al.[68] contained a large number of related sequences, whereas the synthetic probe used by McLaughlin et al.[69] was a homogeneous probe having a unique sequence (which, however, contained both the five-base and the nine-base consensus sequences).

Lambda clones, presumably carrying repetitive *P. falciparum* DNA in*Hin*d III fragments, have also been isolated.[39,70] Subcloning of a 7-kb insert containing repetitive DNA revealed that a similar 21-bp consensus sequence was present in tandem repeats in DNA located at both ends of the insert. The orientation of the repeat in DNA from one end was inverted relative to the other end. This occurs frequently in eukaryotic DNA as indicated by the frequency of stem and loop structures seen in single-stranded DNA by electron microscopy.[46] The repeats in three blocks of DNA appeared slightly more homogeneous within each block than in comparison with the other two blocks. The consensus sequences for these three blocks were closely related, and each consensus sequence contained the same blocks of nine and five bases (as well as a common single base) as that reported by Aslund et al.[71]

Under standard conditions of high stringency, base pairing between the sequence of nine bases and its complement would be stable at 24°C,[18] or slightly above room temperature. While the nine-base sequence is possibly usable by itself as a probe, in general, greater stability would be desirable. This could be accomplished by using a different base sequence containing a higher percentage of G and C bases (which form more stable base pairs) or a longer sequence. A probe composed of any one of the consensus sequences would be preferable to the nine-base sequence. In addition to pairing specifically to sequences having the exact complementary sequence, the consensus probe would also pair with sequences having a few mismatches, particularly if the stringency is relaxed. It is also likely to be true that a DNA probe containing several blocks of DNA (such as the nine-base sequence and the five-base sequence), which pair perfectly or nearly so, can be substantially more stable than a probe containing either block separately, even if mismatching occurs between the blocks.[72]

This repeat seems to be a ubiquitous sequence in *P. falciparum,* since it is apparently found on all chromosomes of the parasite[70] and on a large number of restriction fragments, but is absent in *P. yoelli, P. vivax, P. berghei, P. chabaudi, B. bovis,* and *B. microti.* DNA probes with this sequence can be used to identify *P. falciparum* from other species, as was shown initially by Franzen et al.[68]

No sequence data is published for the repeats studied by Guntaka et al.,[52] but the repetitive

DNA, totaling about 2 to 3% of the genome, studied by those workers was similarly composed of multiple long repeats of small repeat sequences.

A specific probe for *P. falciparum* has also been isolated and tested by Barker et al.[17] The probe DNA was isolated from a pBR322 library of *Sau*3A restriction fragments by hybridization with nick-translated total *P. falciparum* DNA. The clones binding the most radioactive probe were purified, and the DNA applied to a Southern blot where their ability to hybridize to human and *P. falciparum* DNA was tested. A clone which hybidized extremely well to nick-translated *P. falciparum* DNA was assumed to do so by virtue of the presence of a sequence highly repeated in the genome. Hybridization of the nick-translated cloned fragment to a genomic DNA blot confirmed the presence of a probe for repetitive DNA. The probe is capable of detecting 6 pg of purified DNA on a dot-blot assay and 120 pg of DNA in a blood sample treated and applied to a filter. The probe is highly species specific upon dot-blot or Southern analysis and promises to be clinically valuable. No sequence data is available for this probe, and the sequence could be closely related to those discussed previously.

For epidemiological purposes, identification of malaria infection in mosquito vectors is important. Dot-blot hybridization of DNA from salivary gland and gut regions of *Anopheles* mosquitos was capable of detecting the presence of *P. falciparum.*[73]

III. ADVANTAGES OF USING PROBES COMPOSED OF SYNTHETIC DNA

A. Reassociation of Probe DNA

Reassociation of probe DNA with itself (or genomic sequences with genomic sequences) is undesirable since it reduces the amount of probe DNA available in the assay. Complementary sequences of probe DNA in solution would reassociate, although DNA immobilized on filters probably does not, so that a reaction which begins with probe excess would eventually become one in which the filter-bound DNA sequences are in excess. The rate of hybridization to the filter decreases as the concentration of single-stranded DNA in solution decreases. Of equal importance is the difficulty of decreasing the time needed for hybridization by increasing the concentration of probe DNA added, since the probe will self-hybridize more rapidly.

B. Probes Composed of Unique-Strand DNA

Various procedures have been devised to prepare nucleic acid probes that are not self-complementary. Two of these include transcription of a unique strand of DNA into RNA and use of the RNA as a probe, and the use of M13-cloned DNA which provides only single-stranded DNA corresponding to one strand. In addition, procedures have been devised which permit synthesis of probe DNA corresponding only to the insert DNA.[74]

The simplest procedure to obtain a probe which does not self-hybridize is chemical synthesis of only one of the two strands. Use of a synthetic unique strand probe permits the probe DNA to be effectively used at high concentrations, diminishing the time needed for hybridization and permitting hybridization to be carried out for extended periods of time to detect weak signals. Analysis of the results is simpler since absence of further hybridization indicates the reaction has been carried to virtual completion.

A chemically synthesized 21-nucleotide probe corresponding to the repetitive sequence published by Franzen et al.[68] was found to have both high sensitivity and specificity.[69] The synthetic probe hybridized effectively with *P. falciparum* DNA, but did not cross-hybridize with human DNA or with species in the phylum Apicomplexa, *P. vivax,* and *Babesia* species. The synthetic probe has a sensitivity comparable to that reported for cloned DNA containing the same sequence and was only about fivefold less sensitive than nick-translated genomic DNA of the same specific activity. The use of synthetic DNA, therefore, eliminates the need for cell culture and plasmid preparation, and eliminates the possibility that the plasmid could undergo deletions or rearrangements. For commercial use, chemical synthesis is simpler and less expensive than plasmid isolation, especially since very large amounts of probe DNA can be easily synthesized in a single batch. In part, the high sensitivity of the synthetic probe is a reflection of the abundance of the particular repeated sequence. Also, short low complexity probes are kinetically almost as effective as longer probes for filter hybridization.[18] Theoretical considerations indicate that low complexity probes in high concentrations result in the shortest hybridization times.[74]

A major consideration in designing an oligonucleotide probe is the presence of RNA homologous to the probe. If the probe sequence detects a transcribed gene, and the oligomer is complementary to the RNA encoded by that gene, a large variability in signal may result. Unacceptable variation may result from RNA degradation due to the physiological state of the harvested cells as well as the sample handling. This can be circumvented by either treating the blood with ribonuclease or, more easily, by hybridizing with an oligomer carrying the sequence of the "sense" strand. Hybridization to RNA might be advantageous under some circumstances, but it may lead to unacceptable sample variation in routine diagnostic assays.

IV. *BABESIA*

Babesia, like *Plasmodium*, is an intraerythrocytic protozoan parasite, similar in life cycle to *Plasmodium*, except that it does not undergo development in the liver at any time and is classified as a piroplasm.[75] The disease produced by the organism is characterized by fever, hemoglobinuria, and anemia,[76] and largely affects domestic livestock, particularly cattle; an exception to this has been noted in the last 30 years in the northeastern U.S., during which time an outbreak of *B. microti* was observed in the human population.[4]

The major strains associated with disease in cattle include *B. bovis, B. bigemina, B. divergans,* and *B. major*. While the disease produced by any of these is fairly mild in calves, it is quite severe when introduced into adult populations (50 to 90% fatality). Even low parasitemias, as judged by peripheral blood smear, can result in fatality.[76] The incubation period is 8 to 15 days. Diagnosis of these diseases relies heavily on detection of parasites in thick blood smears of infected individuals; this method requires a skilled technician and is quite time consuming, especially in cases of extremely low parasitemia. A rapid diagnostic method for these parasites is essential, since drug treatment of exposed cattle may be employed prophylactically.[77]

The major agent of babesiosis in humans is *B. microti,* although two cases of *B. divergans* have been diagnosed.[4,78] Most cases involved splenectomized individuals, although *B. microti* may cause disease in nonsplenectomized humans.[79] This particular parasite is endemic in the wild rodent population and may be transmitted to domestic pets.

Current diagnostic techniques with *Babesia* species rely on microscopic examination of thick blood films and have the same drawbacks previously mentioned for *Plasmodium* diagnostics. As with *Plasmodium,* attempts to identify infected individuals by monitoring antibody levels or the presence of parasite antigen has met with limited success.[80]

A. *Babesia* DNA

Little information on *Babesia* genetics is currently available. Estimates of DNA content based on parasite number must rely on genome complexities of closely related protozoans.[40] However, an extensive analysis of DNA polymorphism involving the *Bab*R locus has been carried out by Cowman et al.[22] The data indicate substantial DNA rearrangements associated with various geographical isolates of *B. bovis*. Rearrangements are also detected concomitant with a transition from virulent to avirulent phenotype and accompany changes of transcription in the rearranged gene. A probe derived from an avirulent isolate transcript is capable of distinguishing between geographical isolates as well as between virulent and avirulent strains when Southern blot analysis is employed. Areas of the *Bab*R locus, which differ in DNA sequence between isolates, appear to be mutational hotspots; the use of these sequences for distinguishing between the isolates by dot-blot hybridization, therefore, seems unlikely. These type of analyses suggest that caution must be exercised in selecting a species-specific diagnostic probe in organisms that are capable of frequent or extensive genomic rearrangement.

B. Diagnostic Probe for *Babesia*

A diagnostic probe for *Babesia bovis* has been developed from cloned *B. bovis* DNA by McLaughlin et al.[81] The probe was derived by cloning total genomic DNA digested with *Bam*HI and *Cla*I and subsequently inserting fragments into pBR322. Clones were then screened with nick-translated total *B. bovis* DNA, and the colonies of greatest intensity were selected. The individual isolates were then tested for their ability to distinguish between DNAs purified from *B. bovis*, *B. microti*, *B. bigemina*, and bovine DNA upon dot-blot hybridization analysis. Further analysis of the plasmids, carried out through Southern blotting, confirmed the genomic repetition of sequences contained within the plasmids. When nick translated, the plasmid pCB2 is capable of detecting 100 pg of purified *B. bovis* DNA even in the presence of 1 μg of bovine DNA in an overnight exposure. Further experiments are needed to locate a sequence within the plasmid which can be chemically synthesized and implemented as a practical probe. The limit of parasite detection in blood directly applied to nitrocellulose has not been determined for this probe.

V. DETECTION OF HEMOTROPHIC BACTERIA

Bacterial DNA, unlike eukaryotic DNA, does not contain the short highly repeated DNA characteristic of eukaryotic chromosomes. Because of its relatively small size, however, it is possible to achieve reasonable levels of hybridization in short times by hybridizing against unique copy DNA.

The simplest probe is a genomic probe prepared by nick translation. Because all sequences are present in the probe, maximal levels of hybridization would be assured if excess probe was added. However, related strains of bacteria may have substantial cross-hybridization and might not be readily distinguished.

Nick-translated total *Bartonella bacilliformis* DNA easily detects between 0.003 and 0.03 ng of *B. bacilliformis* DNA on a dot blot in an overnight exposure, corresponding to 10^7 to 10^8 bacteria (Figure 1).[84] There is little or no cross-hybridization to as much as 30 ng of human DNA, which corresponds approximately to the number of leukocytes in 10 ml of blood. Thus, removal of white cells from the blood prior to hybridization would be unnecessary.

The probe used in this experiment was not of sufficiently high specific activity to detect bacteria in an infection of low parasitemia without using rather large amounts of blood. However, the specific activity of the probe could easily be increased.

No cross-hybridization was detected with several gram-negative bacteria (*Escherichia coli*, *Salmonella typhi*, *Shigella sonnei*), or with a gram-positive bacteria (*Streptococcus pyogenes*), or with either *P. falciparum* or *P. knowlesi*. A low level of hybridization to *Rickettsia prowaszeki* was observed at the level of a few percent of that observed for *B. bacilliformis*.

Since the presence of *B. bacilliformis* in erythrocytes can be demonstrated either with Giemsa or with ethidium bromide,[82] the primary use for such a probe, at present, probably would be in research rather than in routine diagnosis. Although many instances of animal infections characterized by erythrocyte-associated bacteria are known, few well-characterized infections are known in man, except for those caused by *B. bacilliformis*. Dooley[83] has summarized the available literature on infections apparently caused by erythrocyte-associated microorganisms in man. In one case, the bacteria were gram-positive and so would seem not to be *B. bacilliformis*. Another case from the Sudan was more suggestive of bartonellosis. More readily than any other

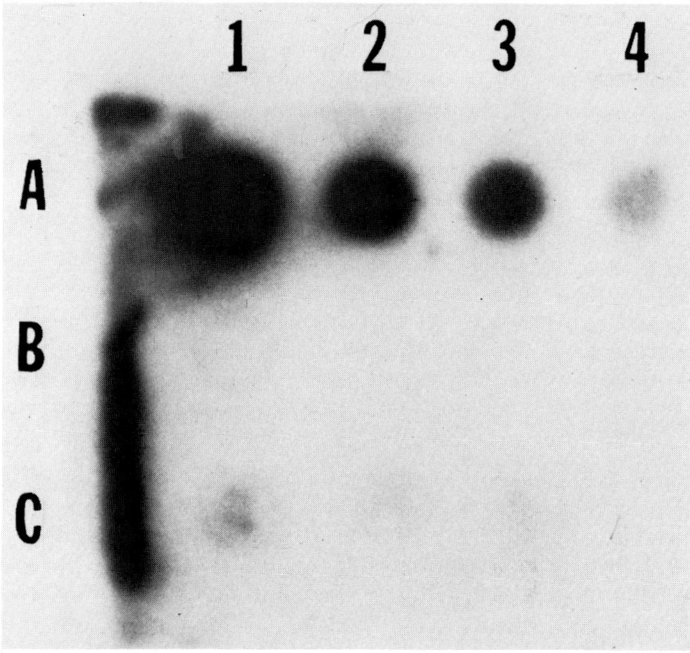

FIGURE 1. *Bartonella bacilliformis* probe. Total genomic DNA (1 μg), labeled by nick translation to 1.7×10^6 cpm/μg, was hybridized to *B. bacilliformis* DNA in row A (A1 = 30 ng; A2 = 3 ng; A3 = 0.3 ng; A4 = 0.03 ng). There was no cross-hybridization to 30 ng of DNA from several other bacteria shown in row B (B1 = *E. coli*; B2 = *Salmonella typhi*; B3 = *Streptococcus pyogenes*; B4 = *Shigella sonnei*). There was a small amount of hybridization to 30 ng of human DNA (C1) and essentially no hybridization to 30 ng of *Plasmodium falciparum* (C2) or 30 ng of *Plasmodium knowlesi* (C3).

procedure, a DNA-based hybridization assay could establish whether the bacteria are *B. bacilliformis* or whether they are related to *B. bacilliformis*.

VI. SUMMARY

Clinical diagnosis and species identification of hemotrophic parasites including *Plasmodium, Babesia,* and bacteria are readily accomplished by DNA hybridization. For eukaryotic parasites, the DNA probe should contain sequences repeated in the genome to maximize sensitivity. Species identification can be achieved by employing species-specific repeated sequences. Chemically synthesized DNA corresponding to repeated parasite sequences has certain advantages over parasite DNA cloned into plasmids. These advantages include the relative ease of preparation of large quantities of inexpensive DNA probe in a single batch which is not subject to sequence variations that might occur with plasmids maintained biologically; also, the synthesized DNA need not be self-complementary, and, therefore, it can be used as a probe in very high concentrations to minimize hybridization times.

ACKNOWLEDGMENTS

We thank Gerald McLaughlin, Siddhartha Kar, Christine Tetzlaff, and Cindy Tripp for helpful discussions. This work was supported by a grant from the Texas Advance Technology Research Program to Garret Ihler and NIH grant AI22713-01A1 to Allison Rice-Ficht.

REFERENCES

1. **Moulder, J.,** Comparative biology of intracellular parasitism, *Microbiol. Rev.,* 49, 298, 1985.
2. **Rowin, K. S., Tanowitz, H. B., Rubinstein, A., Kunkel, M., and Wittner, M.,** Babesiosis in asplenic hosts, *Trans. R. Soc. Trop. Med. Hyg.,* 78, 442, 1984.
3. **Mathewson, H. O., Anderson, A. E., and Hazard, G. W.,** Self-limited babesiosis in a splenectomized child, *Pediatr. Infect. Dis.,* 3, 148, 1984.
4. **Rosner, F., Zarrabi, M. H., Benach, J. L., and Habicht, G. S.,** Babesiosis in splenectomized adults. Review of 22 reported cases, *Am. J. Med.,* 76, 696, 1984.
5. **Weinman, D.,** Infectious anemias due to *Bartonella* and related red cell parasites, *Trans. Am. Philos. Soc.,* 33, 243, 1944.
6. **Bruce-Chwatt, L. J.,** Transfusion associated parasitic infections, *Prog. Clin. Biol. Res.,* 182, 101, 1985.
7. **Perlmann, H., Berzins, K., Wahlgren, M., Carlsson, J., Bjorkman, A., Patarroyo, M., and Perlmann, P.,** Antibodies in malarial sera to parasite antigens in the membrane of erythrocytes infected with early asexual stages of *Plasmodium falciparum, J. Exp. Med.* 159, 1686, 1984.
8. **Whaun, J. M., Rittershaus, C., and Ip, S. H.,** Rapid identification and detection of parasitized human red cells by automated flow cytometry, *Cytometry,* 4, 117, 1983.
9. **Jackson, P. R., Winkler, D. G., Kimzey, S. L., and Fisher, F. M., Jr.,** Cytofluorograf detection of *Plasmodium yoelii, Trypanosoma gambiense,* and *Trypanosoma equiperdum* by laser excited fluorescence of stained rodent blood, *J. Parasitol.,* 63, 593, 1977.
10. **Howard, R. J., Battye, F. L., and Mitchell, G. F.,** *Plasmodium*-infected blood cells analyzed and sorted by flow fluorimetry with the deoxyribonucleic acid binding dye 33258 Hoechst, *J. Histochem. Cytochem.,* 27, 803, 1979.
11. **Avraham, H., Golenser, J., Gazitt, Y., Spira, D. T., and Sulitzeanu, D.,** A highly sensitive solid-phase radioimmunoassay for the assay of *Plasmodium falciparum* antigens and antibodies, *J. Immunol. Methods,* 53, 61, 1982.
12. **McBride, J. S., Walliker, D., and Morgan, G.,** Antigenic diversity in the human malaria parasite Plasmodium falciparum, *Science,* 217, 254, 1982.
13. **Hall, R., McBride, J., Morgan, G., Tait, A., Zolg, J. W., Walker, D., and Scaife, J.,** Antigens of the erythrocytic stages of the human malaria parasite *Plasmodium falciparum* detected by monoclonal antibodies, *Mol. Biochem. Parasitol.,* 7, 247, 1983.
14. **Carter, R. and Voller, A.,** The distribution of enzyme variation in populations of *Plasmodium falciparum* in Africa, *Trans. R. Soc. Trop. Med. Hyg.,* 69, 371, 1975.
15. **Tait, A.,** Analysis of protein variation in *Plasmodium falciparum* by two-dimensional electrophoresis, *Mol. Biochem. Parasitol.,* 2, 205, 1981.
16. **Brown, G. V., Anders, R. F., and Knowles, G.,** Differential effect of immunoglobin on the *in vitro* growth of several isolates of *Plasmodium falciparum, Infect. Immun.,* 39, 1228, 1982.
17. **Barker, R. H., Jr., Suebsaeng, L., Rooney, W., Alecrim, G. C., Dourado, H. V., and Wirth, D. F.,** Specific DNA probe for the diagnosis of *Plasmodium falciparum* malaria, *Science,* 231, 1434, 1986.
18. **Meinkoth, J. and Wahl, G.,** Hybridization of nucleic acids immobilized on solid supports, *Anal. Biochem.,* 138, 267, 1984.
19. **Bresser, J., Hubbell, H. R., and Gillespie, D.,** Biological activity of mRNA immobilized on nitrocellulose in NaI, *Proc. Natl. Acad. Sci. U.S.A.,* 80, 6523, 1983.
20. **Walker, T. K., Rollinson, D., and Simpson, A. J. G.,** Differentiation of *Schistosoma haematobium* from related species using cloned ribosomal RNA gene probes, *Mol. Biochem. Parasitol.,* 20, 123, 1986.
21. **Coppel, R. L., Saint, R. B., Stahl, H. D., Langford, C. J., Brown, G. V., Anders, R. F., and Kemp, D. J.,** *Plasmodium falciparum:* differentiation of isolates with DNA hybridization using antigen gene probes, *Exp. Parasitol.,* 60, 82, 1985.
22. **Cowman, A. F., Bernard, O., Stewart, N., and Kemp, D. J.,** Genes of the protozoan parasite *Babesia bovis* that rearrange to produce RNA species with different sequences, *Cell,* 37, 653, 1984.
23. **Beverly, S. M., Coderre, J. A., Santi, D. V., and Schimke, R. T.,** Unstable DNA amplifications in methotrexate-resistant *Leishmania* consist of extrachromosomal circles which relocalize during stabilization, *Cell,* 38, 431, 1984.
24. **Rieckmann, K. H.,** Falciparum malaria: the urgent need for safe and effective drugs, *Annu. Rev. Med.,* 34, 321, 1983.
25. **Wallach, M., Cully, D. F., Haas, L. O., Trager, W., and Cross, G. A.,** Histidine-rich protein genes and their transcripts in *Plasmodium falciparum* and *P. lophurae, Mol. Biochem. Parasitol.,* 12, 85, 1984.
26. **Hollingdale, M.,** Malaria and liver, *Hepatology,* 5(2), 327, 1985.
27. **Cornelissen, A. W., Langsey, G., and Walliker, D.,** Ribosomal DNA sequences detected in malaria parasites by cytochemical hybridization, *Biol. Cell.,* 49, 87, 1983.

28. **Gritzmacher, C. A. and Reese, R. T.,** Protein and nucleic acid synthesis during synchronized growth of *Plasmodium falciparum, J. Bacteriol.,* 160, 1165, 1984.
29. **Pollack, Y., Metzger, S., Shemer, R., Landau, D., Spira, D. T., and Golenser, J.,** Detection of *Plasmodium falciparum* in blood using DNA hybridization, *Am. J. Trop. Med. Hyg.,* 34, 663, 1985.
30. **Laskey, R. A. and Mills, A. D.,** Enhanced autoradiographic detection of ^{32}P and ^{125}I using intensifying screens and hypersensitized film, *FEBS Lett.,* 82, 314, 1977.
31. **Feinberg, A. P. and Vogelstein, B.,** A technique for radiolabeling DNA restriction endonuclease fragments to high specific activity, *Anal. Biochem.,* 132, 6, 1983.
32. **Schwartz, D. C. and Cantor, C. R.,** Separation of yeast chromosome-sized DNAs by pulsed field gradient electrophoresis, *Cell,* 37, 67, 1984.
33. **Van der Ploeg, L. H. T., Schwartz, D. C., Cantor, C. R., and Borst, P.,** Antigenic variation in *Trypanosoma brucei* analyzed by electrophoretic separation of chromosome-sized DNA molecules, *Cell,* 37, 77, 1984.
34. **Van der Ploeg, L. H., Smits, M., Ponnudurai, T., Vermeulen, A., Meuwissen, J. H., and Langsley, G.,** Chromosome-sized DNA molecules of *Plasmodium falciparum, Science,* 229, 658, 1985.
35. **Kemp, D. J., Coppel, R. L., Stahl, H. D., Bianco, A. E., Corcoran, L. M., McIntyre, P., Langford, C. J., Favaloro, J. M., and Crewther, P. E.,** Genes for antigens of *Plasmodium falciparum, Parasitology,* 91, S83, 1986.
36. **Corcoran, L. M., Forsyth, K. P., Bianco, A. E., Brown, G. V., and Kemp, D. J.,** Chromosome size polymorphisms in *Plasmodium falciparum* can involve deletions and are frequent in natural parasite populations, *Cell,* 44, 87, 1986.
37. **Kemp, D. J., Corcoran, L. M., Coppel, R. L., Stahl, H. D., Bianco, A. E., Brown, G. V., and Anders, R. F.,** Size variation in chromosomes from independent cultured isolates of *Plasmodium falciparum, Nature (London),* 315, 347, 1985.
38. **Giannini, S. H., Schittini, M., Keithly, J. S., Warburton, P. W., Cantor, C. R., and Van der Ploeg, L. H. T.,** Karyotype analysis of *Leishmania* species and its use in classification and clinical diagnosis, *Science,* 232, 762, 1986.
39. **Goman, M., Langsley, G., Hyde, J. E., Yankovsky, N. K., Zolg, J. W., and Scaife, J. G.,** The establishment of genomic DNA libraries for the human malaria parasite *Plasmodium falciparum* and identification of individual clones by hybridisation, *Mol. Biochem. Parasitol.,* 5, 391, 1982.
40. **Pollack, Y., Katzen, A. L., Spira, D. T., and Golenser, J.,** The genome of *Plasmodium falciparum.* I. DNA base composition, *Nucleic Acids Res.,* 10, 539, 1982.
41. **Hough-Evans, B. R. and Howard, J.,** Genome size and DNA complexity of *Plasmodium falciparum, Biochim. Biophys. Acta,* 698, 56, 1982.
42. **Dore, E., Birago, C., Frontali, C., and Battaglia, P. A.,** Kinetic complexity and repetivity of *Plasmodium berghei* DNA, *Mol. Biochem. Parasitol.,* 7, 199, 1980.
43. **Gutteridge, W. E. and Trigg, P. I.,** Incorporation of radioactive precursors into DNA and RNA of *Plasmodium knowlesi in vitro, J. Protozool.,* 17, 89, 1970.
44. **Gutteridge, W. E., Trigg, P. I., and Williamson, D. H.,** Base compositions of DNA from some malarial parasites, *Nature (London),* 224, 1210, 1969.
45. **Schmid, C. W., Manning, J. E., and Davidson, N.,** Inverted repeat sequences in the *Drosophila* genome, *Cell,* 5, 159, 1975.
46. **Wilson, D. A. and Thomas, C. A., Jr.,** Palindromes in chromosomes, *J. Mol. Biol.,* 84, 115, 1974.
47. **Harbers, K., Harbers, B., and Spencer, J. H.,** Nucleotide clusters in DNAs X sequences of the pyrimidine oligonucleotides of mouse L cell satellite DNA, *Biochem. Biophys. Res. Commun.,* 58, 814, 1974.
48. **Biro, P. A., Carr-Brown, A., Southern, E. M., and Walker, P. M. B.,** Partial sequence analysis of mouse satellite DNA: evidence for short-range periodicities, *J. Mol. Biol.,* 94, 71, 1975.
49. **Schmid, C. W. and Jelinek, W. R.,** The Alu family of dispersed repetitive sequences, *Science,* 216, 1065, 1982.
50. **Shafit-Zagardo, B., Maio, J. J., and Brown, F. L.,** KpnI families of long, interspersed repetitive DNAs in human and other primate genomes, *Nucleic Acids Res.,* 10, 3175, 1982.
51. **Gebhard, W., Meitinger, T., Hochtl, J., and Zachau, H. G.,** A new family of interspersed repetitive DNA sequences in the mouse genome, *J. Mol. Biol.,* 157, 453, 1982.
52. **Guntaka, R. V., Gowda, S., Rao, A. S., and Green, T. J.,** Organization of *Plasmodium falciparum* genome. I. Evidence for a highly repeated DNA sequence, *Nucleic Acids Res.,* 13, 1965, 1985.
53. **Blackburn, E. H., Budart, M. L., Challoner, P. B., Cherry, J. M., Howard, E. A., Katzen, A. L., Pan, W. C., and Ryan, T.,** DNA termini in ciliate macronuclei, *Cold Spring Harbor Symp. Quant. Biol.,* 4, 1195, 1983.
54. **Ponzi, M., Pac, T., Dore, E., and Frotali, C.,** Identification of a telomeric DNA sequence in *Plasmodium berghei, EMBO J.,* 4, 2991, 1985.
55. **Blackburn, E. H. and Szostak, J. W.,** The molecular structure of centromeres and telomeres, *Annu. Rev. Biochem.,* 53, 163, 1984.

56. **Van der Ploeg, L. H. T., Liu, A. Y. C., and Borst, P.,** Structure of the growing telomeres of trypanosomes, *Cell,* 36, 459, 1984.
57. **Long, E. O. and Dawid, I. B.,** Repeated genes in eukaryotes, *Annu. Rev. Biochem.,* 49, 727, 1980.
58. **Cornelissen, A. W., Langsley, G., Walliker, D., and Scaife, J. G.,** Gametocytogenesis and ribosomal tRNA gene organisation in the rodent malarias *Plasmodium chabaudi* and *Plasmodium berghei, Mol. Biochem. Parasitol.,* 14, 165, 1985.
59. **Borst, P., Tabak, H. F., and Grivell, L. A.,** Extranuclear genes, in *Eucaryotic Genes, Their Structure, Activity and Regulation,* MacLean, N., Gregory, S. P., and Flavell, R. A., Eds., Butterworths, London, 1983, 71.
60. **Langsley, G., Hyde, J. E., Goman, M., and Scaife, J. G.,** Cloning and characterisation of the rRNA genes from the human malaria parasite *Plasmodium falciparum, Nucleic Acids Res.,* 11, 8703, 1983.
61. **Dame, J. B., Sullivan, M., and McCutchan, T. F.,** Two major sequence classes of ribosomal RNA genes in *Plasmodium berghei, Nucleic Acids Res.,* 12, 5943, 1984.
62. **Dame, J. B. and McCutchan, T. F.,** Identification of 5 S and 5.8 S ribosomal RNA molecules and their genes in *Plasmodium berghei, Mol. Biochem. Parasitol.,* 11, 301, 1984.
63. **Unnasch, T. R. and Wirth, D. F.,** The cloned rRNA genes of *P. lophurae*: a novel rDNA structure, *Nucleic Acids Res.,* 11, 8461, 1983.
64. **Dover, G. A. and Flavell, R. B.,** Molecular coevolution: DNA divergence and the maintenance of function, *Cell,* 38, 622, 1984.
65. **Dame, J. B., Williams, J. L., McCutchan, T. F., Weber, J. L., Writz, R. A., Hockmeyer, W. T., Maloy, W. L., Haynes, J. D., Schneider, I., Roberts, D., Sanders, G. S., Reddy, E. P., Diggs, C. L., and Miller, L. H.,** Structure of the gene encoding the immunodominant surface antigen on the sporozoite of the human malaria parasite *Plasmodium falciparum, Science,* 225, 593, 1984.
66. **Godson, G. N., Ellis, J., Svec, P., Schlesinger, D. H., and Nussenzweig, V.,** Identification and chemical synthesis of a tandemly repeated immunogenic region of *Plasmodium knowlesi* circumsporozoite protein, *Nature (London),* 305, 29, 1983.
67. **Coppel, R. L., Cowman, A. F., Lingelbach, K. R., Brown, G. V., Saint, R. B., Kemp, D. J., and Anders, R. F.,** Isolates specific S-antigen of *P. falciparum* contains a repeated sequence of eleven amino acids, *Nature (London),* 306, 751, 1983.
68. **Franzen, L., Shabo, R., Perlmann, H., Wigzell, H., Westin, G., Aslund, L., Persson, T., and Pettersson, U.,** Analysis of clinical specimens by hybridisation with probe containing repetitive DNA from *Plasmodium falciparum, Lancet,* 1, 525, 1984.
69. **McLaughlin, G. L., Edlind, T. D., Campbell, G. H., Eller, R. F., and Ihler, G. M.,** Detection of *Plasmodium falciparum* using a synthetic DNA probe, *Am. J. Trop. Anim. Health,* 34, 837, 1985.
70. **Oquendo, P., Goman, M., Mackay, M., Langsley, G., Walliker, D., and Scaife, J.,** Characterisation of a repetitive DNA sequence from the malaria parasite, *Plasmodium falciparum, Mol. Biochem. Parasitol.,* 18, 89, 1986.
71. **Aslund, L., Franzen, L., Westin, G., Persson, T., Wigzell, H., and Pettersson, U.,** Highly reiterated non-coding sequence in the genome of *Plasmodium falciparum* is composed of 21 base-pair tandem repeats, *J. Mol. Biol.,* 185, 509, 1985.
72. **Edlind, T. and Ihler, G.,** Electron microscopic mapping of complementary sequences on single strands of bacteriophage lambda DNA, J. M*ol. Biol.,* 160, 23, 1982.
73. **Wilairat, N., Tirawanchai, C., Intapruk, S., Tungpradabkul, S., Sertsrivanich, R., Panyim, S., and Yuthavong, Y.,** Recombinant DNA techniques as potential diagnostic means, *Ann. Inst. Super. Sanita,* 21, 299, 1985.
74. **Ashley, P. L. and MacDonald, R. J.,** Synthesis of single-stranded hybridization probes from reusable DNA templates bound to solid support, *Anal. Biochem.,* 140, 95, 1984.
75. **Callow, L. L.,** in *Immunity to Blood Parasites in Animals and Man,* Miller, L. H., Pino, L. H., and McKelvey, J. J., Jr., Eds., Plenum Press, New York, 1977.
76. **Mahoney, D. F.,** *Babesia* of domestic animals, in *Parasitic Protozoa,* Krier, J. P., Ed., Academic Press, New York, 1977, 1.
77. **Kuttler, K. L. and Johnson, L. W.,** Chemoprophylactic activity of imidocarb, diminazene and oxytetracycline against *B. bovis* and *B. bigemina, Vet. Pathol.,* 21, 107, 1986.
78. **Gombert, M. E., Goldstein, E. J., Benach, J. L., Tenebaum, M. J., Grunwaldt, E., Kaplan, M. H., and Evelerd, L. K.,** Human babesiosis: clinical and therapeutic considerations, *J. Am. Med. Assoc.,* 248, 3005, 1982.
79. **Benach, J. L., Habicht, G. S., and Hamburger, M. I.,** Immunoresponsiveness in acute Babesiosis in humans, *J. Infect. Dis.,* 146, 369, 1982.
80. **Timms, P., Stewart, N. P., Rodwell, B. J., and Barry, D. N.,** Immune responses of cattle following vaccination with living and nonliving *B. bovis* antigens, *Vet. Parastol.,* 16, 243, 1984.
81. **McLaughlin, G. L., Edlind, T. D., and Ihler, G. M.,** Detection of *Babesia bovis* using DNA hybridization, *J. Protozool.,* 33, 125, 1986.

82. **Benson, L. A., Kar, S., McLaughlin, G., and Ihler, G. M.,** Entry of *Bartonella bacilliformis* into erythrocytes, *Infect. Immun.*, 54(2), 347, 1986.
83. **Dooley, J. R.,** Haemotropic bacteria in man, *Lancet*, 2, 1237, 1980.
84. **Kar, S. and Ihler, G.,** unpublished data, 1986.

Chapter 19

LEISHMANIASIS: DNA PROBES FOR DIAGNOSIS

Dyann F. Wirth, William O. Rogers, Heitor Dourado, and Bernadino Albuquerque

TABLE OF CONTENTS

I. INTRODUCTION

Parasitic diseases are still prevalent in many parts of the world, causing both human suffering and economic loss. The World Health Organization estimates that over 300 million people are infected with some parasitic disease. Major efforts to control and eradicate parasitic diseases have met with some success, but in many developing tropical countries, parasitic diseases remain major health problems. Parasitic diseases, such as malaria, schistosomiasis, filariasis, trypanosomiasis, and leishmaniasis, can pose a significant barrier to economic development, and their control is an important goal for improved world health. Intensive research is being devoted to the development of new control measures for many parasitic diseases. These control measures include development of vaccines and new chemotherapeutic agents, as well as improved vector control strategies. Previous experience has demonstrated the need for extensive baseline information on the epidemiology of a disease, including its prevalence, incidence, and transmission, before the introduction of any control program, and the need for continued monitoring of the control program in order to assess its effectiveness. The agents of these diseases have complex life cycles involving both vectors of transmission and often intermediate hosts, both of which have an impact on the transmission of disease and can affect the outcome of any control measure. For example, in the case of leishmaniasis, there are several potential intermediate hosts, including the sloth and opossum. Control measures should allow for the potential switch in the primary reservoir when environmental conditions have changed.

Recent developments in biotechnology, including the use of monoclonal antibodies and recombinant DNA, have provided new tools for the collection of information on parasitic diseases and have the potential for providing more extensive and detailed information on the parasite in the infected human and in insect vectors. This article focuses on these new methodologies for epidemiological studies of the parasitic disease leishmaniasis for which new methods of detection, both in man and insect vectors, have been developed.

II. LEISHMANIASIS: THE DISEASE

Human leishmaniasis is caused by at least 14 different species and subspecies of the genus *Leishmania*. The clinical manifestations of the disease depend in part on the infecting *Leishmania* organism and fall into three general categories: simple cutaneous disease, which is often self-limiting; mucocutaneous disease, which involves the destruction of nasal tissue; and visceral disease, a systemic infection that is often fatal if untreated. In simple cutaneous leishmaniasis, a cutaneous ulcer develops at the site of the infected sandfly bite and heals spontaneously after a period of several months to a year. In mucocutaneous leishmaniasis, metastases from a primary cutaneous lesion spread to the mucosa of the nose and pharynx, causing extensive destruction and, eventually, death from secondary infection or respiratory impairment. Diffuse cutaneous leishmaniasis resembles lepromatous leprosy; there is a failure of the cellular immune response to leishmanial antigens, and the patient develops diffuse, nonulcerated lesions teeming with infected macrophages. Finally, certain species of *Leishmania* are capable of infecting tissue macrophages of the liver and spleen, causing a protracted and, ultimately, fatal febrile illness referred to as visceral leishmaniasis. In fact, the diseases share relatively few properties except that they are caused by organisms of the same genus and that, in humans, the parasite grows in the phagolysosomal vesicles of macrophages. Epidemiological studies over the last 25 years have shown that leishmaniasis is a zoonotic disease; the parasite is transmitted to man from a reservoir mammalian host by a phlebotomus sandfly vector during a blood meal. The reservoir hosts are different in different geographic locations. Domestic and wild dogs are implicated as reservoir hosts for visceral leishmaniasis caused by *L. chagasi* in northern Brazil, while sloths are implicated as the major reservoir host for cutaneous disease caused by *L. braziliensis* in the

Table 1
DISTRIBUTION OF *LEISHMANIA* SPECIES

Disease	Species	Primary geographic location[a]
Cutaneous leishmaniasis	*L. mexicana* ssp. *mexicana*	Mexico, Central America
	L. mexicana ssp. *amazonensis*	Brazil, Amazon region
	L. mexicana ssp. *pifoni*	Venezuela
	L. major	Southern U.S.S.R., Middle East
	L. tropica	Asia, southern Europe, northern and western Africa
	L. braziliensis ssp. *guyanensis*	Northern South America
	L. braziliensis ssp. *panamenensis*	Central America
	L. braziliensis ssp. *peruviana*	Peru
Diffuse cutaneous	*L. mexicana* ssp. *amazonensis*	Brazil, Amazon region
	L. aethiopica	Ethiopia and Kenya
Mucocutaneous	*L. braziliensis* ssp. *braziliensis*	Western and northern South America
Visceral disease	*L. donovani*	India
	L. donovani ssp. *infantum*	Mediterranean area
	L. chagasi	Northern South America

[a] Based on Marinkelle, C. J.[10]

Amazon region of Brazil. There is limited data available on reservoir hosts because of the difficulty of isolating and identifying parasites from captured wild animals. Only a subset of the numerous *Leishmania* species that infect mammals infect and cause disease in humans. Presumably, this selectivity is a combination of factors including the intrinsic susceptibility of humans and the feeding habits of the sandfly vector. Enormous effort has been devoted to the isolation and characterization of *Leishmania* organisms that infect man and to the identification of the principal mammalian reservoirs and species of phlebotomus vector.[1] The result of many such studies has been the correlation of particular clinical manifestations with certain species or subspecies of the parasite (Table 1).

III. IDENTIFICATION OF *LEISHMANIA* SPECIES

The identification of *Leishmania* species is based on a variety of ecological, biological, biochemical, and immunological criteria.[2] These include growth in culture, isoenzyme analysis,[3,4] monoclonal antibody identification,[5] kinetoplast DNA density,[6] kinetoplast DNA restriction length polymorphism, and molecular karyotype analysis.[7] Each cultured isolate of the parasite has been analyzed by the use of one or more of these criteria and categorized as to species and subspecies. This work is done for research purposes, but is not routinely performed in the clinical laboratory. There remain certain controversies as to whether organisms isolated in distant geographic locations, but sharing certain common properties, belong to the same or distinct subspecies of *Leishmania*. Whether these organisms represent strains of the same subspecies or distinct subspecies cannot be resolved because there is no single generally accepted method for species identification in the genus *Leishmania*. This uncertainty complicates comparison of the disease epidemiology in distinct geographic locations and represents a potential limitation on the transfer of control measures from one geographic location to another. The World Health Organization has addressed this problem by establishing a set of reference strains for the various species and subspecies to be used for comparison and classification of new isolates.

IV. CONTROL PROGRAMS

The diverse nature of leishmaniasis clearly requires a diverse control program with specific targets for each focus of the disease. For example, a control program for cutaneous leishmaniasis in an urban setting near Manaus, Brazil will be different than one for visceral leishmaniasis in India. This will require more extensive collection of baseline data with regard to the infecting *Leishmania* species in humans, the relevant sandfly vector, and the principal mammalian reservoir of the particular species or subspecies. These data must be collected at each focus for proper implementation of any control program directed at either the sandfly vector or the mammalian reservoir. For control measures that involve identification and treatment of patients, accurate, rapid diagnosis of leishmaniasis must be achieved before treatment with relatively toxic chemotherapeutic agents is begun.[8] Current chemotherapy is with pentavalent antimony compounds or arsenic compounds. New drugs, including allopurinol riboside, are being tested, but are not yet in clinical use.

One of the major problems in further analysis of the ecology and epidemiology of the disease is the extremely laborious task of identifying the parasite. All of the work documenting the species and subspecies of *Leishmania* is dependent on the isolation of the organism either directly in culture or after passage through a susceptible laboratory animal, most commonly the hamster. This has several limitations, namely, the number of samples that can be handled at any one time, the time it takes to grow the parasites and subsequently type them (3 to 4 weeks), and perhaps more important, the selection from an otherwise mixed population of parasites of those that grow either in vitro or in experimental animals. No data are available with regard to the prevalence of mixed infections, since all previous identification has been based on parasites which have been cultured. A method of direct identification of *Leishmania* parasites from lesions, sandfly vectors, and intermediate mammalian reservoirs is necessary if broader epidemiological studies involving large numbers of samples are to be initiated.

V. PROBLEMS WITH DIAGNOSIS

Current diagnosis of leishmaniasis is achieved either by direct examination of a tissue biopsy as a Giemsa-stained touch preparation, or after standard histopathology preparation, or by means of a delayed-type hypersensitivity test referred to as the Montenegro test.[9] Neither of these methods is able to distinguish *Leishmania* species or subspecies, and the Montenegro test cannot distinguish current from previous infection. For certain forms of leishmaniasis, the most effective control measures may be the direct treatment of infected patients[10] and will require specific diagnosis of *Leishmania* species or subspecies from lesion material in order to design treatment regimens that minimize morbidity and mortality.[11] For example, cutaneous infection with *L. braziliensis* is often associated with subsequent mucocutaneous disease, and early diagnosis could facilitate treatment and perhaps reduce the frequency of mucocutaneous disease.

VI. DNA PROBES IN THE DIAGNOSIS OF LEISHMANIASIS

A new methodology, based on DNA probes specific for the kinetoplast DNA of various *Leishmania* species, was developed to provide a direct diagnosis of patients with leishmaniasis and to eliminate the need for culturing parasites before species identification.[12] This touch-blot procedure, diagrammed in Figure 1, is a modification of standard diagnostic techniques. A punch biopsy is taken from the raised edge of the cutaneous lesion and applied directly to a solid support filter, making triplicate impressions on the filter. In all the studies for validation of the probe, a portion of the biopsy was removed and placed directly in culture medium. The filter (nitrocellulose in earlier studies and Gene Screen Plus in later studies) was then divided into

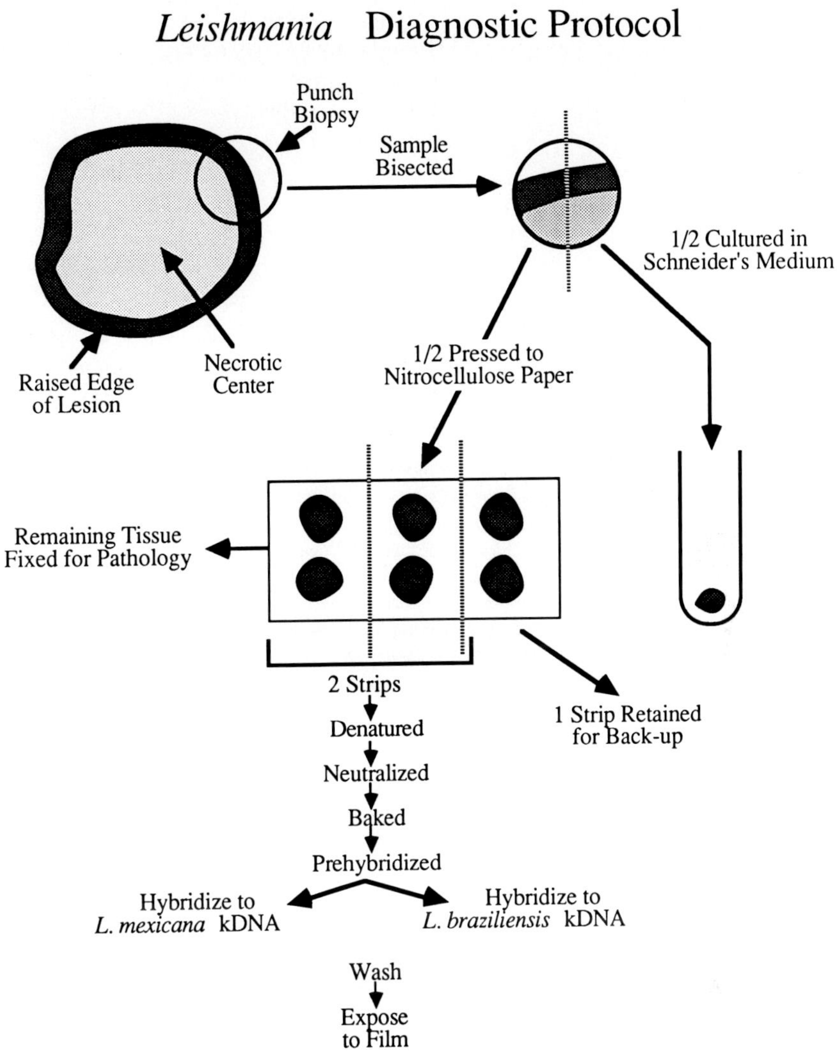

FIGURE 1. Tissue touch blot method. Shown is a diagram of the tissue touch procedure developed for the detection of *Leishmania* spp. in tissue biopsies of cutaneous lesions.

three identical strips, and processed for hybridization. The DNA was denatured *in situ* in the presence of base and subsequently hybridized with DNA probes specific for *L. mexicana* or *L. braziliensis*.

This methodology allows direct diagnosis from lesion material without requiring isolation of the parasite. Such direct diagnosis of *Leishmania* species, which had not been possible with any of the existing methodologies, provides the basis for the clinical management of this disease. In the initial studies, the DNA probes, which were based on total kinetoplast DNA (kDNA), could differentiate the major species complexes in the new world, *L. mexicana* and *L. braziliensis*. Subsequent experiments, in which recombinant DNA methodologies were used, resulted in the development of DNA probes that can differentiate species, subspecies, and even distinct isolates of the parasite.[12-17]

The basis for the DNA probes is the minicircle, which is a highly repeated small circular DNA molecule found within the mitochondria of the parasite. It has no known function and has an apparent high rate of DNA sequence divergence as measured by restriction site polymorphism and DNA hybridization studies.[18-22] In the new world *Leishmania* spp., the kDNA minicircles

Table 2
COMPARISON OF *LEISHMANIA* IDENTIFICATION METHODS

Test result	No. of patients
Montenegro positive	43
Hybridization positive	32
L. mexicana	2
L. braziliensis	30
Culture positive	27
Histopathology	17

Note: Results from 57 patients examined at the *Leishmania* clinic, Instituto de Medicina, Tropical de Manaus Amazonas, Brazil during January and Feburary 1984. Each patient was given a Montenegro test and those with positive results in the test were given further diagnostic tests which included a tissue biopsy (4 mm) taken directly from the lesion. This biopsy was divided in half, and one half was placed directly in Schneider's *Drosophila* medium and the other half touched to nitrocellulose. The remainder was processed for histopathology. The hybridization was performed by standard procedures as described.[12] The *L. mexicana* probe was a combination of DNA extracted from strain WR303 *(L. mexicana* ssp. *amazonensis)* and L11 *(L. mexicana* ssp. *mexicana)*. The L. *braziliensis* probe was isolated from strain WR2903. All hybridization results were read independently by two people.

isolated from *L. mexicana* do not share any sequence homology with those isolated from *L. braziliensis*.[12,13] These differences in DNA sequence have provided the basis for a DNA probe that can distinguish the two *Leishmania* species directly when material obtained from a lesion is applied to nitrocellulose.

These DNA probes have been used to diagnose leishmaniasis in patients from the Instituto de Medicina, Tropical de Manaus[23] (Table 2, Figure 2). In each case, the results of the DNA probe were compared to standard diagnostic tests, including the Montenegro hypersensitivity test, histopathology, and culturing and subsequent characterization of the parasite. As can be seen in Table 2, the DNA probe detected the disease in 32 of the 43 patients who had positive results in the Montenegro test. This observation may mean that in some patients, the parasite density was below the detection limit of the assay, which, based on laboratory experiments, is as few as 50 organisms in a single spot on nitrocellulose. There are cases in which organisms were cultured from lesions that were originally negative. Another possibility is that the delayed-type hypersensitivity test detected either a previous infection or a cross-reacting antigen from another type of infection, for example, *Trypanosoma cruzi,* the causative agent in Chagas' disease.[24] A third possibility is that the parasite in the lesion is of a type not recognized by the kDNA probe; however, every parasite that has been isolated from a lesion has reacted with either the *L. braziliensis* or *L. mexicana* kDNA probes.

The DNA probe clearly detects infections in more patients than either histopathology or culturing. Although this probably results from the greater sensitivity of the DNA probe in comparison to these other methods, it may also represent false positives in the DNA probe method. Because there is no single "gold standard" for the detection of parasites, the exact determination of false positives is not possible. However, in ten cases in which histopathology failed to detect parasites and the result of the DNA probe was positive, the culturing of parasites clearly demonstrated their presence. The question of the false positive reaction with DNA probes has been tested experimentally. These kDNA probes did not hybridize with touch preparations

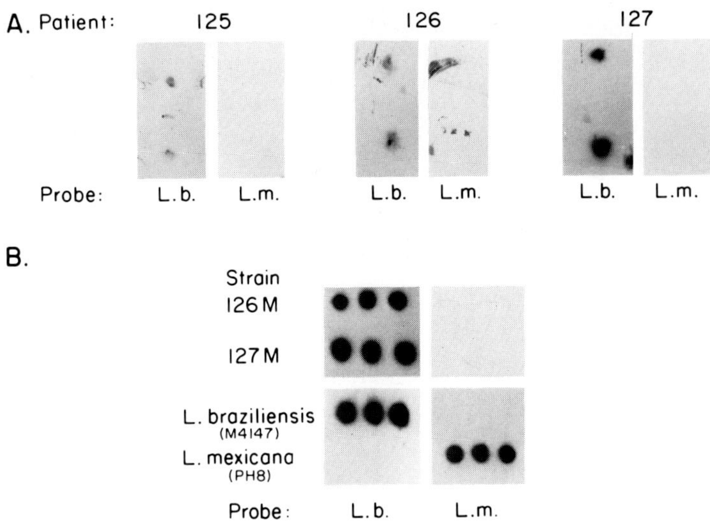

FIGURE 2. DNA probe detects *Leishmania* in tissue biopsies. Shown is an autoradiogram of tissue touch preparations from three patients with cutaneous leishmaniasis (A) or cultured promastigotes (B) which have been hybridized with radiolabeled kinetoplast DNA probes. The left filters in each pair were probed with labeled *L. braziliensis* kDNA (L.b.) and the right filters were probed with *L. mexicana* (L.m.).

of uninfected tissue from several animals and from a limited number of human lesions that subsequently proved not to be due to leishmaniasis. In addition, these DNA probes did not react with *T. cruzi, Plasmodium falciparum*, or *Escherichia coli*. Further evidence that the number of false positives with the kDNA probes is relatively low was obtained by testing duplicate touch preparations of each lesion with kDNA from both *L. mexicana* and *L. braziliensis*. In every case in which there was a reaction with the kDNA probe, it was specific for either *L. mexicana* or *L. braziliensis,* and the duplicate lesion showed no reaction above background. When parasites were isolated from these lesions and subsequently tested by kDNA hybridization, the original identification was confirmed. An alternative method of limiting the number of false positive reactions, due to nonspecific binding of labeled DNA to tissue or blood, is to perform *in situ* hybridization and examine each preparation under the microscope, as suggested by Barker and co-workers.[15] This is a very time-consuming and expert process not easily adapted to large numbers of samples.

One of the limitations of the kDNA minicircle as a hybridization probe is that, although it can distinguish the major species complexes of new world cutaneous leishmaniasis, the kDNA minicircle from each complex is homologous to all the subspecies, and it is, therefore, impossible to distinguish subspecies.[12,13] This identification of subspecies is important clinically because certain manifestations of the disease are specific to the parasite subspecies and is also important in the description of any intermediate host or insect vector carrying a particular subspecies. For example, mucocutaneous leishmaniasis is associated with only one subspecies of *L. braziliensis, L. braziliensis* ssp. *braziliensis,* and early identification of this subspecies could be important in designing treatment regimens. In addition, our work and the work of others have indicated that in the old world *Leishmania* species causing cutaneous and visceral disease, kDNA sequence homology occurs among different species.[12,16,17,24-27] For example, kDNA isolated from *L. major* hybridizes with both of the other cutaneous species, *L. tropica* and *L. aethiopica*, and with the visceral strain *L. donovani*.[17]

Therefore, a new approach is required for the development of DNA probes that can distinguish subspecies in new world cutaneous leishmaniasis and differentiate the species complexes of the old world isolates. Several groups have used recombinant DNA methods to

Table 3
DNA PROBE STUDIES TO DIFFERENTIATE *LEISHMANIA* SPECIES AND SUBSPECIES

Source	Specificities	Ref.
L. donovani	Species	28
L. infantum	Species	8, 28
L. donovani	Visceral complex	16
L. chagas	Non-Indian visceral	16
	Complex isolate	16
L. major	Isolate	17
L. mexicana ssp. *amazonensis*	Species	30
	Subspecies	30
	Isolate	30
L. major	Isolate	14
L. tropica	Isolate	14
L. aethiopica	Isolate	14

develop such DNA probes with narrower specificities[14-17,26,28-30] (Table 3). The method has been to clone restriction fragments of a kDNA minicircle and to use these cloned subfragments as more specific probes for species, subspecies, and even isolates. The general observation from this body of work is that within the minicircle population there are DNA sequences that have undergone rapid sequence divergence and can thus serve to differentiate even closely related organisms.[28] For example, we have cloned subfragments of minicircle DNA isolated from *L. mexicana* ssp. *amazonensis* which hybridize with only a small subset of all *L. mexicana* ssp. *amazonensis* strains tested. In addition, we have shown that this sequence divergence can occur within a single minicircle. A nested set of deletions of a single cloned minicircle fragment from *L. mexicana* ssp. *amazonensis* was generated and then tested for hybridization specificity. The full-length minicircle had a similar specificity to total kDNA, and two deletions demonstrated species and isolate specific hybridization patterns[30] (Figure 2).

The next step for the utilization of DNA probes will be in the detection of infected insect vectors and intermediate hosts. Preliminary laboratory experiments show that parasites can be detected in infected sandflies that have been squashed directly on nitrocellulose.[31] This approach must now be tested in the field. Similarly, laboratory-based experiments have demonstrated that parasite infections can be detected in tissue touch preparations from animals experimentally infected;[12] however, both the intensity of infection and the target tissue in natural hosts will be different, and thus the DNA probes must be tested directly in field-extracted material.

The DNA probe methodology should be readily adaptable to field situations. Once the tissue biopsy or sandfly vector is obtained, it is applied directly to nitrocellulose or other solid supports and is stable in this form indefinitely. Thus, samples could be collected from distant sites and returned for processing. The major disadvantage of this methodology is the requirement for a radioisotope. Alternative methods of labeling DNA, which are being developed and tested for such biological specimens, should be useful in field situations.

REFERENCES

1. **Lainson, R. and Shaw, J. J.,** Epidemiology of Leishmaniasis, *Nature (London),* 273, 595, 1978.
2. **Chance, M. L. and Walton, B. C., Eds.,** *Biochemical Characterization of Leishmania,* UNDP/World Bank/ World Health Organization, Geneva, 1982.
3. **Gardner, P. J., Chance, M. L., and Peters, W.,** Biochemical taxonomy of *Leishmania.* II. Electrophoretic variation of malate dehydrogenase, *Ann. Trop. Med. Parasitol.,* 68, 317, 1974.
4. **Kreutzer, R. D. and Christensen, H. A.,** Characterization of *Leishmania* spp. by isoenzyme electrophoresis, *Am. J. Trop. Med. Hyg.,* 29, 199, 1980.
5. **McMahon-Pratt, D. and David, J. R.,** Monoclonal antibodies that distinguish New World species of *Leishmania, Nature (London),* 291, 581, 1981.
6. **Chance, M. L., Peters, W., and Shchory, L.,** Biochemical taxonomy of *Leishmania.* I. Observations on DNA, *Ann. Trop. Med. Parasitol.,* 68, 307, 1974.
7. **Giannini, S. H., Schittini, M., Keithly, J. S., and Van der Ploeg, L. H. T.,** Karyotype analysis of *Leishmania* species and its use in classification and clinical diagnosis, *Science,* 232, 762, 1986.
8. **Peters, W., Evans, D. A., and Lanham, S. M.,** Importance of identification in cases of leishmaniasis, *J. R. Soc. Med.,* 76, 540, 1983.
9. **Montenegro, J.,** Cutaneous reaction in leishmaniasis, *Arch. Dermatol. Syphilol.,* 13, 187, 1926.
10. **Marinkelle, C. J.,** The control of leishmaniasis, *Bull. W.H.O.,* 58, 807, 1980.
11. **Deane, L. M. and Grimaldi, G.,** *Leishmaniasis,* Chang, K. P. and Bray, R. S., Eds., Elsevier, Amsterdam, 1986.
12. **Wirth, D. F. and McMahon-Pratt, D.,** Rapid identification of *Leishmania* species by specific hybridization of kinetoplast DNA in cutaneous lesions, *Proc. Natl. Acad. Sci. U.S.A.,* 79, 6999, 1982.
13. **Barker, D. C. and Butcher, J.,** The use of DNA probes in the identification of leishmanias: discrimination between isolates of the *mexicana* and *braziliensis* complexes, *Trans. R. Soc. Trop. Med. Hyg.,* 77, 285, 1983.
14. **Kennedy, W. P. K.,** Novel identification of differences in the kinetoplast DNA of *Leishmania* isolates by recombinant DNA techniques and in situ hybridization, *Mol. Biochem. Parasitol.,* 12, 313, 1984.
15. **Barker, D. C., Gibson, L. J., Kennedy, W. P. K., Nasser, A. A. A. A., and Williams, R. H.,** The potential of using recombinant DNA species-specific probes for the identification of tropical *Leishmania, Parasitology,* 91, S139, 1986.
16. **Lopes, U. G. and Wirth, D. F.,** Identification of visceral *Leishmania* species with cloned sequences of kinetoplast DNA, *Mol. Biochem. Parasitol.,* 20, 77, 1986.
17. **Wirth, D. F. and Rogers, W. O.,** Rapid identification of *Leishmania* species using specific hybridization of kDNA probes, in *Rapid Detection and Identification of Infectious Agents,* Kingsbury, D. T. and Falkow, S., Eds., Academic Press, New York, 1985, 127.
18. **Englund, P. T.,** Free minicircles of kinetoplast DNA in *Crithidia fasciculata, J. Biol. Chem.,* 254, 4895, 1979.
19. **Kidane, G. Z., Hughes, D., and Simpson, L.,** Sequence heterogeneity and anomalous electrophoretic mobility of kinetoplast minicircle DNA from *Leishmania tarentolae, Gene,* 27, 265, 1984.
20. **Jackson, P. R., Wohlhieter, J. A., Jackson, J. E., Sayles, P., Carter, L. D., and Hockmeyer, W. T.,** Restriction endonuclease analysis of *Leishmania* kinetoplast DNA characterizes parasites responsible for visceral and cutaneous disease, *Am. J. Trop. Med. Hyg.,* 33, 808, 1984.
21. **Steinert, M., Van Assel, S., Borst, P., and Newton, B. A.,** *Evolution of Kinetoplast DNA,* Saccone, C. and Kroon, A. M., Eds., North-Holland, Amsterdam, 1976, 71.
22. **Lopes, U. G., Momen, H., Grimaldi, G., Marzochi, M. C. A., Pacheco, R. S., and Morel, C. M.,** Schizodeme and zymodeme characterization of *Leishmania* in the investigation of foci of visceral and cutaneous leishmaniasis, *J. Parasitol.,* 70, 89, 1984.
23. **Wirth, D. F., Rogers, W. O., Barker, R., Dourado, H., Suesebang, L., and Albuquerque, B.,** Leishmaniasis and malaria: new tools for epidemiological analysis, *Science,* 234, 975, 1986.
24. **Aston, D. L. and Thorley, A. P.,** Leishmaniasis in central Brazil: results of a Montenegro skin test survey among Amerindians in the Xingu National Park, *Trans. R. Soc. Trop. Med. Hyg.,* 64, 671, 1970.
25. **Arnot, D. E. and Barker, D. C.,** Biochemical identification of cutaneous leishmaniasis by analysis of kinetoplast DNA. II. Sequence homologies in *Leishmania* kDNA, *Mol. Biochem. Parasitol.,* 3, 47, 1981.
26. **Barker, D. C., Arnot, D. E., and Butcher, J.,** DNA characterization as a taxonomic tool for identification of kinetoplast flagellate protozoans, in *Biochemical Identification of Leishmania,* Chance, M. L. and Walton, B. C., Eds., World Health Organization, Geneva, 1982, 139.
27. **Spithill, T. W. and Grumont, R. J.,** Identification of species, strains and clones of *Leishmania* by characterization of kinetoplast DNA minicircles, *Mol. Biochem. Parasitol.,* 12, 217, 1984.
28. **Lawrie, J. M., Jackson, P. R., Stiteler, J., and Hochmeyer, W.,** Identification of pathogenic *Leishmania* promastigotes by DNA: DNA hybridization with kinetoplast DNA cloned into *E. coli* plasmids, *Am. J. Trop. Med. Hyg.,* 34, 257, 1985.

29. **Frasch, A. C. C., Goijman, S. G., Cazzulo, J. J., and Stoppani, A. O. M.,** Constant and variable regions in DNA minicircles from *Trypanosoma cruzi* and *Trypanosoma rangeli:* approaches to species and stock differentiation, *Mol. Biochem. Parasitol.,* 4, 163, 1981.
30. **Rogers, W. O. and Wirth, D. F.,** Kinetoplast DNA minicircles: regions of extensive sequence divergence, *Proc. Natl. Acad. Sci. U.S.A.,* 84, 565, 1987.
31. **Rogers, W. O. and Wirth, D. F.,** Use of kinetoplast DNA probes in the detection of infected sandflies, submitted, 1988.

Part VII: Future Directions

Chapter 20

DNA PROBES AND ANTIMICROBIAL SUSCEPTIBILITY TESTING

Fred C. Tenover

TABLE OF CONTENTS

I. INTRODUCTION

New diagnostic reagents utilizing DNA hybridization technology are now becoming available for the direct detection and identification of pathogenic microorganisms in clinical specimens. Together with the monoclonal antibody-based tests, these reagents provide a highly specific means of rapidly identifying infectious agents, which, in many cases, will obviate the need for culture, the traditional first step in bacterial identification.

As the use of these new technologies becomes more widespread and the desire to discontinue concomitant bacterial cultures grows, the need for alternate means of determining the antimicrobial susceptibility of those infectious agents identified by noncultural means will become more pronounced. Several investigators have suggested adopting a statistical approach to susceptibility testing, deriving the probability of susceptibility by computer, using the previous experience of the medical center with that organism and antimicrobial agent combination. Standardized susceptibility results, carried out on organisms isolated in pure culture, which would still be the majority of tests performed by the laboratory, would be used to continually update the database of the system.

A second approach is to determine the genetic potential of an organism for resistance to an antimicrobial agent using a DNA probe. Probes could be used to directly analyze a clinical specimen for resistance genes, obviating the need for isolating the pathogen in pure culture for traditional testing. Such an approach would allow simultaneous identification and susceptibility testing of microorganisms within clinical specimens in a matter of hours. Clearly, however, the task of assembling the appropriate probes for the many bug-drug combinations will be a monumental task. Yet, the growing interest in using DNA probes in the clinical laboratory now makes such an undertaking feasible.

Herein, I will review the reasons for determining the antimicrobial susceptibility pattern of a microorganism and the advantages and disadvantages of using DNA hybridization techniques for this purpose.

II. PURPOSE OF ANTIMICROBIAL SUSCEPTIBILITY TESTING

Antimicrobial susceptibility testing of microorganisms can serve several purposes. First, and most commonly, it provides clinicians with the data necessary to choose the most appropriate antimicrobial agent or combination of agents with which to treat an infectious process. This is particularly important when the susceptibility of an organism to an agent cannot be predicted on the basis of previous experience with other strains of the same species. Second, antimicrobial susceptibility results can be used to aid in the identification of microorganisms. For example, gram-negative bacilli that are resistant to ampicillin and carbenicillin, but susceptible to first-generation cephalosporins, are often *Klebsiella* species. Verifying the identification of a bacterium with its susceptibility pattern is an important aspect of quality control for the clinical laboratory, particularly with the growing use of automated systems that use unusual substrates for identification. Third, the results of susceptibility testing can be used as epidemiologic markers to trace the movement of either a particular species or even a specific resistance gene within a hospital or community. Such epidemiologic studies can also be undertaken to determine the prevalence of particular resistance determinants in a given geographic area. Finally, susceptibility testing can be used to determine the spectrum and activity of new antimicrobial agents.[1]

While DNA probes directed against resistance genes will not be useful as a means of culture confirmation, they can provide data about the susceptibility of an organism to a given class of antimicrobials, and they have already proven to be excellent tools for epidemiologic studies.[2] Thus, probes hold promise as an alternate means of rapid susceptibility testing.

III. KEY ISSUES TO BE ADDRESSED

A. Defining the Term "Resistant Microorganism"

Traditionally, a microorganism is said to be resistant to an antimicrobial agent when the concentration of that antimicrobial achievable in the human body is below that necessary to inhibit the growth of that microorganism. Thus, "resistance" is a phenotypic trait and a reflection of the mode of action of the agent, its pharmacokinetics, and its interactions with that particular microorganism. The goal of performing antimicrobial susceptibility testing and designating microorganisms as susceptible or resistant to antimicrobial agents is to provide data for clinical action. Thus, before DNA hybridization can be initiated as a means of susceptibility testing, criteria for interpreting the results must be established.

A positive hybridization result is evidence that an organism carries a resistance gene. This is presumptive evidence that the organism is resistant to that antimicrobial. In some cases, however, probes will detect the presence of resistance determinants in microorganisms that are not expressed; thus, the organism clinically would respond to that drug. Should microbiologists take this possibility into account and report results as "presumably resistant by DNA probe assay" instead of just "resistant"? One can imagine the confusion that such a report would generate. Thus, if probes are to be used for susceptibility testing, it should be only in those instances where both microbiologists and clinicians are willing to agree that a positive hybridization test will be interpreted as equivalent to a resistant phenotype.

B. Classes of Antimicrobial Resistance Genes Amenable to Probe Analysis

Antimicrobial resistance genes may be located on the bacterial chromosome, plasmids, or on transposable elements. Resistance can be the result of random genetic events on the bacterial chromosome, such as a spontaneous mutation within the gene that alters the site of action of the drug,[3,4] derepression of a cryptic gene, as with the inducible cephalosporinases of several species of *Enterobacter* and *Pseudomonas*,[5-8] or rearrangements in the regulatory region of a resistance gene.[9] Attempting to monitor all spontaneous mutations on the chromosome of an organism with gene probes would be virtually impossible. Thus, only certain classes of antimicrobial resistance, i.e., those with identifiable structural genes, will be amenable to probe analysis. These would include the plasmid-borne β-lactamase, tetracycline, aminoglycoside, chloramphenicol, trimethoprim, and macrolide-lincosamide-streptogramin (MLS) resistance genes.[10-12]

C. Types of Clinical Specimens Amenable to Probe Analysis

Ideally, all types of clinical specimens could be examined for resistance genes using DNA probes. However, many of the same mechanisms of resistance seen in pathogens are present in members of the normal flora of humans. This is particularly true for the β-lactamase genes present in *Haemophilus influenzae* and *Neisseria gonorrhoeae*, which have their origins in the gram-negative flora of the intestine.[13-16] Thus, the best way of avoiding false positive hybridization results would be to use DNA probes only to examine clinical specimens from normally sterile body sites (and hence free from commensal flora). Such sites as cerebral spinal fluid, urethral exudates from males, and pleural fluid usually contain a single pathogen type which decreases the problem of positive hybridization results that reflect the resistance genes carried by commensal organisms rather than those of the pathogen. Alternately, specimens containing mixed flora could be screened for resistance genes to the empiric therapeutic regimens commonly used in the hospital.

An additional problem is the large number of probes that would be required to test for even the most common resistance phenotypes in enteric organisms if each resistance gene required its own probe. For example, there are at least 21 different β-lactamase genes that can result in penicillin or ampicillin resistance.[17] Thus, a series of DNA probes capable of detecting more than one resistance gene are needed. Data presented by Marshall et al. for tetracycline resistance genes suggest that this would be feasible for the common enteric genes.[18]

D. A β-Lactamase Gene Probe

Perine et al.[19] conducted a study to determine the feasibility of screening males for penicillinase-producing *Neisseria gonorrhoeae* (PPNG) by using a dual probe system directly on urethral exudate material. Gonococci were detected with labeled gonococcal plasmid DNA, and a β-lactamase gene probe was used to predict the susceptibility of those organisms to ampicillin. The probe results were compared to a standard culture method, and the presence of β-lactamase was determined by the chromogenic cephalosporin method on the isolated *N. gonorrhoeae* colonies. The β-lactamase probe exhibited 93% sensitivity and 96% specificity when compared to chromogenic cephalosporin tests. The results suggest that a combined probe system for some specimen types is feasible.

E. Extrapolation of Results

Caution must be exercised, however, when trying to generalize these results. First, chromosomally mediated penicillin resistance has been reported in *N. gonorrhoeae*,[20] and the probe described by Perine et al. would not detect these resistant strains. Second, specimens collected from the female genital tract may be contaminated with gram-negative bacilli harboring β-lactamase genes that could produce false positive results. As noted previously, β-lactamase genes are known to be widely distributed on plasmids of enteric bacteria. Thus, the conclusion one may draw from these results is that DNA probes can be used effectively as an alternate means of susceptibility testing for therapeutic prediction of resistance if specimens are obtained from a normally sterile body site in which a single pathogen, with a known mechanism of resistance, is likely to occur, although some other forms of resistance may be missed. The ability to successfully probe more complex specimens remains to be determined.

IV. CRITERIA FOR INTERPRETING RESULTS OF PROBE TESTS

Antimicrobial resistant isolates are those organisms whose growth is unaffected by the presence of a defined concentration of antibiotic. As noted above, it is a phenotypic determination, and one that tries to account for the concentration of antimicrobial that can be achieved in the blood or tissues, the pharmacokinetics of the drug, and the degree of protein binding. Susceptibility testing, while not an exact science, represents the laboratory's best estimate of the response of an organism to a drug.[21]

With DNA probes, a different set of criteria must be applied. A DNA probe signals the genetic potential of an organism to manifest resistance to an antimicrobial. It is a different way of looking at resistance, but not necessarily a less effective means. Both probes and traditional methods have certain advantages and disadvantages. Traditional tests determine the phenotypic susceptibility of an organism at the time, or shortly after the time, it was isolated from the patient. They cannot accurately predict, as probes can, the derepression of genes that may lead to clinical failure during therapy, nor can they be used directly on a patient's specimen for a rapid assessment of a therapeutic modality. While it is difficult to imagine the replacement of the traditional disc diffusion[22] or automated minimal inhibitory concentration (MIC) methods[23] by DNA hybridization methods, probes do offer some advantages over these traditional tests. For instance, such problems as inoculum effects[24,25] and tolerance,[26,27] that often make MIC results difficult to interpret, are not problematic with probe tests in which the results provided are simply the presence or absence of a gene.

Table 1
STUDIES OF ANTIMICROBIAL RESISTANCE GENES USING DNA PROBES

Antimicrobial agent	Class/type/gene	Ref.
Aminoglycoside	Adenyltransferase	28, 29, 30
	Phosphotransferase	31, 32
β-lactam	*amp* C	33
	TEM, OXA-2	34
	TEM	19, 35
Chloramphenicol	CAT	36
Macrolides	MLS	37—40
Mercury	R100	41
Tetracycline	Enteric A—D	18, 42
	Streptococcal L—N	43
	tetM	44, 45
Trimethoprim	DHFR I and II	46, 47
	Tn7	48

V. EXPERIENCE WITH PROBES FOR DETERMINING THE PRESENCE OF ANTIMICROBIAL RESISTANCE GENES

With the exception of the study cited above, there are no clinical trials of DNA probes as a method of susceptibility testing to either encourage or discourage our pursuit of this approach. There are, however, a large number of DNA probes that have been described and used to answer a variety of questions concerning the distribution of antimicrobial resistance determinants both in hospitals and among the bacterial flora of animals. Table 1 lists several examples of DNA probes that have been developed. The results of these studies will play a central role in the types of probes chosen to test in the clinical laboratory. For example, Marshall et al. have examined the distribution of five classes of tetracycline resistance genes found in enteric bacilli in the fecal flora of animals and humans.[18] Their data suggest widespread dissemination of these resistance determinants, particularly of Class B determinants. Thus, a hybridization method for screening samples for tetracycline resistance must include at least a probe that recognizes the Class B determinant. They also noted similarities between the genes encoding the proteins of Classes A, B, and C.[18] These data could be used to construct a probe that could detect the genes encoding all three classes of tetracycline resistance instead of just a single class.

Tenover et al. have developed probes to aminoglycoside resistance genes and used them to study a cohort of patients at the Seattle Veterans Administration Medical Center in Washington who have carried aminoglycoside resistant organisms over extended periods of time.[29] These studies suggest that the ANT(2″) resistance gene is the most common mechanism of resistance in this hospital. Thus, screening of isolates for aminoglycoside resistance could focus at least initially on this gene.

Many other studies have focused on the epidemiology of resistance genes in hospitals as well as in community settings.[28-48] Such epidemiologic studies using probes will continue to be of value in providing data on which resistance genes to screen in clinical samples.

VI. ADDITIONAL STUDIES REQUIRED BEFORE IMPLEMENTATION OF DNA PROBES FOR ANTIMICROBIAL SUSCEPTIBILITY TESTING

How often are resistance genes harbored by microorganisms but not expressed? How often are cryptic resistance genes derepressed during therapy? The answers to these questions, while unknown, are critical to the success of probes as a means of susceptibility testing. Thus, there is a need to conduct such studies and develop this information base.

There are some data available from the research literature to guide our investigations. For example, all isolates of *Serratia marcescens* are thought to carry a chromosomal gene for an aminoglycoside *N*-acetyltransferase enzyme, AAC(6′)-I, that mediates resistance to amikacin, kanamycin, neomycin, and tobramycin.[49] However, only a small fraction of *S. marcescens* isolates exhibit resistance to these aminoglycosides, even though enzymatic activity can be demonstrated in many of these strains.[50,51] The results of hybridization tests using a probe to the AAC(6′)-I gene would invariably be positive. Thus, although phenotypically susceptible, these organisms would be interpreted as resistant to most aminoglycosides.

Clearly, much more information about the distribution of resistance genes in various genera will be required before probes can be used with confidence in the clinical laboratory in all but a few well-defined situations.

VII. DIRECT SUSCEPTIBILITY TESTING OF SPECIMENS

Using DNA probes for direct susceptibility testing of patient specimens seems to be a novel idea. However, direct susceptibility testing of patient specimens prior to the isolation of an organism in pure culture is commonly undertaken in microbiology laboratories, particularly for the examination of positive blood cultures.[52] Using the Gram stain of the positive blood broth to determine the types of antimicrobials to be tested, susceptibility tests are often initiated using the blood broth as the inoculum. These results are then confirmed by standard methods when the organism is isolated in pure culture.[52]

On the other hand, there are several reasons why direct susceptibility tests have not been performed on urine specimens or have been limited to procedures incorporating automated equipment.[53] First, when compared to positive blood cultures, there is a greater likelihood urine will contain multiple organisms. The presence of two or more organisms in a susceptibility test can lead to erroneous results due to (1) inhibition of the growth of one organism by another, (2) transfer of genetic material between organisms altering the antibiogram, and (3) more rapid growth of one organism obscuring the zones of inhibition of another. It is conceivable that, in the future, susceptibility of microorganisms to the antimicrobials used for the treatment of urinary tract infections may be determined by probe assay if a specific genetic determinant is found to mediate resistance to agents, such as norfloxacin or other quinolones, that can be administered orally and that are effective against a wide range of microorganisms.

VIII. USING DNA PROBES FOR SUSCEPTIBILITY TESTING OF ORGANISMS GROWN IN VITRO

There are three instances where DNA probes may prove of value for determining the susceptibility of bacterial isolates already grown in vitro. The first is for screening fastidious microorganisms for the presence of a given resistance gene. Some organisms do not grow well on standard Mueller Hinton agar or do not grow well in many automated susceptibility systems, thus posing problems for the laboratory in determining resistance. If the mechanism of resistance is known and is mediated by a recognizable DNA sequence (as opposed to a point mutation), then probes would be useful for susceptibility testing. Probes may be used to test organisms such as *Bordetella pertussis, Brucella species, Mycobacterium tuberculosis,* or even *Chlamydia trachomatis*.

The second use for probes would be to test organisms known to carry derepressable antimicrobial resistant genes, such as *Enterobacter cloacae*, for cephalosporinases active against third generation cephalosporins. Here, a gene to a regulatory mutant may be possible.

The last use for probes would be to test for resistance genes known to be involved in heteroresistance, such as methicillin resistance in staphylococci,[54] where only a portion of the bacterial population may be resistant to the drug, but that subpopulation could well lead to treatment failure.

IX. PROBES FOR EPIDEMIOLOGIC STUDIES OF ANTIMICROBIAL RESISTANCE

The majority of probes to antimicrobial resistance genes developed to date have been used to study the movement of drug resistance genes in hospitals. These studies are valuable in determining trends in antimicrobial resistance and should continue. Clinical laboratories may wish to be involved in these studies since the majority of isolates originate in such laboratories. Using biotinylated probes and colony hybridization techniques, such as the spot-blot method described by Gootz et al.,[55] these studies would not be difficult undertakings for a clinical laboratory and would yield valuable information, particularly concerning the frequency of expression of different classes of resistance genes in various bacterial genera.

X. PROBES TO OTHER ANTIMICROBIAL RESISTANCE GENES

The mechanisms of resistance to antimicrobials manifested by organisms other than bacteria are not well understood. It is possible that, in the future, probes may be used to determine the resistance of *Plasmodium falciparum* to chloroquine and fansidar, *Candida albicans* to amphotericin B, and herpes simplex virus to acyclovir.

XI. CONCLUSIONS

DNA probes may be helpful for determining the antimicrobial susceptibility of a highly select group of organisms to specific antimicrobials. These would include organisms that are either difficult to propagate in the laboratory, those that have well-defined modes of antimicrobial resistance and are often present in normally sterile body sites, and the direct testing of clinical isolates for specific resistance genes. The latter category of tests could only be undertaken if the organism is growing in a normally sterile body site, free from other flora, and has a known mechanism of antimicrobial resistance. The testing of urethral exudates from males suspected of having gonorrheae with a β-lactamase probe would be an example of the latter category.

One major drawback associated with the use of probes for more widespread determinations of antimicrobial susceptibility is the unknown distribution of antimicrobial resistance genes in nature. The number of organisms harboring silent genes remains problematic. By using probes, however, we may discover why a given percentage of patients fail therapy for "unknown reasons".

ACKNOWLEDGMENTS

I wish to thank Drs. Arnold Smith and Fritz Schoenknecht for their valuable contributions to this manuscript.

REFERENCES

1. **Barry, A. L.,** *The Antimicrobial Susceptibility Test: Principles and Practices,* Lea & Febiger, Philadelphia, 1976.
2. **Tenover, F. C.,** Studies of antimicrobial resistance genes using DNA probes, *Antimicrob. Agents Chemother.,* 29, 721, 1986.
3. **Sugarman, B. and Pesanti, E.,** Treatment failures secondary to *in vivo* development of drug resistance by microorganisms, *Rev. Infect. Dis.,* 2, 153, 1980.
4. **Mendelman, P. M., Chaffin, D. O., Stull, T. L., Rubens, C. E., Mack, K. D., and Smith, A. L.,** Characterization of non-β-lactamase-mediated ampicillin resistance in *Haemophilus influenzae, Antimicrob. Agents Chemother.,* 26, 235, 1984.
5. **Preheim, L. C., Penn, R. G., Sanders, C. C., Goering, R. V., and Giger, D. K.,** Emergence of resistance to β-lactam and aminoglycoside antibiotics during moxalactam therapy of *Pseudomonas aeruginosa* infections, *Antimicrob. Agents Chemother.,* 22, 1037, 1982.
6. **Seeberg, A. H., Tolxdorff-Neutzling, R. M., and Wiedemann, B.,** Chromosomal β-lactamases of *Enterobacter cloacae* are responsible for resistance to third-generation cephalosporins, *Antimicrob. Agents Chemother.,* 23, 918, 1983.
7. **Gootz, T. D., Sanders, C. C., and Sanders, W. E., Jr.,** Resistance to cephamandole: derepression of β-lactamases by cefoxitin and mutation in *Enterobacter cloacae, J. Infect. Dis.,* 146, 34, 1982.
8. **Sanders, C. C.,** Inducible β-lactamases and non-hydrolytic resistance mechanisms, *J. Antimicrob. Chemother.,* 13, 1, 1984.
9. **Yamamoto, T., Yamagata, S., Horu, K., and Yamagishi, S.,** Comparison of transcription of β-lactamase genes specified by various ampicillin transposons, *J. Bacteriol.,* 150, 269, 1982.
10. **Foster, T. J.,** Plasmid-determined resistance to antimicrobial drugs and toxic metal ions in bacteria, *Microbiol. Rev.,* 47, 361, 1983.
11. **Davies, J. and Smith, D. L.,** Plasmid-determined resistance to antimicrobial agents, *Annu. Rev. Microbiol.,* 32, 469, 1978.
12. **Bryan, L. E.,** *Antimicrobial Drug Resistance,* Academic Press, Orlando, Fla., 1984.
13. **Anderson, E. S.,** The ecology of transferable drug resistance in the *Enterobacteriaceae, Annu. Rev. Microbiol.,* 22, 132, 1968.
14. **Falkow, S.,** *Infectious Multiple Drug Resistance,* Pion, London, 1975.
15. **de Graaff, J., Elwell, L. P., and Falkow, S.,** Molecular nature of two beta-lactamase-specifying plasmids isolated from *Haemophilus influenzae* type b, *J. Bacteriol.,* 126, 439, 1976.
16. **Elwell, L. P., Roberts, M., Mayer, L. W., and Falkow, S.,** Plasmid-mediated beta-lactamase production in *Neisseria gonorrhoeae, Antimicrob. Agents Chemother.,* 11, 528, 1977.
17. **Levesque, R. C., Medeiros, A. A., and Jacoby, G. A.,** Molecular cloning and DNA homology of plasmid-mediated β-lactamase genes, *Mol. Gen. Genet.,* 206, 252, 1987.
18. **Marshall, B., Tachinbana, C., and Levy, S. B.,** Frequency of tetracycline resistance determinant classes among lactose-fermenting coliforms, *Antimicrob. Agents. Chemother.,* 24, 835, 1983.
19. **Perine, P. L., Totten, P. A., Holmes, K. K., Eng, E. H., Ratnam, A. V., Widy-Wersky, R., Nsanze, H., Habte-Gabr, E., and Westbrook, W. G.,** Evaluation of DNA-hybridization method for detection of African and Asian strains of *Neisseria gonorrhoeae* in men with urethritis, *J. Infect. Dis.,* 152, 59, 1985.
20. **Faruki, H., Kohmescher, R. N., McKinney, W. P., and Sparling, P. F.,** A community-based outbreak of infection with penicillin-resistant *Neisseria gonorrhoeae* not producing penicillinase (chromosomally mediated resistance), *N. Engl. J. Med.,* 313, 607, 1985.
21. **Jones, R. N.,** Antimicrobial susceptibility testing (AST): a review of changing trends, quality control guidelines, test accuracy, and recommendation for the testing of β-lactam drugs, *Diagn. Microbiol. Infect. Dis.,* 1, 1, 1983.
22. **Bauer, A. W., Kirby, W. M. M., Sherris, J. C., and Turck, M.,** Antibiotic susceptibility testing by a standardized single disk method, *Am. J. Clin. Pathol.,* 45, 493, 1966.
23. **Sherris, J. C. and Ryan, K. J.,** Evaluation of automated and rapid methods, in *Rapid Methods and Automation in Microbiology — 1981,* Tilton, R. C., Ed., American Society for Microbiology, Washington, D.C., 1982, 1.
24. **Sabath, L. D., Garner, C., Wilcox, C., and Finland, M.,** Effect of inoculum and beta-lactamase on the antistaphylococcal activity of thirteen penicillins and cephalosporins, *Antimicrob. Agents Chemother.,* 8, 344, 1975.
25. **Eng, R. H. K., Smith, S. M., and Cherubin, C.,** Inoculum effect of new β-lactam antibiotics on *Pseudomonas aeruginosa, Antimicrob. Agents Chemother.,* 26, 42, 1984.
26. **Sherris, J. C.,** Problems in in vitro determination of antibiotic tolerance in clinical isolates, *Antimicrob. Agents Chemother.,* 30, 633, 1986.
27. **Tuomanen, E., Durack, D. T., and Tomasz, A.,** Antibiotic tolerance among clinical isolates of bacteria, *Antimicrob. Agents Chemother.,* 30, 521, 1986.

28. **Tenover, F. C., Gootz, T. D., Gordon, K. P., Tompkins, L. S., Young, S. A., and Plorde, J. J.,** Development of a DNA probe for the structural gene of the 2″-*O*-adenyltransferase aminoglycoside-modifying enzyme, *J. Infect. Dis.*, 150, 678, 1984.
29. **Tenover, F. C. and Plorde, J. J.,** DNA probes for epidemiologic studies of aminoglycoside-resistant bacteria, in *Microbiology — 1986*, Schlessinger, D., Ed., American Society for Microbiology, Washington, D.C., 1986, 125.
30. **Groot Obbink, D. J., Ritchie, L. J., Cameron, F. H., Mattick, J. S., and Ackerman, V. P.,** Construction of a gentamicin resistance gene probe for epidemiologic studies, *Antimicrob. Agents Chemother.*, 28, 96, 1985.
31. **Young, S. A., Tenover, F. C., Gootz, T. D., Gordon, K. P., and Plorde, J. J.,** Development of two DNA probes for differentiating the structural genes of subclasses I and II of the aminoglycoside-modifying enzyme 3′-aminoglycoside phosphotransferase, *Antimicrob. Agents Chemother.*, 27, 739, 1985.
32. **Lambert, T., Gerbaud, G., Trieu-Cuot, P., and Courvalin, P.,** Structural relationship between the genes encoding 3′-aminoglycoside phosphotransferases in *Campylobacter* and in gram positive cocci, *Ann. Inst. Pasteur/Microbiol.*, 136B, 135, 1985.
33. **Bergstrom, S., Olsson, O., and Normark, S.,** Common evolutionary origin of chormosomal beta-lactamase genes in enterobacteria, *J. Bacteriol.*, 150, 528, 1982.
34. **Boissinot, M., Mercier, J., and Levesque, R. C.,** Development of natural and synthetic DNA probes for OXA-2 and TEM-1 β-lactamases, *Antimicrob. Agents Chemother.*, 31, 728, 1987.
35. **Jouvenot, M., Deschaseaux, M. L., Royez, M., Mougin, C., Cooksey, R. C., Michel-Briand, Y., and Adessi, G. L.,** Molecular hybridization versus isoelectric focusing to determine TEM-type β-lactamases in gram negative bacteria, *Antimicrob. Agents Chemother.*, 31, 300, 1987.
36. **Roberts, M. C., Actis, L. A., and Crosa, J. H.,** Molecular characterization of chloramphenicol-resistant *Haemophilus parainfluenzae* and *Haemophilus ducreyi*, *Antimicrob. Agents Chemother.*, 28, 176, 1985.
37. **Christie, P. J. and Dunney, G. M.,** Antibiotic selection pressure resulting in multiple antibiotic resistance and localization of resistance determinants to conjugative plasmids in streptococci, *J. Infect. Dis.*, 149, 74, 1984.
38. **Arthur, M. and Courvalin, P.,** Contribution of two different mechanisms to erythromycin resistance in *Escherichia coli*, *Antimicrob. Agents Chemother.*, 30, 694, 1986.
39. **Arthur, M., Andremont, A., and Courvalin, P.,** Heterogeneity of genes conferring high-level resistance to erythromycin by inactivation in enterobacteria, *Ann. Inst. Pasteur/Microbiol.*, 137A, 125, 1986.
40. **Thakker-Varia, S., Jenssen, W. D., Moon-McDermott, L., Weinstein, M. P., and Dubin, D. T.,** Molecular epidemiology of macrolides-lincosamides-streptogramin B resistance in *Staphylococcus aureus* and coagulase-negative staphylococci, *Antimicrob. Agents Chemother.*, 31, 735, 1987.
41. **Barkay, T. D., Fouts, D. L., and Olsen, B. H.,** Preparation of a DNA gene probe for detection of mercury resistance genes in gram negative bacterial communities, *Appl. Environ. Microbiol.*, 49, 686, 1985.
42. **Mendez, B., Tachibana, C., and Levy, S. B.,** Heterogeneity of tetracycline resistance determinants, *Plasmid*, 3, 99, 1980.
43. **Burdett, V., Inamine, J., and Rajagopalan, S.,** Heterogeneity of tetracycline resistance determinants in *Streptococcus*, *J Bacteriol.*, 149, 995, 1982.
44. **Roberts, M. C., Koutsky, L. A., Holmes, K. K., LeBlanc, D. J., and Kenny, G. E.,** Tetracycline-resistant *Mycoplasma hominis* strains contain streptococcal *tetM* sequences, *Antimicrob. Agents Chemother.*, 28, 141, 1985.
45. **Morse, S. A., Johnson, S. R., Biddle, J. W., and Roberts, M. C.,** High-level tetracycline resistance in *Neisseria gonorrhoeae* is result of acquisition of streptococcal *tetM* determinant, *Antimicrob. Agents Chemother.*, 30, 664, 1986.
46. **Fling, M. E., Walton, L., and Elwell, L. P.,** Monitoring of plasmid encoded, trimethoprim-resistant dihydrofolate reductase genes: detection of a new resistance enzyme, *Antimicrob. Agents Chemother.*, 22, 882, 1982.
47. **Rudy, R. P. and Murray, B. E.,** Evidence for an epidemic trimethoprim-resistance plasmid in fecal isolates of *Escherichia coli* from citizens of the United States studying in Mexico, *J. Infect. Dis.*, 150, 25, 1984.
48. **Pulkkinen, L., Houvinen, P., Vuorio, E., and Toivanen, P.,** Characterization of trimethoprim resistance by use of probes specific for transposon Tn7, *Antimicrob. Agents Chemother.*, 26, 82, 1984.
49. **Miller, G. H., Sabatelli, F. J., Hare, R. S., and Waitz, J. A.,** Survey of aminoglycoside resistance patterns, *Dev. Ind. Microbiol.*, 21, 91, 1980.
50. **Larson, T. A., Garrett, C. R., and Gerding, D. N.,** Frequency of aminoglycoside 6′-*N*-acetyltransferase among *Serratia* species during increased use of amikacin in the hospital, *Antimicrob. Agents Chemother.*, 30, 176, 1986.
51. **Katoh-Kanno, R., Kimura, M., Ikeda, T., and Kimura, S.,** Survey of modifying enzymes and plasmids in amikacin resistant *Serratia marcescens*, *Microbiol. Immunol.*, 30, 509, 1986.
52. **Coyle, M. B., McGonagle, L. A., Plorde, J. J., Clausen, C. R., and Schoenknecht, F. D.,** Rapid antimicrobial susceptibility testing of isolates from blood cultures by direct inoculation and early reading of disk diffusion tests, *J. Clin. Microbiol.*, 20, 473, 1984.

53. **Wellstood, S. A.,** Direct identification and susceptibility testing by the AutoMicrobic system for gram negative bacilli from urine specimens, *J. Clin. Microbiol.*, 23, 1068, 1986.
54. **Beck, W. D., Berger-Bachi, B., and Kayser, F. H.,** Additional DNA in methicillin-resistant *Staphylococcus aureus* and molecular cloning of *mec*-specific DNA, *J. Bacteriol.*, 165, 378, 1986.
55. **Gootz, T. D., Tenover, F. C., Young, S. A., Gordon, K. P., and Plorde, J. J.,** Comparison of three DNA hybridization methods for detection of the aminoglycoside 2″-*O*-adenylyltransferase gene in clinical isolates of bacteria, *Antimicrob. Agents Chemother.*, 28, 69, 1985.

Index

INDEX

A

B

C

D

E

F

G

I

N

O

R

S

U

V

W

X

Y

Z